Wild Cultures: A Comparison Between Chimpanzee and Human Cultures

How do chimpanzees say, "I want to have sex with you?" By clipping a leaf or knocking on a tree trunk? How do they eat live aggressive ants? By using a short stick with one hand or a long stick with both? Ivorian and Tanzanian chimpanzees answer these questions differently, as would humans from France and China if asked how they eat rice. Christophe Boesch takes readers into the lives of chimpanzees from different African regions, highlighting the debate about culture. His ethnography reveals how simple techniques have evolved into complex ones, how teaching styles differ, how material culture widens access to new food sources, and how youngsters learn culture. This journey reveals many parallels between humans and chimpanzees and points to striking differences. Written in a vivid and accessible style, *Wild Cultures* places the reader in social and ecological contexts that shed light on our cultures, which will help us refine human nature and its multitude of facets.

Christophe Boesch is Professor and Director of the Department of Primatology at the Max-Planck Institute of Evolutionary Anthropology in Germany. He has studied the chimpanzees of Taï National Park in Côte d'Ivoire for the last 33 years and those of Loango National Park in Gabon for 6 years. The author of two published books and the founding president of the Wild Chimpanzee Foundation, he fights for a better future for the remaining wild ape populations at a grassroots level.

Wild Cultures

A Comparison Between Chimpanzee and Human Cultures

CHRISTOPHE BOESCH

CAMBRIDGE UNIVERSITY PRESS
Cambridge, New York, Melbourne, Madrid, Cape Town,
Singapore, São Paulo, Delhi, Mexico City

Cambridge University Press
32 Avenue of the Americas, New York, NY 10013-2473, USA

www.cambridge.org
Information on this title: www.cambridge.org/9781107025370

First published 2012

Printed and Bound in the United Kingdom by the MPG Books Group

A catalog record for this publication is available from the British Library.

Library of Congress Cataloging in Publication data
Boesch, Christophe.
Wild cultures : a comparison between chimpanzee and
human cultures / Christophe Boesch.
pages cm
Includes bibliographical references and index.
ISBN 978-1-107-02537-0
1. Chimpanzees–Behavior. 2. Animal behavior. I. Title.
QL737.P96B5955 2012
599.885–dc23 2012018828

ISBN 978-1-107-02537-0 Hardback

To all – people and chimpanzees – whose cultures are so fascinating and make life on Earth so rich,

To Lukas and Léonore for accompanying us through this adventure,

To Brutus and Fédora for coaching us through the secrets of their Culture

To Grégoire and Moda for sharing the intimacy of their Culture

Contents

Acknowledgments

Wild Cultures is about differences within and between species and such comparisons were only possible thanks to the generosity, tolerance, and cooperation of many others. When Jane Goodall first invited my wife Hedwige and I to visit Gombe in 1983, we were naturally very impressed by the Gombe chimpanzees, given our still very slow progress in habituating the Taï chimpanzees at that time. But we were just as impressed by how dramatically different the Gombe habitat was compared to the Taï forest. And as we progressed in our observations of the Taï chimpanzees, we rapidly felt that they did not fit perfectly into what we had read about the chimpanzees of Gombe and Mahale. That was the beginning of this book.

It was not only a great honor then when both Jane Goodall and Toshisada Nishida invited me to their study sites for longer stays and encouraged me to make comparisons between the Taï, Gombe, and Mahale chimpanzees. This was not only a wonderful opportunity, it was also an eye-opener to what population differences in this species really mean. I am not only very grateful to these two pioneers of chimpanzee research for inviting and supporting me; I am equally grateful that they made my stays in their camps and my learning experiences there such a pleasure. I want to thank my teachers in Gombe who so patiently taught me about cultural differences there: Frodo, Evered, and Prof introduced me to their hunting behavior, Fanny and Fred showed me leaf grooming; Fifi, Fax, and Gremlin introduced me to termite fishing; and Goblin and Wilkie introduced me to their complex interactions with red colobus monkeys. I would also like to thank my teachers in Mahale: Alufo and Bonobo's refined leaf folding allowed me to understand this behavior; Masudi, Caliope, Opal, and Gwekulo introduced me to skilled ant fishing; and Darwin and Carter ably demonstrated their specific way of leaf clipping. Without these teachers, I would not have been convinced about all these differences and an important aspect of chimpanzee life would have remained obscured.

After my illuminating visits to Gombe and Mahale, I became very curious about the breadth and richness of cultural differences in chimpanzees and was lucky enough to have generous colleagues who graciously invited me to visit a variety of chimpanzee groups living in a variety of ecological conditions. I am thus very grateful to John Mitani, David Watts and the Ngogo chimpanzees, to Vernon Reynolds and the Budongo chimpanzees, and to Richard Wrangham and the Kanyawara chimpanzees for accepting me at their sites, for guiding my vision into the fascinating individuals of these study communities, and for graciously answering all my questions as I wanted to know whether my impressions gained on their study groups corresponded to theirs. Here, I would also like to thank Crickette Sanz and David Morgan for introducing me to the Goulalougo chimpanzees, and David for memorably showing me how the chimpanzees use sequential tools to access underground termites.

The start of our project in Taï National Park would not have been possible without the trust Hans Kummer placed in us. Hans Kummer's intellectual mix of rigor and keen longing to understand the unknown have greatly influenced my own attitude towards science. My present involvement in primate research and conservation is part of the legacy he left me. In 1991, after visiting the Taï chimpanzee project, Professor Steven Stearns of the University of Basel offered me an academic position at his institute that would allow me to continue my fieldwork in Taï forest. After 12 years in the forest, this generous offer allowed me to touch base with scientific progress and intensify the Taï Chimpanzee Project as well as develop my scientific career. I would also like to express my deepest gratitude to Professor André Aeschlimann from the University of Neuchâtel, head of the Swiss National Foundation and President of the CSRS in Abidjan for many years, whose support of the Taï Chimpanzee Project throughout its early years contributed importantly to the long-term success of this project, now in its 33rd year.

My wife Hedwige Boesch-Achermann was enthusiastic, curious, and open to leave the so-called civilized world. She was by my side from the first day on and for more than 12 years of permanent life in the Taï forest. If the study of the Taï chimpanzees is successful, it is only because she was there with me all those years. This transformed our life into an adventure, a long inquiry into and discovery of some of the most intimate aspects of the lives of our cousins from the forest. Most of the ideas developed in this book matured over the years we spent together in the forest, but also after this time when she continued her work for the chimpanzees in Europe. Trying to understand this elusive species and discovering the many aspects of their life was and will always be a part of our life. Therefore, throughout the text, the use of "I" should often be read as "we" as the Taï experience was a joint experience that moulded so many aspects of my scientific and personal way of thinking.

As the Taï Chimpanzee Project developed, we began to include both students and local field assistants to increase the number of observations done with the chimpanzees as well as to habituate three other chimpanzee communities. Each of our many field assistants have our deepest gratitude. Not only were they willing to spend 12–14 hours per day following the chimpanzees on their often over 5 km-long forays in the forest, but they spent nights in the forest to protect seriously ill chimpanzees from possible leopard attacks. And they were willing to stay with the chimpanzees during some very unstable periods, including the civil wars from 2002 to 2004 and in 2011, thereby actively protecting the study site and the chimpanzees. They have been our ambassadors in the local populations and made the development and continuation of the project possible since it began in 1978. We would especially like to thank the senior field assistants Kpazahi Honora, Bolé

Camille, Oulaï Nicaise Daurid, Bally Louis Bernard, Tahou Mompeho Jonas, Gouyan Bah Nestor, Gnahe Djirian Appolinaire, Yagnon Valentin, Tah Alain Pahi, Sioblo Arsène, Guy Sylvain, Guiro Ferdinand Thia, Kevin Charles Bally, late Blaise Blé Téré, Ignace Dezaï, Camille Dji, Gabriel Gnombouhou Kouya, Gérald Gah, Benjamin Goullaon, Mathias Douosson, and Denis Lia.

At the same time, I would like to thank the many students, far too numerous to be named individually, from all over the world who helped to follow and study the chimpanzees in Taï forest for short periods. I thank all students and colleagues who were involved in the Taï project for longer periods for their dedication and passion for the chimpanzees and their forest, including Pascal Gagneux, Christian Falquet, Miriam Behrens, Gerd Radl, Diane Doran, Martina Funk, Margaret Hoitink, Andy Kurt, Penny Simpson, Brigitte Schmid, Suzanne Pieren, Ulrike Ratkjen, Annemarie Fränkl, Gregory Roduit, Paco Bertolani, Chloe Cipoletta, Dean Anderson, Nick Malone, Ilka Herbinger, Roman Wittig, Myriam Sele, Cathy Crockford, Steven Bada, Nicola Paterson, Pola Abaza Daniel Hanus, Hjalmar Kuehl, Julia Riedel, Janna Rist, Melissa Tauber, Kathleen Beese, Sabrina Locatelli, Lionel Egger, Fabian Leendertz, Sandra Junglen, Rebecca Stumpf, Antoine N'Guessan, Zoro Goné Bi, Anja Blankenburg, Svenja Schenk, Siva Aina Jensen, Célestin Kouakou, Simone Ban Dagui, Nadine Eckhardt, Lydia Luncz, Livia Wittiger, Sonja Metzger, Ammie Kalan, Karline Jaanmart, Serge Soiret, and Danielle Spitzer.

In any long-term project, some difficult periods may occur that might have dramatic consequences for the chimpanzees we have habituated to human presence. Thanks to the courage and altruism of some individuals, in most cases we were able to prevent the worst – the disappearance for unknown reason and/or the killing of the chimpanzees. My warmest thanks therefore go to Claudia Steiner and Franca Donati, who were present during the emotionally wrenching *Ebola* outbreaks in the Taï chimpanzees in 1992 and 1994, to Thomas Pfluger who stayed near the camp when, in 1991, unrest from Liberia spilled over into the Taï region and also to Emmanuelle Normand, Yasmin Moebius, Tobias Deschner, and Cristina Gomes for having decisively contributed to keeping the project going during some of the most difficult times of civil unrest in Côte d'Ivoire during 2002–2004. Finally, my warmest thanks go to Livia Wittiger, Lydia Luncz, Nadine Eckhardt, Sonja Metzger, and Danielle Pfitzer for keeping the project alive during the terrible disease outbreak of Christmas 2009 and the following periods of unrest, as well as to Serge Keagnon Soiret and Simone Ban Dagui during the war of early 2011. These individuals' contributions made it possible for the Taï Chimpanzee Project to remain active, which will undoubtedly contribute to our understanding of chimpanzees in the years to come.

I am most grateful to Carolyn Rowney for enthusiastically and efficiently correcting the writing in this book. A special thank you to my office assistant Claudia Nebel, our IT expert Andreas Walther, and our multimedia expert Silvio Tüpke for their great continuous support, which makes it possible for me to spend time writing books! And I thank Hedwige Boesch, Cleve Hicks, Claudia Borchers, Tobias Deschner, Linda Vigilant, Geraldine Fahy, Livia Wittiger, Josephine Head, Erin

Wessling, Claudio Tennie, Richard Moore, and the other members of my department who enthusiastically read and discussed all draft chapters of this book.

Our long-term study of the chimpanzees in Taï National Park was only possible because of the constant and amicable support of the Ivorian Authorities. Thirty-three years is a long time and by far surpasses the normal life of any government, but nevertheless, the support from the Ivorian government remained inflexible. We wholeheartedly admire the constancy of the support we received from all the different people working in and heading the "Ministère des Eaux et Forêts," as well as within the "Ministère de la Recherche Scientifique," and in particular those working and heading the "Direction de la Protection de la Faune" of the "Office Ivoirien des Parcs et Réserves" and of the "Direction du Parc national de Taï," as well as the numerous park agents, especially those of the Taï Sous-Préfecture. Throughout the years, Taï National Park has been subject to attacks from various directions, mainly in the shape of logging, poaching, civil unrest, and farming, and this has threatened the survival of the forest and its fauna, including the chimpanzees. The Ivorian authorities have always undertaken the steps necessary to guarantee the survival of this precious park and unfailingly supported the continuity of the project. The Centre Suisse de Recherches Scientifiques en Côte d'Ivoire (CSRS) has been our base in Abidjan since our very first visit to the country in 1976, and they have remained a key partner in the project ever since. We thank all the successive directors, Jean-Francois Graf, Marianne Knecht, Felix Naef, Michel Gremaut, Peter Lehman, Guéladio Cissé, and Bassirou Bonfou for going out of their way to support us. In Taï, we profited a great deal from the support of the directors of the Station d'Ecologie Tropicale and their staff, Denis Vivet, the late Théo Tiépkan Zoroa, and Paul Zouhou, respectively. A special thanks also to my colleagues at the University of Cocody and Abobo in Abidjan for supporting the collaboration between our institutions.

Our new project in Loango National Park in Gabon has profited from the unfailing support of the Gabonese authorities, l'Agence Nationale des Parcs Nationaux (ANPN), and of the Centre National de Recherches Scientifique et Techniques (CENAREST). This project is a collaboration among and financially supported by Société pour la Conservation et le Développement (SCD), Wildlife Conservation Society (WCS), and the Max Planck Society. We are very grateful to Rombout Swanborn for his financial and logistical support of the Loango project, and Tomo Nishihara, Lee White, and Edward Truter for their constant logistical support. I am especially grateful to the students who helped right from the start and contributed to the development of the project under sometimes quite strenuous conditions: Josephine Head was with us from the very first day and she was later joined by Luisa Rabanal, Nikki Tagg, Beke Graw, Emilie Fairet, Edward Wright, Lyndsay Rankin, Melanie Gregoire, Christopher Orbell, and Carolyn Sanguinetti. We are indebted to Loïc Mackaga, Erick Reteno Guizard, Kharl Remanda, Aimé Relonga, Herve Ngoundou, Ulrich Moussouami, Yannick Nkoma, Jhon Essongue, Martial Pambo, and Pierre Boukoussou for their invaluable tracking assistance in the forest and contributing to the long-term data collection.

No scientific project can survive without financial support and here I wish to stress the continuous, generous, and unwavering support we have received from the Swiss National Science Foundation for the first 19 years and from the Max Planck Society for the past 13 years. Very few public funding organizations dare to support a long-term field project for such a length of time and we are immensely grateful to them for seeing the value in such a study. We would also like to thank additional funding sources, including the Messerli Foundation, the Leakey Foundation, the Schultz Stiftung, the Jane Goodall Institute, the Wenner Gren Foundation, the Roche Foundation, and the Freie Akademische Gesellschaft.

How dull would our life have been in the forest without our encounters with the wonderful, friendly, helpful, curious, and tolerant African people near and in our forest camp, who joyfully shared their knowledge with us? What would our life have been like over the years without our friends and family from abroad who visited, sent presents, letters, food, and books, and stayed in contact while we were in Africa and back in Europe? Thank you to all of you.

Last but not least, we would like to thank all the chimpanzees who have tolerated our presence for so many years, allowed us to share so much of their lives with us, and made us wonder on a daily basis about our own species. During the process of habituating and following them, we also inadvertently contributed to their confrontation with threats such as diseases and made them more vulnerable to poaching. This has been a constant worry for us from the beginning and I hope that by increasing the status of chimpanzees in the scientific community and the public at large, this book will allow us to pay back some of our debts to them.

It is also in the spirit of "paying back" a debt to the chimpanzees that my wife and I have created the "Wild Chimpanzee Foundation" (www.wildchimps.org). We are most grateful for the wonderful support, motivation, and involvement, be it financial or technical, we have experienced in this undertaking from so many of the team at the Max Planck Institute. We could not do it without them.

Introduction

Humans descended from the apes! My dear, let us hope that it is not true, but if it is, let us pray that it will not become generally known.

The wife of the Bishop of Worcester, after learning about Darwin's theory of evolution (1860).

A Man without culture resembles a zebra without stripes

African Proverb

In the 1600s, the "civilized world" was horrified to learn that there were barbarians in the Americas practicing cannibalism. However, the French philosopher Michel de Montaigne said: "There is nothing barbarian and savage in this nation, according to what I have been told, except that each of us call barbarian what is not in our habits; to be honest, it seems that we have no other reference of truth and reason than the example and idea of the opinions and habits of the country we come from. That country always has perfect religion, the perfect police, the perfect and accomplished usage of things."[1] In other words, Montaigne was reminding his contemporaries that western values are not universal and that we need to understand other cultures before judging them. He also provocatively suggested that it might be less barbarous to eat someone who was already dead than to dismember someone who was still alive (a practice that westerners had been doing for centuries in the name of civilized justice).

Humans are the only animals on Earth that have composed musical pieces like "La Traviata" and "Don Giovanni," painted works of art like the "Joconda," cooked food using a variety of energy sources, invented computers, constructed skyscrapers, and travelled to the moon. At the same time, different groups of humans also wear very different clothing, have developed very different eating habits, and use many different languages and types of social greeting. This must not be forgotten in our discussion about the evolution of culture. Culture here is defined as all socially learned behavior that distinguishes individuals belonging to one group from individuals belonging to other groups. At the same time, let us be candid for a moment; I have never done anything even remotely close to composing "Don Giovanni,"

[1] « *Il n'y a rien de barbare et de sauvage en cette nation, à ce qu'on m'en a rapporté, sinon que chacun appelle barbarie ce qui n'est pas de son usage ; comme de vrai, il semble que nous n'avons autre mire de la vérité et de la raison que l'exemple et idée des opinions et usages du pays où nous sommes. Là est toujours la parfaite religion, la parfaite police, le parfait et accompli usage de toutes choses.* » ("Des Cannibales", Essais, 1st volume, chapter 30, 1595.)

I have never painted a work of art, and I have no idea what is inside the computer I am writing this on. And the same is probably true for most of us. Hence, a more realistic way to describe the achievements of human culture would be to say that a few human geniuses invented computers and even fewer composed Don Giovanni, while the rest of us are happy to be able to drive a car and cook food on a gas stove and speak the languages and adopt the social habits of the group we belong to. For most of us, the world is made up of products that were invented and manufactured by other humans.

Culture has always been considered as one of the most important human characteristics. Researchers from many academic disciplines, including anthropology, archaeology, and much of psychology, are dedicated to studying this uniquely human ability. How can you not be impressed by the power of culture in humans when you see, for example, the unbelievable cave paintings in the "Grotte de Chauvet" in the South of France? The groups of elegant horses, lions, and antelopes were painted on the cave walls some 32 000 years ago by individuals who possessed skills that are comparable to Picasso's and Braque's. Similarly, when I play Chopin's "Nocturnes" on the piano, I share a cultural inclination and imagine myself giving life to Chopin's intentions by interpreting those pieces with the level of feeling he might have intended them to be played with some 160 years ago. These cultural skills, which humans have shared for centuries, are clearly hallmarks of humanity and many have thought them to be uniquely human.

But then the chimpanzee came along. The major finding that science has garnered about chimpanzees over the past 25 years is that each population in Africa differs to a surprisingly high degree from any other in many behavioral and social domains. The puzzle was that such behavioral differences could not be predicted directly from the chimpanzees' environments. In addition, these population differences are often strikingly arbitrary; in some populations, males initiate a sexual encounter with a female by knocking with their knuckles on the trunks of trees, while in others they do so by ripping leaves with their fingers. In yet another population, males rip leaves with their teeth to invite a female for sex, while in another this would signal the start of a play session. Some chimpanzee populations crack open nuts with hammers while others ignore nuts entirely. But all these differences sounded uncannily similar to culture to us and, thus, we were forced to rethink our ideas about culture being uniquely human.

Now that the notion of culture in animals was no longer heresy, examples of population-specific differences in other animal species also began to emerge. Observers working with Sumatran orangutans noted that some populations used tools to open a certain variety of fruit, while populations on the other side of the river ate the same fruit but never used tools. There were reports of dolphin populations producing "clicks" in different, group-specific ways in the Atlantic compared to the Pacific Ocean. Capuchin monkeys in Central America have been observed to develop astounding group-specific play techniques, like one individual poking a finger in the eye of another or biting hair from the partner's shoulder, and, like fashion trends, these behavioral patterns spread rather quickly to most other group members. One population of New Caledonian crows was seen to make and use simple tools to fish insects out of branches, while a subpopulation invented a more efficient hooked tool. Baboons and bonobos have also been proposed to belong to the culture club and you can be sure that the list will grow as researchers observe and compare more populations of animal species.

And the culture war started ...

The entire history of our quest to understand "what makes us human" has been punctuated with debates. Following the remarks by the Bishop of Worcester's wife, one such debate began on whether or not humans are a class of living beings separated by divine intervention from all other living beings. Luckily, it quickly included some scientific questions. Is tool use uniquely human, as is suggested by the name of our famous ancestor "*Homo faber*," or are tool-using abilities shared with chimpanzees? This debate has been ongoing since the Gombe chimpanzees were first reported to use tools in the early '60s. This is a particularly heated debate as, even today, the association with tools remains one of the most important criteria for deciding if a fossil belongs to our genus, *Homo*. So it should not come as a surprise that observations of tool use in animals have immediately been questioned. Like Desmond Morris said, "our climb to the top has been a get-rich-quick story, and, like all *nouveaux riches*, we are very sensitive about our background."[2]

Similarly, does hunting define humanity as the label "Man the hunter" would suggest, or are other primates hunters as well? Evidence of our hunting past is abundant; however, hunting has now been observed in all chimpanzee populations in Africa, which undoubtedly brings the value of using hunting to define humanity into question. Maybe it is something in the way we hunt. For example, could cooperative hunting be a marker of humanity, or are the cooperative hunts seen in lions and chimpanzees similar in kind? Since it was suggested that cooperation requires sophisticated social intelligence, the debate about cooperation has grown in visibility and became a very lively contest in which observations from the field are pitted against experiments with captive animals, and vice versa.

Life was much simpler for anthropologists 50 years ago. Without any observations of wild animals for comparison, it was easy to define what was strictly human. When culture was thought to be a uniquely human ability, defining it was very straightforward and all scientists happily embraced definitions that incorporated a direct reference to human or uniquely human abilities like language and writing. Now that we know the situation is not so simple, however, there is an urgent need to better understand what culture means in nature. Nevertheless, some minimal agreement prevails in the sense that all would agree, at the very least, that culture is a distinctive collective practice that is passed on to group members, and, at the insist-

[2] *The naked ape* by Desmond Morris (2nd edition, 1983, p. 241).

ence of some culture anthropologists, includes elements based on shared meanings between members of the same group or society.[3]

The factions of the culture debate can be summarized as follows: The "human uniqueness" faction holds that only humans have culture and that what we observe in animals is something different altogether and therefore must be named differently. The "human superiority" faction holds that what is observed in other animal species could be conceded to be culture but regardless, it is of a lower form, and higher forms of the culture phenomenon, like imitation, teaching, and sharing goals and intentions, are strictly human. The "ape–human clade uniqueness" faction proposes that all great apes, including humans, share such abilities and they are based on some cognitive abilities that are unique to them. And finally, the "biological" faction proposes that culture is widespread and we should look at it in a comparative way. Culture, from this point of view, is expected to be especially important in species with enough learning abilities and in environments where flexible solutions are required.

Chimpanzee ethnography to uncover culture

As a field worker who has now spent 30 years observing chimpanzees in many parts of tropical Africa, I have been lucky enough to see for myself most of the behavioral differences that I discuss in this book. My aim is to present culture from a different perspective than previous works have. First, I will focus on the different properties of the culture phenomenon as observed in different species. I will not discuss whether or not culture exists in species other than humans as 40 years of studies on whether or not chimpanzees and other animal species *have culture* has rendered this last question a little outdated. Second, I will concentrate on animal culture rather than on human culture and by doing so, I will highlight the different levels of cultural complexity that can be reached in animal species, with a special emphasis on chimpanzees. Third, I will guide the reader through the ever-increasing diversity of cultural behaviors that we are discovering in chimpanzees. Over the years, it has become apparent that each chimpanzee population possesses a distinct behavioral repertoire. In fact, these differences are so clear that if you tell me how a chimpanzee behaves, I can tell you where he or she comes from. This is akin to a behavioral fingerprint! Fourth, using this knowledge, I will try to identify the specifics of the culture phenomenon in humans. However, to answer the question of "what makes us human," we should not forget the great sociologist Pierre Bourdieu's advice to concentrate on what humans do in practice and not what they say they do.

To better understand what culture is and how it is acquired, I will distinguish between three main cultural domains. The first domain, *material culture*, includes all aspects of culture that are concerned with the external material world that

[3] Some examples of discussions about human cultures can be found in Kuper (1999), Barnard (2000), and many other publications.

individuals encounter daily. Most of the work that has been published so far on animal culture has focused on this domain and through such studies it has been possible to use artifacts such as tools to document cultural diversity in a variety of species, including chimpanzees, orangutans, and New Caledonian crows.[4] The second domain, *social culture*, includes all social aspects that allow individuals to develop and benefit from the advantages provided by living in social groups. The third domain, *symbolic culture*, encompasses all means of communication, in a very general sense, between individuals during their social interactions. These last two domains have rarely been considered in primates, although they have been more systematically discussed for sea mammals, like whales and dolphins, and for birds.[5] These three domains are, of course, not mutually exclusive; for example, a symbolic cultural behavior may include an object of the environment or a social partner. However, all three domains encompass primarily different substrates – tools, social partners, or arbitrary meanings – and rely upon different cognitive abilities, and, most importantly, they are acquired with the help of different learning mechanisms.

As a general rule, the more we learn about animals and their behavior, the more we are forced to redefine "what makes us human." In this book, I hope to familiarize the reader with some of the chimpanzees that I have known over the last 30 years and through them, share how culture affects the way they live. Indeed, early propositions of what defined humanity, such as "Man the hunter," "*Homo faber*" – the tool user, "Cultured Man," "*Homo reciprocans*" – the sharer of food, the Cooperator, and the Mind Reader[6] have all been challenged by observations of wild animals. But our knowledge of animals is still limited in comparison with our knowledge of humans and thus, one essential approach to learning what really makes humans unique is to study animals in the wild and consider all the evidence they present us with.

Cultural biases and scientific progress

All of us, myself included, are influenced by our personal experiences. Those of us who had a strongly religious upbringing might have difficulty with the idea that, for millions and millions of years, we shared a common ancestor with chimpanzees and many other mammals. On the other hand, those of us trained as biologists sometimes have difficulty imagining that human behavioral patterns might not have a precursor form in one of our closest animal relatives. In a similar

[4] For chimpanzees, see McGrew (1992, 2004), for orangutans, see van Schaik and Knott (2001) and van Schaik *et al.* (2003) and for New Caledonian crows, see Hunt (1996) and Hunt and Gray (2003).

[5] The main exception for primates has been the work of Sapolsky (2006) on baboons. For whales and dolphins, see Whitehead (1998) and Rendell and Whitehead (2001).

[6] "Man the Hunter" (Lee and DeVore 1968), *Homo faber* (Leakey 1961, Mithen 1996, Wolpert 2007), *Homo reciprocans* (Bowles and Gintis 2002), Cooperator (Henrich *et al.* 2006), Mind Reader (Meltzoff 1996, Tomasello *et al.* 2005).

way, experimental psychologists have been trained extensively to think in terms of experiments and they remain skeptical of any observations that cannot be duplicated in an experiment with controlled conditions. With a parallel bias, mathematicians tend not to accept a conclusion that is based on observations if it is not supported by a mathematical model confirming the causal link between a parameter and an effect. Those who observe animals in the wild, in contrast, tend to be skeptical of captive observations if they are not supported by natural observations, and stress the limitations of captive studies as being unnatural and therefore difficult to generalize. All approaches have their benefits and limitations and the key to making progress is to be aware of them and combine the beneficial aspects of each approach.

Our brain should see through our eyes. As a fieldworker, I tried to accept what I observed about the behavior of an animal species in the wild without being influenced by my education, experiences, and beliefs. What determines how chimpanzees behave is not what we think they are capable of doing; we should document what we see in a detailed way without preconception. The challenges encountered in the dense rainforests of Africa are very different from those encountered in the open savannas, but this only became apparent after observers compared their observations of chimpanzees finding fruit or a driver ant nest or hunting for monkeys in both environments. To me, a city boy without any such prior experience, this was a lengthy education.

However, you see only what you know! This learning to see is a painstaking experience that can move one away from what you had previously learned to be "true." Once during my early days in the Taï Forest, I was following Brutus and Falstaff, the two oldest male chimpanzees of the study community, on a hunt when they suddenly passed behind me and disappeared in the dense undergrowth. I was totally astounded; why were they running in the seemingly wrong direction? It took me many months and many more hunts until I realized that they were anticipating what would happen during the hunt and knew that under the pressure of the other hunters high up in the tree, the prey would try to escape in a particular direction, which they would then head towards to close the trap. I was lucky enough to study a chimpanzee population that lived in a habitat where group hunting was required and that included two old, gifted hunters like Falstaff and Brutus, who would repeatedly perform complicated anticipatory hunting movements for me to learn to see and understand them. Descriptions of the cooperative hunts by Taï chimpanzees can read like a foreign language if you have not experienced them firsthand. Furthermore, some might dismiss such descriptions as purely anthropomorphic[7]

[7] "Anthropomorphism," defined as the attribution of mental states or properties to nonhuman animals, has been criticized by some as being a non-scientific explanation of animal behavior (Keeley 2004, Goodrich and Allen 2007, Wynne 2007). It has then been used to dismiss the work of animal advocates, who they believe are too emotional about and believe they see human feelings in animals. As such, anthropomorphism would prevent observers who have spent too much time observing animals from making rational judgments. On the other hand, some argue that a certain level of anthropomorphism, based on empathy and intuition, are essential to deciphering animal behavior patterns that

and not realize that they might be the ones wearing the anthropocentric glasses that lead them to place humans as the reference for all comparisons.

Similarly, too many still believe that the western way of life is representative of the entire human species, but this is simply not true. In my opinion, scientists who hold such ethnocentric beliefs are in no way helping to improve our knowledge of "what makes us human." Only without these glasses can we see, for example, that humans throughout the world have very different ways of understanding distances and directions and have developed very original solutions to orient themselves under different socio-economical situations. Similarly, humans belonging to different cultural groups exhibit an incredible diversity of responses to tests for altruistic propensity and willingness to enforce rules through punishment. As we will see later in this book, there is an abundance of persuasive examples of cultural solutions in different cognitive domains in humans. In many instances, it is actually our western way of life that is different from the way humans think throughout the rest of the world. No one way of life should be considered more valuable than any other, as all are responses to the socio-ecological conditions faced by each of these populations. Leaving ethnocentrism aside, what is so fascinating about humans is how much more diverse our range of possible solutions seem to be compared to chimpanzees and any other species (we will come back to this point later).

Finally, we see only what we want to see. It requires a special effort to put aside our anthropocentric and ethnocentric glasses – after all, we have "always known" that our food tastes better, that our grass is greener, that domestic cars are more reliable, and that our music is the most harmonious – and open our eyes without being judgmental to the richness of human cultural differences and to the behavior of animals. Humans are the judges when it comes to comparisons with other animal species and this makes us prone to hasty or unjustified claims. In this book, it is my intention to consider all our knowledge on chimpanzees and to show that, with the help of all observations, a productive distinction can be reached that pinpoints the unique cultural abilities of humans and chimpanzees.

To address these issues, I propose to take you on a trip into the tropics and into the lives of the chimpanzees, where we will follow some individuals as they develop in their respective cultures. I will first review some of the different approaches to animal culture, starting with my personal experiences when I worked in Africa (Chapter 1), and then try to explain why "culture" is such a hotly debated concept and why the resistance of some to accept it in nonhuman animals is so great (Chapter 2). I will then present some evidence for what chimpanzees have achieved in the material culture (Chapter 3), social culture (Chapter 4), and symbolic culture (Chapter 5) domains. To finish this review, I will concentrate on how cultural traits are taught and acquired in different chimpanzee and human groups (Chapter 6).

are new and unexpected (de Waal 2001). Finally, critiques of anthropomorphism often come from people who are anthropocentric themselves and therefore extremely reluctant to concede any so-called "human prerogatives" to nonhuman species and too often anthropomorphism is used as a last-resort argument to avoid considering the evidence.

Next, I will discuss how chimpanzees react to death and what this indicates about their feelings (Chapter 7). Then, through a direct comparison between chimpanzees and humans, I will explore the relationship between culture and cognition in both species (Chapter 8). Finally, I will revisit the question of human and chimpanzee cultural uniqueness (Chapter 9).

1 Studying culture in the wild

Every circumstance concurs in demonstrating that the brutes are actuated by appetite only, and that man is influenced by a superior principle. The only doubt that remains is the difficulty of conceiving how appetite only should produce in animals effects so similar to those produced in men by intelligence

George de Buffon (Natural History, History of Man, vol. 3, 1748, English translation 1812)

In the 4th century, Diogène asked Plato what a man is. Plato said, "It is a bipedal with a naked skin." Diogène left the room and came back a bit later with a featherless chicken and told the audience, "Here, Plato's man!"

In 1979, just a short time after beginning my PhD research on the chimpanzees of Taï National Park, I experienced my first culture shock. Our friends, the Sangbé, a Baoulé family who lived near the park boundary, were very supportive of us living in the middle of the forest and "working" with wild chimpanzees. The Sangbés came to live there because it had become impossible to sustain themselves in their home village, which was located in the progressively drying savannah region in the middle of Côte d'Ivoire. Rains were much more abundant close to the large Taï forest and they felt privileged to be able to grow cocoa, yams, bananas, and Raphia trees, which are used to make Bangui wine. Their hospitality remains one of our greatest memories of life deep in traditional Africa. I remember us gathering around the petrol lamp under the starlit sky in their village and dancing to the music of their second-born son, N'Guessan, with his homemade guitar. Mama Sangbé held our son in a "boubou" tied around her back, and N'Goran, their first-born son, drummed on an empty plastic gasoline jug that held the Bangui wine we drank until very late in the night. Without ever asking us, they knew that we did not eat primates and they declared them to be our totem. N'Dri, their third son, often gave me the best parts of the meat in honor of my status as a guest, so I received the heads of the duikers, the long hairless tail of the big Gambian rats, the crest of the chickens, and the eyes of the goats. I was always jealous of Hedwige, my wife, who received the lesser-esteemed parts like goat and duiker legs, or the white meat of the chicken. It took me many years before I realized that I was not forced to eat these prized pieces as long as I recognized the honor they were bestowing upon me and would sacrifice some of my Bangui wine to the ancestors by pouring the last drops from my gourd onto the floor. Once it had gotten late and we were ready to take our leave, following the Baoulé tradition, we asked "for the road" so that we would find our way home. More Bangui wine would then appear from the dark and our hosts would offer us only half the road so that we would be sure to visit again.

Once in the forest, another culture shock awaited me. When I first saw chimpanzees, I found it impossible to distinguish between all the furry black shadows in the dark forest and this made me panic; how on earth would I be able to study their social interactions and follow their behavior? Hedwige shared my dismay at this challenge. As if this was not enough, the Taï forest was so dark and so dense that after taking 20 steps in, you had no idea where you were and could very easily get lost. Valentin, a local Guéré hunter who was our first guide, felt very confident in the forest and continually saw interesting plants and animals. However, he was totally unconcerned about bringing us back to our camp, and would happily bring us to his village, a good 6-hour walk away, instead. It was not that he had a poor sense of orientation; in fact, his amazing navigational skills allowed him to walk in a perfectly straight line for kilometers without the help of any technological devices as long as the sun was out, a feat no city boy like me could even dream of. It was possible that the time he spent as a child in the forest with his father had taught him about moving in the forest and he acquired a sense of orientation that, in comparison to my own, made me feel like a broken weathervane.

Understanding cultural diversity first requires modesty. Fresh after graduation from Geneva University, enrolled in a PhD program in one of the best Swiss Universities, and armed with a three-year grant from the highly competitive Swiss Science Foundation, I had the perfect plan to understand the nut-cracking behavior of the Taï chimpanzees. They were one of the rare African chimpanzee populations that were believed to use stone hammers to crack nuts. There were no eyewitness reports of this behavior, only accounts from local people who had attributed signs of nut-cracking in the forest to the chimpanzees. I had come to Côte d'Ivoire to elucidate this. But when I arrived in the middle of Africa, in the extraordinary Taï forest, all my knowledge about mathematics, statistics, chemistry, and evolutionary theory did not seem to be of great use. I could draw a perfect map of the park from satellite images, but would be irreparably lost in it the minute I lost my compass.

To observe chimpanzees, one first has to find them, and due to the dramatic increase in the number of humans in Africa, all of them hunting in one form or another for meat, they survive only in the most remote places on the continent. To reach such places, I needed the help of local guides who knew the area. We went looking for information about intact populations of chimpanzees near the village of Keïbly, north of Taï. The villagers all agreed that they could be found near their village, but when I asked where exactly, it was difficult to gain a clear answer. Incurably optimistic, Hedwige and I decided to try to find them. Our guide, Valentin, a former hunter and game warden, knew the region and pointed to the south as the place where we needed to go, but we first headed east and then more northeast, between cocoa fields intertwined with banana and yam plantations. We greeted some farmers as we crossed their "campements," which consisted of a couple of huts they used during the week, and each time, Valentin stopped to accept a drink of the palm wine they had offered us. Time passed quickly, which naturally concerned us city folk, but Valentin insisted that it would be rude to refuse their hospitality. As we continued walking for hours in different directions, Valentin, who was invariably confident about our route, happily pointed in what my compass showed to be a different direction each time. The passing time had me worried. Eventually, I confronted Valentin and implied that he was playing tricks on us and did not actually know anything about the chimpanzees. He was clearly hurt by my suspicions. "Just one more hour and we'll be there," he assured me. Not wanting to start a fight in the middle of nowhere, I agreed, but I remained nervous as he did not even have a wristwatch. We saw many more small forest patches intermixed with cocoa fields but no traces of chimpanzees anywhere. Four hours later, Valentin was still as optimistic as he had been in the morning but by then, I was fully convinced he was tricking us.

It took me many years to realize that, to Valentin, like the other people in the village, it was more important to please me by agreeing that chimpanzees were nearby than to tell me the bleak reality – that only farmland remained around the village. In their culture, hospitality is imperative, especially when dealing with a visitor from afar. In addition, traditional farmers like Valentin have a notion of space that is topographic and therefore based on their knowledge of the fields surrounding their village, and they do not extract direction precisely even though they are perfectly able to recall all the types of fields and campgrounds along each path between the village and whatever point you name. Such a notion of space works perfectly well for their needs, but as it is not based on a system of maps and compasses as we had learned to use at school, I was not prepared to understand it. Finally, as he had never had a watch, he was not measuring time in hours like I was used to doing, but instead he used the position of the sun and was generally much more relaxed about time than I was trained to be. I was completely naïve about these three simple cultural differences and this transformed my interactions with locals into a long string of miscommunications.

Cross-cultural communication remained challenging. Over the years, I realized that the chimpanzees were susceptible to a variety of diseases. In May 1999, 12 adult and infant chimpanzees died from an unknown disease. To prevent the possible transmission of such disease outbreaks from the human observers to the chimpanzees, we introduced the use of face masks to prevent aerosols, which are air droplets that we expel as we cough, from infecting the chimpanzees. Furthermore, we wanted to avoid leaving human-introduced worms in the forest, which could also potentially infect the chimpanzees; to speak plainly, this meant that if a member of the chimpanzee project had to relieve him- or herself in the forest, they had to collect their feces and bring it with them when they returned to camp. It was not easy to explain to my 18 local field assistants that diseases are not caused by divine curses, but instead the result of contact with pathogens and that we could inadvertently infect the chimpanzees with the pathogens we carry even if we are not yet ill. But when I told them that they would have to collect and carry their feces back to camp, I was met with dead silence. Finally, after a lengthy discussion, one of them explained to me that one of his uncles had died after someone had secretly placed feces under his bed. A second told me a similar story about a cousin who had died because some of his feces had been stolen. In other words, in their minds, I was asking them to risk their lives to do something that might not even benefit the chimpanzees! We discussed this matter for days and although they still did not believe me, in the end they understood that my request was serious and, out of respect for me, they decided to change their diets so that they would not have to relieve themselves during the time they spent in the forest.

To study animals, all you need is love

During our first two months in Taï forest, we would rush towards any chimpanzee calls we heard, but we never saw them as they would run away if we made even the slightest sound. I had known it would be hard to see wild chimpanzees but I had not realized it would be so frustrating. Friends back in Zürich had warned me that it was crazy to try to do a PhD on wild chimpanzees and that I would never finish it, but Hedwige and I loved to be in this magnificent forest and we were happy to try to decipher the minds of the chimpanzees, particularly when it came to their special nut-cracking behavior. The question we initially set out to resolve was how these chimpanzees were so consistently able to choose the most efficient hammers for cracking each of the five nut species they ate.

The Nobel Prize-winner Konrad Lorenz said that only those who love animals are willing and able to endure the prodigious amount of observation time that is the necessary basis for understanding animals.[1] We did not realize it at the time, but by going out into the forest every day to track these black shadows, we were a live illustration of this. At the same time, Hedwige and I felt like two "Sherlock Holmes" in the forest; we found only remnants of cracked nuts and anvils, and occasionally hammers, wooden clubs, and stones, but we knew that if we looked at them long and hard enough, we could open a window into the minds of the chimpanzees who had been feeding on those nuts. Did chimpanzees know that they needed to select heavier, harder hammers to crack the harder nuts? How did they choose which hammers to transport to nut-laden trees? Could they really have some mental memory of tree and hammer locations in such a dense forest? By carefully reading the traces and fresh leftovers, we felt sure that we would be able to find the answers. Nut-cracking turned out to be a miraculous help to us. During the 4-month long nut-cracking season of the abundant *Coula* nuts, we used the hammering sounds as our guide. When a chimpanzee uses a 1–4 kg hammer to crack one of these hard nuts, it produces a loud pounding sound that a careful ear can hear from almost a kilometer away. An individual typically has to hit a nut five to seven times before it cracks open, and we would use these time lapses to move forward unnoticed and then freeze while the chimpanzee ate the nut. We would move forward again when

[1] Konrad Lorenz's book, *King Solomon's ring* (1952), was an inspiration for young naturalists of my generation.

the chimpanzee started hammering the next nut until, finally, we would reach them. Frustratingly, during our first three years there, they always ran away. But as the individuals fled, we were able to sex them thanks to the striking pink skin that surrounds the genitals on the females and age them from their stature. We then looked at how many hits were needed to open a nut and how many nuts were eaten per minute by individuals of each sex and age and this was what turned out to be the basis of my PhD thesis – a study about chimpanzee backsides! This was not really what I had envisioned myself studying back when I was in university.

Chimpanzees, like many other species, have been exterminated in many parts of Africa, first, as a result of humans' destroying their habitat, and second, because they are regularly hunted for food. So it is not surprising that they are wary of humans approaching them, even if that human comes all the way from Switzerland. When faced with a gun, an individual does not have much time to learn and chimpanzees understandably have a hard time distinguishing between human friends and enemies. So, the adventure of trying to understand our close cousins from the forest always starts with many frustrating months and years of trying to overcome their natural fear of humans. This process of habituating chimpanzees to human observers took us five years, but we were lucky; some of my colleagues were unsuccessful even after trying for seven years or longer. Such a long investment, which must be made before you can even start to observe the animals, is part of what Konrad Lorenz referred to when he said you need to love the animals you study. Thanks to our newly acquired detective skills, we were quickly able to identify and name the fleeing chimpanzees: Falstaff was the only male with a whitish left eye and Brutus was the big male with the tops of both ears bent backwards. Saphir had a small sexual swelling that seemed irresistible to the males while Salomé had a huge swelling. Tosca had a large, lateral cut on her right ear and a small son, Tarzan. It took us more than three years to realize that the community we were following, which was led by Brutus at that time, was 80 chimpanzees strong.

Once the chimpanzees were habituated to our presence, they set our working hours. They rise at dawn every day of the year and remain active until sunset. They can be completely silent for days, so we needed to be under their nests before they leave them in the early morning or we might spend the day alone in the forest. They forage for 12 hours and cover anywhere from 4 to 8 km of their 20 km^2 territory in a day and we followed them until they constructed a new nest for the night. After nesting them, we usually had a good hour-long walk back to camp at night through the forest, always wary of twigs that could metamorphose into snakes. Back in our small home, we cleaned up and typed out our data, ate dinner, and relaxed for an hour before falling asleep, only to awaken well before dawn the next day and rush out to meet the chimpanzees before they left their nests. This might seem like a rough life, but once we were able to follow the chimpanzees at close range, we shared our days with them and felt privileged to discover their special nut-cracking behavior, their sophisticated cooperation during hunts and intergroup encounters, and their many stunning acts of altruism. Even now, after 30 years of studying these chimpanzees, we still find ourselves surprised by new and unexpected behaviors.

First steps towards chimpanzee culture

However, my real first steps toward understanding cultural diversity in chimpanzees happened when I was lucky enough to visit other chimpanzee populations. I had read about the behavior of the chimpanzees of Gombe and Mahale in Tanzania, but reading alone does not sufficiently show you exactly how differently groups behave. I was privileged to see those behaviors with my own eyes and evaluate exactly how much they differ from the behavior of the Taï chimpanzees. When I contacted Jane Goodall, she was very supportive and invited me without hesitation to follow the chimpanzees of Gombe to study how their hunting strategies differed from those I had seen in Taï.[2] This first cross-population experience really opened my eyes to the cultural dimension of chimpanzee behavior.

Gombe National Park, Tanzania, May 1992

Prof and Frodo are wandering on a ridge on the high slopes in the center of their territory. Far below, the sun is transforming Lake Tanganyika into a silvery mirror, on which the small white triangular sails of the fishermens' boats are scattered. Suddenly, the silence is broken by the high-pitched screams of a group of red colobus monkeys in a woodland forest patch below. Exchanging a glance, the two males move silently down the slope and enter the forest. Arriving under the group of colobus monkeys, they look for the part of the group with the most females and babies, their favorite prey. Then, Frodo, the best hunter in the group, climbs a 20 m tall tree, which causes the panicking monkeys to disperse downhill in all directions. Frodo rapidly concentrates his efforts on two females, each of whom carried a baby on their belly, while I see Prof speed downhill and quickly climb a tree in pursuit of a juvenile monkey. Frodo, who is not at all frightened by the females' threats, dives and grabs the baby from the belly of the closest one, sits close by on a branch and silently eats it.

I run down to look for Prof just in time to see him descending the tree empty-handed. He continues downhill towards the river in the valley without a glance at Frodo, who is still up in a tree eating the baby. A few minutes later, he arrives at a fully active nest of driver ants, which live in colonies of many millions with soldiers with powerful biting mandibles. Standing on a rock above the entrance, he breaks off a 120 cm long stick and inserts one end into the swarm of ants at the nest entrance. They immediately attack the stick and start climbing up it. Once the ants have climbed 25 cm up the stick, Prof swiftly pulls it out of the entrance, turns it upwards with his left hand and pulls the stick through his right fist, thereby collecting the ants in a ball. Moving quickly, he brings the lump of ants into his mouth and chews them vigorously.

[2] In the early 1960s, Jane Goodall started her pioneering project on this population of chimpanzees in a park bordering Lake Tanganyika in western Tanzania. This developed into the longest-running research project in the wild and has produced many skilled students. Jane Goodall was supportive of my research from very early on and invited me to visit her research site three times to conduct a very instructive comparative study of the Gombe and Taï chimpanzees.

Taï National Park, Côte d'Ivoire, October 1989

Brutus, Falstaff, and a few younger adult males are moving through the forest when we hear the calls of a group of red colobus monkeys further ahead. When the chimpanzees reach the monkeys, which are high up in several 40–60-m high emergent trees, Snoopy, one of the youngest chimpanzees in the group, climbs very slowly directly under the monkeys. When the monkeys discover Snoopy approaching, they begin to move away while Snoopy continues climbing until he is about 15 m below them. Another male joins Snoopy and, together, they walk out onto the high branches of the tree, which pressure the monkeys to move in one constant direction. These two are moving far too slowly to successfully capture the monkeys or force them to fall. But Brutus and Falstaff, who have been intently watching what is going on above, start to run in the direction of the colobus' escape path and remain ahead of them. Brutus continues in the same direction while Falstaff veers off slightly to the south and climbs a tree ahead of some monkeys that have fled in that same direction. Once high enough, Falstaff sits on a branch and stares at the monkeys. His presence blocks the possibility of an escape for those who have been herded by Snoopy. Suddenly, Brutus dashes away in the same direction. As I follow him, I see him climb a vine on a large tree, being very careful to avoid moving any branches. When Brutus emerges into the canopy, the first monkeys who have been heading away from Falstaff and Snoopy arrive, and they now find themselves blocked by Brutus, who dashes towards them to close the trap. However, the trap fails as the monkeys manage to escape on the thinnest, highest branches above Brutus.

The hunters give up, descend, and start to move northwards, where they encounter a driver ant nest. Brutus uses two fingers to open a small hole in the loose soil and the soldier ants swarm slowly out. He breaks a 35 cm long stick off a nearby sapling and inserts one end directly into the nest. Once the ants have climbed 5 cm up the stick, he swiftly puts the stick into his mouth, closes his teeth tightly, pulls the stick out, and chews vigorously. He repeats this for 15 minutes and Falstaff and Snoopy join him, using the same technique, for 8 minutes.

It struck me how similar and yet completely different the Gombe and Taï chimpanzee hunting techniques were. Seeing Frodo and Prof hunt in 1992, after I had followed Brutus and Co. on so many hunts in the Taï forest, showed me how flexible chimpanzee behavior can be.

The differences in the ant collection techniques, also called ant-dipping, of the Gombe and Taï chimpanzees are also very instructive because chimpanzees at both sites eat driver ants. I have personally tested both the long and short stick techniques at both sites and I was not able to determine why chimpanzees at one site would prefer one technique over the other. Admittedly, there were some painful practice attempts before I could successfully eat the large mandible-equipped ants without them first biting me, but young chimpanzees must also learn to put the ants well inside their mouths so that the ants do not bite the soft skin on their lips and cheeks, then chew really fast while moving the ants around in their mouths so that they do not have time to bite them. Once they have learned these techniques, they can enjoy the taste, which, for me, is comparable to tasty raw shrimp. My point is that I was able to use both Gombe and Taï techniques at either site and could not find any differences that would preclude the use of one of the methods over the other. However, the two techniques are not equivalent; in fact, the Gombe technique allows a chimpanzee to eat four times as many ants per minute.[3] So the question is not, why do the Gombe chimpanzees collect ants with a long stick, but, instead, why do the Taï chimpanzees use the less-efficient method? If these chimpanzees were simply looking for the optimal ant-dipping technique, they would all use the Gombe technique to catch ants, but they do not.

The Gombe chimpanzees' way of hunting monkeys and catching ants differs from the Taï chimpanzees' way. In both cases, the solutions the chimpanzees came up with while trying to access the same prey are only partly influenced by ecological differences. In its simplest form, this is what culture is about: population-specific behavior patterns that are socially acquired. Large parts of the culture debate have concentrated on determining whether a behavior is really socially learned or whether it could be explained by ecological differences, even if they are insignificant or presently absent. When one considers how many human cultural habits are so obviously affected by ecological conditions – for example, clothing habits reflect climatic conditions and cooking traditions are influenced by which plants and animals are available – it seems strange that this should critically challenge the presence of culture in animals. A skeptic could argue that Prof had no choice but to use a long stick if the nest structure of the ants was much deeper underground in Gombe than in Taï. For that reason, I tested both short and long sticks in both locations and was able to reject this possibility in both cases.

About animals, ignorance, and anthropocentrism

Aristotle and Descartes were the two most famous proponents of the idea that animals are mere machines, while humans are thinking beings with souls. One must concede that *neither of them* had information about the behavior of animals in the wild and we can only guess what they would have concluded if they had known then all we know now about the sophisticated behavior of many animal species. Classical thinkers could only speculate about the true abilities of animals and as they lived during a time when the dominance of humans over all other life forms was dogma, not the least because of religious beliefs, we can understand how they might have come up with such propositions. For centuries, intellectuals developed their ideas about animals on the basis of the very sparse observations made by explorers. Knowing how curious humans are about the world around us, this must

[3] When I visited Gombe in 1992, I was impressed by the number of ants that held onto the sticks while the Gombe chimpanzees dipped them and I began to count the number of ants the chimpanzees were able to eat per minute of activity using the two different techniques (Boesch and Boesch-Achermann 2000).

have been a frustrating and unsatisfying situation. As such, one would expect that scientists and thinkers would have greeted any reliable information about wild animals with enthusiasm, but this was overwhelmingly not the case. For example, Jane Goodall's first observations in the early 1960s of Gombe chimpanzees hunting for meat were challenged and some suggested that such behavior was abnormal and the result of Goodall provisioning the chimpanzees with bananas early on in her work there.[4] Similarly, her observations of chimpanzees spontaneously using tools were rapidly dismissed by many. After all, at that time, tool use and hunting were still thought to be uniquely human characteristics.

Why is the idea of human dominance still so strong? To me, having immersed myself for so many years in the life of another species, it is sometimes shocking how careless scientists can be. Many eminent psychologists, philosophers, and economists who compare humans with other animal species in a quest to find what makes us unique have not worked with the animal species they write about. One would expect that they would, at the very least, read the relevant original literature before drawing any conclusions but, unfortunately, this is rarely the case. When talking about those who doubt the mental capabilities of animals, Frans de Waal, the famous primatologist, said, "What makes such critics unfathomable to me is their total absence of humility when faced with a group of animals they have never worked with."[5] I can only share his reaction and be mystified at some colleagues' lack of curiosity. At a conference about the human mind in Denmark, a speaker was explaining to us that, unlike chimpanzees, humans have a "unique sense of fairness," an idea she further developed in the rest of her talk. At the end of her talk, I asked which data she based these claims about chimpanzees on. She admitted that she had no idea, but had heard someone else say this before. On another occasion, during an international psychology congress in Germany, I was in a debate with a colleague in front of 400 delegates and he claimed that recent data showed that the differences in ant-dipping behavior in chimpanzees had now been proven to be due to ecological differences in ant behavior. Problematic to his argument was that I was a co-author on one of the papers he cited and associated with a second paper and each clearly stated that some differences observed between chimpanzee populations could not be explained by ecological differences (see also next chapter).

After 40 years of reports of tool use by chimpanzees from all populations throughout tropical Africa, it is mystifying to read that some still deny them such abilities. One such skeptic wrote, "It is not that chimpanzees lack visual imagination or are unable to learn complex tasks by trial and error, but they do not reason about things. They have, for example, no concept of force, and even worse, no concept of

[4] Some authors have argued that Jane Goodall artificially provisioning the Gombe chimpanzees with bananas led to a nutritional imbalance, which the chimpanzees had to compensate for by hunting for meat (Power 1991, Clark 2002, Hart and Sussman 2005). This idea stuck around for many years, until observations of hunting behavior in populations of chimpanzees that had not been artificially provisioned, such as the Taï and Ngogo chimpanzees, were published (Boesch and Boesch 1989, Mitani *et al.* 2002).

[5] Citation is from Frans de Waal (2001).

causality … One may illustrate the differences in chimpanzee and human thinking with the claim that an ape seeing the wind blowing and shaking a branch till the fruit falls would never learn from this to shake the branch to get the fruit".[6] Wow, talk about a sweeping claim. We have had many human visitors in Taï who have tried to crack nuts and, if they succeeded at all, they often ended up smashing the nut into pulp. A female chimpanzee, on the other hand, is perfectly able to crack open nuts and leave the kernel perfectly intact. Doing this, she can eat two nuts per minute for hours at a time, which leaves me wondering who lacks the concept of force. The skeptic's final comment about the blowing wind was a citation from another scientist with equally little experience with animals in the wild. In fact, in the Taï forest, we have often observed chimpanzees shaking small trees to make caterpillars fall onto the ground, where they collect and eat them in bulk. Similarly, during the most intense moment of pursuit in a hunt, hunters regularly forcefully shake the branches their prey is on to either prevent them from jumping away or to make them misstep and fall to the ground, where other chimpanzees are waiting. Misrepresenting facts or perpetuating hearsay is too often done for the sake of a scientific argument guided by personal convictions.

Luckily, not all philosophers and scientists are anthropocentrics. The most famous exception, Charles Darwin, noted in 1871 that "the more the habits of any particular animal are studied by a naturalist, the more he attributes to reason and the less to unlearnt instinct." A century earlier, the French naturalist Charles-George Le Roy wrote: "You know, Madam, I held that it comes to the hunters to appreciate the intelligence of animals. To know them well, one has to live in society with them; and most philosophers are blind to that. To know animals, you should not consider isolate cases. What is important to consider are their daily activities; it is the whole of the actions modified by the circumstances."[7] Over 300 years ago, some had already realized the importance of natural observations and of observing animals that are as undisturbed as possible. Fortunately, in the last 50 years, many have taken up this challenge, and for the first time in centuries, we are in a situation to base our discussions about differences between humans and other animal species, including chimpanzees, on detailed and well-documented facts. In this book, I make full use of this luxury.

[6] Another extract from this "scientific" work: "Animals do not understand the causal relation between their acts and the outcome they experience. Apes for example, cannot select an appropriate tool for a simple physical manipulation without extensive teaching … Animals have very limited capacity for refining and combining objects to make better tools … Another important difference is that chimpanzees are slow to pick up skills from other animals. In essence, chimpanzees lack the technical intelligence needed for manipulating and transforming physical objects" (Lewis Wolpert, 2007, pp. 168–171). Note that for all these claims, we have direct observations of chimpanzees contradicting them, but the author makes no effort to consider such literature.

[7] « Vous savez, Madame, que je soutiens qu'il appartient qu'aux chasseurs d'apprécier l'intelligence des bêtes. Pour bien les connaître, il faut avoir vécu en société avec elles ; et la plupart des philosophes n'y entendent rien. » « … pour faire connaître les animaux, il ne faut pas tenir compte des faits isolés. Ce qu'il est important d'examiner, c'est leur conduite journalière ; c'est l'ensemble des actions modifiées par les circonstances, … » (*Lettres sur les animaux*, 1761–1781).

Nowadays, anthropocentrism can also take subtler forms, as is often shown in the one-sided attitude of some psychologists searching for human cognitive abilities in animals. Such an attitude is terribly reductive for we know the importance of ecology and experience on each individual and, therefore, need an approach to animals that does not place humans at its comparative center. The questions should focus on what it means to be a living being of any kind, rather than immediately restricted to some comparison with humans. Anthropocentrism, whether positive or negative, needs to be acknowledged and contained because it denies animals their own voice. Psychologists tend to ask, "Can animals do what people do, and if so, how?", while biologists tend to ask, "How and why do animals do what they do in the wild?" Thus, the contrast between traditional psychological and biological approaches is one between anthropocentric and ecological approaches.[8] For example, we should try to determine if memory as a cognitive mechanism is available to other species, not whether other animals have a specific human memory. It should be obvious that they do not since they are not human, and it is inappropriate to define memory as a phenomenon of the modern human mind rather than in terms of cognitive abilities.

Time to realize that Descartes got it wrong

Both Prof's and Brutus' methods for feeding on driver ants are essential to the culture debate. However, neither Gombe nor Taï chimpanzees are representative of all chimpanzees. For example, studies have shown that the Bossou chimpanzees in Guinea use another method entirely to collect ants, and that the tendency to hunt in groups at Ngogo in Uganda is dramatically different than in both Taï and Gombe. The main point is that we have to get rid of this tendency to associate an entire species with one "standard" population. The Gombe population is often regarded as the chimpanzee standard. This makes sense, as Gombe was the first of the ongoing long-term chimpanzee study sites and, along with its founder, Jane Goodall, this group has become world famous and produced more literature than any other site. But despite all of this, the tendency of the Gombe chimpanzees to hunt solitarily does not mean that this is how all chimpanzees hunt. As we will see in more detail in later chapters, this is only one of the ways chimpanzees hunt. This argument can be generalized to other species, like the orangutans, where some behavioral elements from the population living in Suaq Balimbing, Sumatra, differ from those of Ketambe, Sumatra, or from different regions of Kalimantan. Similarly, behaviors and calls of Atlantic dolphins differ from those of Pacific dolphins. René Descartes' simplistic views about animals are wrong and it is time to incorporate this clearly into how we think about animals.

[8] For more on this specific dimension of anthropocentrism, see Allen (2004), Shettleworth (1998), and Barrett *et al.* (2007).

To reject Descartes' ideas about animals also implies that we cannot simply consider animals living in captivity as equal to those living in the tropical rainforests of Gabon or the savannahs of Senegal. Captivity represents a set of specific ecological and social conditions that are dramatically different from many conditions in the wild. First, captivity presents the animals with a very *passive* environment, where everything is provided to them in a very restricted space and any challenges, dangers, and diversities that are found in the wild are suppressed. Second, captivity presents the animals with an extremely *impoverished* social life, where social challenges and stimulations are very limited. However, these socio-ecological conditions are real to the captive individuals and, as such, could reveal some of the species' potentials when living under special conditions. In addition, we should be aware that captive conditions also vary and they should also not all be considered equal. Most animals who live in biomedical facilities are kept in either total isolation or in very small groups. In contrast, most who live in sanctuaries and behavioral research facilities are kept in small peer groups, although some are kept in larger multi-aged groups, and a very small minority are kept in groups with more than one adult male. In contrast, wild chimpanzee groups comprise between 30 and 175 individuals of all ages, with many adult males and females, and live in territories ranging from 15 to over 30 km^2 with all individuals rarely, if ever, together at the same time. These are dramatically different social settings that play a crucial role in the development of social abilities (see below).

Recently, Duane Rumbaugh, Sue Savage-Rumbaugh and colleagues, who have specialized in working with captive chimpanzees, came to a very similar conclusion: "Our main thesis is that there is no average or universal chimpanzee. It is for this single reason that different laboratories are reporting very contradictory conclusions as to whether or not chimpanzees imitate or emulate the behaviors of others. This presents a very serious problem and a source of unwarranted controversy to no constructive end. The view that chimpanzees are chimpanzees and that one may take any of them and reach valid population-wide conclusions regarding the species is simply wrong."[9]

Human societies are also extremely diverse. They range from the Inuit in very cold environments to the Bwa pygmies in tropical rainforests. There are too many to list, but the important aspect to consider is that, too often, a deterministic attitude about humans prevails, where data collected from only one human group are systematically generalized to the whole species without any consideration for the important differences found in different groups. It is striking that 99% of the human subjects used in psychological studies come from western industrialized countries, where only 12% of the world's human population lives.[10] Furthermore, 80% of those human subjects come from undergraduate psychology classes. The rejection of Descartes' ideas about animals should come about with a recognition that, in humans also, learning is a very important force that molds our intelligence

[9] Rumbaugh *et al.* (2008, pp. 101–102).
[10] Arnett (2008), Henrich *et al.* (2010a).

in the face of the socio-economic conditions of our upbringing and adult life. When I consider the human side of the story throughout this book, I will bear such diversity in mind whenever data permit.

Synopsis

- Culture is probably the single most central concept in anthropology and, thus, has long been considered to define humans.
- Culture should be personally experienced or else, at first, it can be very confusing and hard to understand.
- When animals look at humans, they tend to see the enemy. Thus, observing them in their natural environment requires a lot of patience to overcome such a handicap.
- Culture is revealed from comparing the behavior of individuals belonging to different social groups living in their natural world.
- The human tendency to consider ourselves the "Crown of the Evolutionary Chain" has regularly prevented us from looking at animals in an unbiased way and has favored simplistic criteria to define our own superiority.
- Descartes' proposition that animals are mere machines is simply wrong and it is time we start to genuinely look at what animals are capable of.

2 From human culture to wild culture

It is the Man of the past that by force is predominant in us, as the present is only little compared to our long past during which we shaped ourselves and from which we resulted.

Emile Durkheim, *L'évolution pédagogique en France*, 1938.

The Grotte de Chauvet contains the oldest painting in the world, so man not only produces tools but also makes artistic objects.

Ryszard Kapuscinski, 2003.

Eight-year-old Sartre was sitting on a root, facing his mother, Salomé, and watching her precisely positioning *Panda* nuts so that she could hit them exactly along the crack and open the shell without damaging the nuts inside. I could not stop thinking about the numerous drawings I had seen of our naked, hairy ancestors using stones to open nuts or crack long bones to reach the marrow. Sartre and his consorts produced numerous stone flakes while using granite hammers to crack some of the hardest nuts in Africa. I excavated such a chimpanzee site with the help of an archaeologist and, similar to what you would find in sites where humans made and used tools, we found hundreds of stone flakes, most of which were unintentional by-products of the nut-cracking activities. As a *Panda* nut requires a compression force of 1600 kg before it will break, it is no wonder little Sartre needed all his strength; to succeed, he had to stand and lift his 6-kg stone above his head with both hands and use all the force he could muster to both forcefully and precisely hit the nut. In contrast, Salomé, a strong adult, made this exercise look much easier with her more adept use of the same hammer. Sartre put his left hand out in a begging gesture and Salomé handed him half of her nut. Sartre broke a thin branch from a nearby sapling, cut a 12 cm long piece with his teeth, and chewed on one end to sharpen it. He then used the stick like a fork to extract the almond pieces embedded within the broken nutshell. Like the first time I had seen this done by an adult female, Sartre made a tool in front of my eyes and it took him less than one minute to select a twig, make three modifications to it, and start to use it. This all happened with such ease, as if making a tool was the most natural thing in the world to a chimpanzee! I am now used to seeing chimpanzees use tools, but remain impressed by how easily they make them. Here again, my mind wandered off as I imagined our ancestors stone-knapping and flaking while youngsters nearby watched what was going on and progressively learned the technique. Having regularly cracked walnuts and hazelnuts myself in the forests of Switzerland, nut-cracking had seemed such a human activity to me!

Two hours later, I followed Salomé and Sartre as they came upon a huge fallen tree that had opened a nice clearing in the forest canopy. Chimpanzees love to rest and groom each other in such spots for long periods of time, but this time, they had not come here to rest. I noticed that Salomé and Sartre had made tools from saplings before reaching the fallen tree, cutting off leafy twigs, branches, and leaves until they were left with 20 cm long sticks. When they reached the tree with these tools in hand, Salomé rapidly removed a 1-m long piece of bark, and the two of them scratched the sawdust away to expose the bare wood. They were looking for the white dot of an exposed horseshoe beetle shell in which grubs grow and eventually develop into adults before flying away. Once they found one, Salomé and Sartre used their sticks to break open the shell and then dexterously extracted the 4 cm long white grub without harming it. They ate this delicacy together with some fresh leaves they had stripped from a *Coula* sapling. Both searched for grubs for over 45 minutes, peeling the bark away as they worked their way about 20 m along the fallen tree and successfully extracted about 20 grubs. Salomé and Sartre made 12 new tools of the same size while they were eating.

Without Salomé, I never would have realized that large beetles lay eggs that take weeks to grow into adult beetles under the thick bark of some large, dead trees. I can almost hear the philosopher Frederick Engels in his "Dialectics of Nature" (1934) say, "The most that the animal can achieve is to collect; man produces, he prepares the means of life, makes the environment serve his ends, and masters it." Were Salomé and Sartre humans or animals? What seemed obvious to me is that, thanks to their use of tools, both chimpanzees were able to access food that would have not been available to them otherwise. Tools allowed them to shape the environment to serve their needs.

After a short rest, Salomé and Sartre moved on. As we came near a large buttressed tree, Sartre tore away a handful of fresh leaves that he put in his mouth and chewed into a ball. He then ran to one of the largest buttresses on the ground and used the balled up leaves as a sponge in a depression where water had accumulated. He repeatedly put the sponge into the depression and then sucked the water out of it. After his fourth time, he had to rush through the undergrowth to catch up with his mother. Up ahead, Salomé broke off a piece of dead wood that had been hanging from a vine, exposing a round hole from which the sound of buzzing adult wood-boring bees protecting their nest could be heard. She quickly selected a 5-mm thick twig and cut it into a 25 cm long stick, which she inserted into the hole (see Fedora doing the same in Figure 2.1). She made forceful grinding movements with the stick as if she was trying to harm the adult bees inside. Twice, large black injured bees tried to fly out of the nest, but Salomé used her fist to smash them against the ground and ate them both by carefully squishing them between her molars. She then opened the dead branch lengthwise and ate all the grubs that were inside. During this time, Sartre watched her actions, the tools she used, and the bee nest from just 20 cm away.

In his 1762 "Natural History of Quadrupeds," Count George de Buffon wrote that humans differ from other animals in the sense that "the education of the child is

Figure 2.1 **Fedora uses a stick to extract wood-boring bee grubs from their nest.** The juvenile female had broken a dead branch at the level of the entrance to the nest (the round shape of the entrance hole can still be seen on the end of the branch she holds in her left hand), and then inserted the 20 cm long twig she had made to extract the 3 larvae that were inside the nest.

no longer purely individual, since his parents teach him not only what they acquired from nature, but also what they learned from their own parents and from the society to which they belong".[1] This corresponds precisely with what most modern psychologists would say characterizes modern human education. However, in many respects, Sartre is getting a great deal of information from Salomé about the dense forest and about where and how to extract food that, at a first glance, is either invisible or out of reach. In addition, nut-cracking, other use of tools, and many aspects of his life are specific to his group, and his mother is doing something that she learned from her own mother and that has been going on for many generations.[2]

We are touching on the heart of what culture is: a process whereby infants learn some specific behavioral patterns of their population from their parents and other group members. Just as we learn our culture from our families and friends, Sartre

[1] « … *l'éducation* de l'enfant n'est plus une éducation purement individuelle puisque ses parents lui communiquent non seulement ce qu'ils tiennent de la Nature, mais encore ce qu'ils ont reçu de leurs aïeux et de la société dont ils font partie; » (Buffon, 1762, p. 42).

[2] We were able to prove this for the nut-cracking behavior because these chimpanzees regularly use stone hammers, which, unlike wooden hammers or sticks, remain well preserved over time, even when buried. This study, the first archaeological study done with animals, showed that the Taï chimpanzees were nut-cracking as far back as 4300 years ago, which represents about 220 chimpanzee generations (Mercader *et al.* 2002, 2007).

was learning the specifics of the Taï chimpanzee culture while following his mother Salomé or admiring Brutus and Falstaff, the two big males that are his mother's friends. Even though this might seem relatively straightforward, coming up with a simple definition of culture has not been easy and entire books have been written in an attempt to define the concept (see Box 2.1). What can be viewed as minimally agreed upon, though, is the notion that culture is a *population-specific product* that is *learned from social partners* – we distinguish ourselves from others by our different languages, greetings, or cooking methods, for example. What is more difficult to agree upon is the content of these two concepts. For example, how abstract must these population-specific products be – such as greeting gestures, language, religion, myths, or institutions – and what type of social learning is required – such as mimicking, imitation, or specific forms of teaching – for them to qualify as culture? To some, this may seem less relevant than the existence of a population-specific product, but to members of the "human uniqueness side" and the "human superiority side," such nuances are essential and make all the difference.

Box 2.1: What is culture?

Culture is one of the central concepts of anthropology, and as such, attempts by anthropologists to define culture began early on. This has resulted in a deluge of definitions, which has made it difficult to reach a general consensus. I do not intend to review all present concepts of culture, but instead just want to give a sense of the dominant positions within different disciplines interested in culture. Historically, anthropology has been the first to use this concept and one classical definition was presented in 1940 by Robert Murdock (pp. 364–368) and included seven attributes.

- *Culture is learned.* It is not instinctive or innate or transmitted biologically, but is composed of habits acquired by each after birth.
- *Culture is inculcated.* All animals are capable of learning, but man alone seems able, in any considerable measure, to pass on his acquired habits to his offspring.
- *Culture is social.* Habits of cultural order are shared by human beings living in societies and kept relatively uniform by social pressure.
- *Culture is ideational.* To a large extent, cultural habits are conceptualized as ideal norms or patterns of behavior.
- *Culture is gratifying.* Culture always and necessarily satisfies basic biological needs and secondary needs derived therefrom.
- *Culture is adaptive.* Cultures change and the process of change appears to be an adaptive one by which culture tends to become more adjusted to the geographic environment.
- *Culture is integrative.* As one product of the adaptive process, the elements of a given culture tend to form a consistent and integrated whole.

In contrast, psychologists tend to concentrate on the mechanisms sustaining cultural acquisition and transmission. In other words, they focus on the first two of Murdock's seven criteria (Galef 1992, Heyes 1993, Tomasello 1990). From this, a classic psychological definition would include the following criteria.

- The naïve individual learns a specific behavior from social partners with a social learning mechanism granting some conformity in the copy.
- Alternatively, the model individual directly influences the learning of the behavior by the naïve individual (e.g. by teaching interventions).
- Culture is the result of an accumulation of modifications.

Culturally oriented psychologists, however, would add that "culture as a socially interactive process of constructing comprises two main components: shared activity with cultural practices and shared meaning with cultural interpretation." Therefore, "cultural learning is stimulated and adapted to a particular ecocultural niche" (Greenfield *et al.* 2003a).

Finally, biologists have started to show a growing interest in culture because of its independence from biological reproductive events and, therefore, its potential to contribute to more rapid evolutionary changes than a purely biological one (Kummer 1971, Bonner 1980, Wilson 1975, Maynard-Smith and Szathmary 1995). To them, a biological definition of culture would be centered on the following three elements:

- culture is a non-genetic method of transferring information between individuals;
- culture not only allows for vertical transmission (from parents to offspring), but also for oblique and horizontal transmission;
- culture is copied from other individuals.

Therefore, I propose the following *consensus definition* that could unify the concept of culture across these different disciplines.

- Culture is learned from group members.
- Culture is a distinctive collective practice.
- Culture is based on shared meanings within a group or society.

Tradition has sometimes been proposed as an alternative for culture, but with here again many differences between authors about what is understood under "tradition." As in both cases a behavior has to be learned socially and then transmitted to group members, many consider both terms as synonyms (e.g. Galef 1992, Fragaszy 2003, Whiten and van Schaik 2007). For some others, culture requires more elaborate social learning mechanisms, such as imitation and teaching, shows more accumulation of complexity or includes multiple traditions. In the present book, I will use only culture and consider it as including all these possibilities.

One last aspect to consider when defining culture is the difference of opinions about the function of culture. One group argues that cultural traits provide the individual with a direct benefit and that this explains why these traits spread within a population (e.g. Wilson 1975, Alexander 1974, Cavalli-Sforza and Feldman 1981). This has also been called the "gene–culture coevolution" theory, where culturally beneficial traits may lead to positive genetic selection (e.g. Lumsden and Wilson 1981, Boyd and Richerson 1985, Durham 1991). Others argue that beneficial cultural traits represent only a minor part of human cultural traits, and that, by and large, culture presents no direct advantage to the individual and might even be detrimental (e.g. Kroeber and Kluckhohn 1952, Sahlins 1976, Kuper 1999, Barnard 2000).

References for recent discussions of culture: Boesch and Tomasello 1998, Kuper 1999, Barnard 2000, 2011, Rendell and Whitehead 2001, Tuttle 2001, Alvard 2003, Boesch 2003, Fragaszy 2003, Greenfield *et al.* 2003a, Laland and Hoppitt 2003, van Schaik *et al.* 2003, Galef 2004, McGrew 2004, Laland and Janik 2006, Whiten and van Schaik 2007.

Culture and ecology in humans

"Try to imagine a way of life where land, shelter, and food are free, and where there are no leaders, bosses, politics, organised crime, taxes, or laws. Add to this the benefits of being part of a society where everything is shared, where there are no rich people and no poor people ...where there is no evidence that anyone ever made stone tools or weapons ... To this day they have not learned to make fire." With these words, Kevin Duffy introduces the Mbuti pygmies from the Ituri Forest in the Democratic Republic of Congo.[3] As he says, there are many possible reasons why the Mbuti forest people did not become more "advanced" over the last several thousands of years. Such forests often contain stones, but perhaps none of them easily flake into tools. In addition, there was probably no seasonal starvation or ice age to stimulate innovative technical changes. Nevertheless, the Mbuti culture has survived for many thousands of years, while so-called "great" civilizations like ancient Egypt, Greece, Rome, and others rose and fell. The specifics of the forest in which they live have made the Mbuti culture uniquely successful in terms of longevity, but technologically very simple.

Cultural diversity can be seen whenever one travels between the multitudes of different human societies that inhabit the planet. There are differences in all facets of life in these different societies, from differences in how people dress, speak, build houses, and educate children to differences in social norms and religion. As for the Mbuti people, contrary to some naïve and idealistic views that present human culture as being entirely disconnected from nature, many complex aspects of human

[3] *The children of the forest: Africa's Mbuti pygmies* (Duffy, 1984).

culture are related to the changing reality faced by individuals. A striking example comes from two human populations from Indonesia that relied heavily on whales for subsistence hunting. The Lamalera hunters relied on subsistence hunting for toothed whales but had strong taboos against hunting baleen whales, which they said were too large and dangerous. In contrast, their neighbors, the Lamakera hunters, which had moved towards a more mixed economy, were still hunting baleen whales but had strong taboos against hunting toothed whales, which they said were too large and dangerous. It was suggested that this difference emerged when the Lamakera hunters shifted to a more mixed economy, as whaling became less profitable in their fishing waters. Taboos in humans are social norms that limit, among other things, the consumption of certain food types. They often result from a once-hunted species falling out of the optimal diet so that continuing to pursue these species would provide lower returns.[4] Thus, even if pygmies in the Ituri Forest possess many food taboos, this results in only a very limited loss of calories. So, taboos may be cultural responses to changing ecological conditions.

However, culture in humans has led to the construction of cities and the invention of sophisticated travel technologies, refined musical instruments, and computers, and is the basis of many aspects of our internal lives, such as our beliefs, knowledge, myths, and religions. Many of these are very complex and elaborate cultural products that one individual alone could not have invented and that took many generations to become the products they currently are. To account for this reality, many have emphasized the cumulative dimension of culture, by which, over generations, human cultural products are elaborated on and improved and it is these improved products that are transmitted. Cumulative culture is what we observe today and in many cases, we are not aware of all the intermediate steps that happened on the way to products becoming the cultural items we have learned to use and rely on.

Different approaches to culture

Sartre, Salomé, and Brutus are constantly confronted with problems about what they will eat next, how they should behave when they encounter group members, or how they communicate with others. Sartre must learn to behave in a way that others will accept (if he greets another individual in the wrong way or at a bad time, he is very likely to get slapped), in a way that others will understand (so he can successfully attract the attention of the female with the attractive sexual swellings), and in a way that will allow him access to the food he likes (so he can crack nuts with a hammer or access ants despite their formidable jaws). Different scientific disciplines have tended to concentrate on different aspects of the challenges Sartre faces. Anthropologists concentrate on the question of "what" constitutes culture; that is, what the different elements of culture in any specific groups of humans they study

[4] For more on the whaling differences, see Alvard (2003); for more on taboos in humans, see Hill and Hawkes (1983) and Alvard (2003); for the pygmies of the Ituri forest in DRC, see Aunger (1994, 1996).

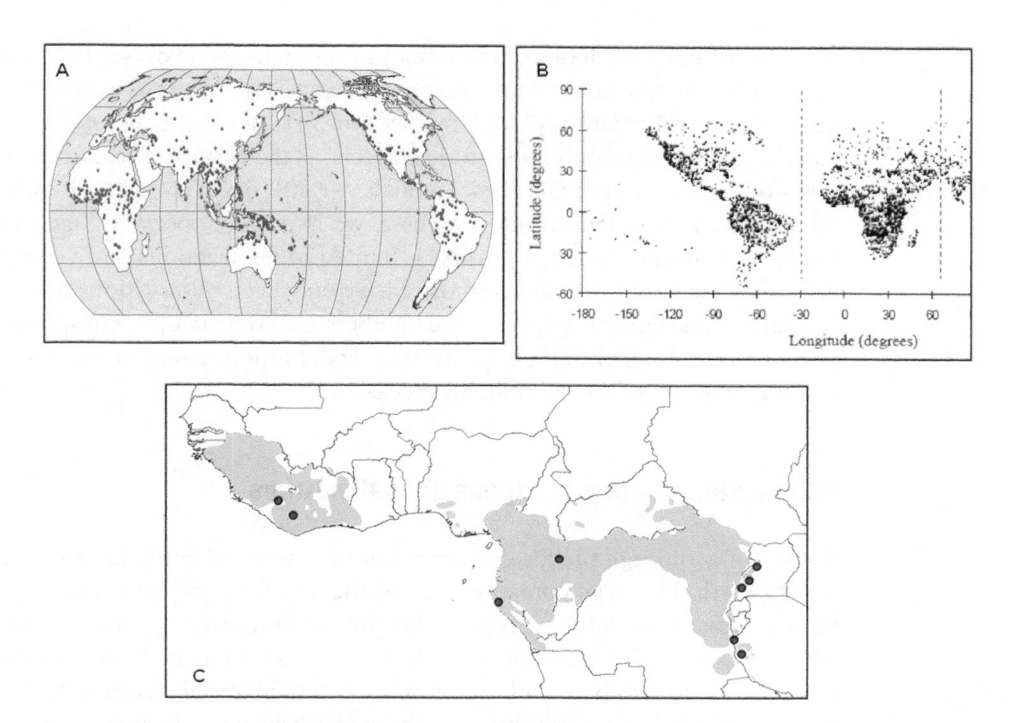

Figure 2.2 Our knowledge of cultural diversity in humans rests on data from hundreds or sometimes even thousands of different populations, while only nine populations of chimpanzees have been studied. (A) presents the 130 populations included in a study about the evolution of human languages (Atkinson 2011), while (B) presents over 2000 human populations that were included in a study about cultural diversity (Collard and Foley 2004). (C) shows the six chimpanzee populations that were included in the first large-scale study of chimpanzee culture (Whiten *et al.* 1999), plus three new populations that are now being studied.

are and how this is influenced by contact between different groups. They see culture as a phenomenon with content. Psychologists, on the other hand, have concentrated on the "how" question; that is, on how a cultural trait is acquired and what cognitive abilities are required. They see culture as a learning process. Finally, biologists concentrate on the question of "why" cultures exist in order to understand the benefits of culture and under which circumstances one would expect cultures to evolve. They see culture as an adaptive process.

If the study of culture in humans is facilitated by the availability of data on many hundreds of different societies, we must keep in mind that this is far from the case for other species. The most studied animal species and, therefore, the one that we have by far the most knowledge of is our own species. Our knowledge of humans is based on centuries of work by explorers, missionaries, traders, and hunters, followed by anthropologists, ethnographers, and psychologists. For example, we know about 2000 different languages, we can compare the warfare strategies of over 400 different human societies, we have access to observations on hundreds of different societies living in different socio-economic systems, and we have information on centuries of cultural changes in hundreds of different societies (see Figure 2.2). Yes, with such vast amounts of information, we can start to make claims about what humans are and what they do. However, *this luxury simply does not exist for any other species*. The second most studied species is the chimpanzee, but even after 50 years of fieldwork studying these animals, we still only have precise observations on 9 different populations and for these, we have only observed 3 generations at the most. Our knowledge of gorillas is limited to observations on five populations with differing levels of detail, and there have only been observations on five groups of orangutans. For other species, the numbers are even smaller. Thus, comparisons of humans with other animal species are, by definition, going to *underestimate* the cultural diversity of the nonhuman species.

Animal ethnography to expose animal cultures

As he walks through the forest, Sartre not only sees the lovely tall trees, the dense undergrowth, his social companions, and the monkeys and birds flocking around, he also sees a material dimension to his surroundings that is going to provide him with food, a social dimension where he has his social partners and in which all of his social interactions unfold, and a more personal social dimension that is symbolic, without direct connection to the real world but which still allows him to communicate with others. To take into account such different dimensions through which all individuals move on a daily basis, I will structure my presentation of culture into three main components. As we will see, the main advantage to doing this is that we can distinguish between different products and substrates of culture and thereby more clearly distinguish the different transmission mechanisms that will support the three culture domains.

The first is *material culture*, which can be seen today in our cars, houses, and computers, but which was already visible in the first stone flakes made by *Homo habilis* in the Oldowan Gorge in Tanzania some 2.5 millions of years ago, followed by the splendid stone hand-axes made by *Homo erectus*, and later by the differently shaped arrows, spears, fishing and digging sticks, huts and clothing found in traditional foraging societies. Material culture encompasses all the interactions individuals have with their environment. It has been suggested that it is this domain of culture that has allowed humans to become independent from the limitations and constraints imposed by their environments. At the same time, it is this domain of culture that is most often proposed to be present in different animal species.

The second is *social culture*, which is key for social living individuals who are faced with many and very flexible social challenges. Many of the social cultural traits are based on the fact that the presence of social partners can allow individuals to reach goals that he or she would not have been able to reach alone. Such social actions depend on an initiator's ability to coordinate or manipulate social partners to his or her advantage, as in cooperation or social exchanges. Different social structures, composition, and organization will influence the emergence of possible social cultural traits adopted within a social group. In contrast to symbolic culture, nothing needs to be arbitrary in social culture, but in contrast to material culture, it does not involve or concern objects of the physical environment.

The third is *symbolic culture*, which expresses itself in the widespread human social norms and rules, institutions, beliefs, myths, and religions. For many of us, symbolic culture seems to be impossible without language, which is a uniquely human attribute. A special attribute of this domain of culture is that by being much less dependent upon the prevailing environmental conditions, the specifics of these cultural traits in a group are often much more arbitrary and based more on shared meanings than is true for material cultural traits. Therefore, the learning of symbolic traits requires different mechanisms to conform to the performance in all group members. Symbolic culture can interact with material culture and there are many clear examples of humans bridging both domains of culture.

Whatever human groups one looks at, one will find abundant examples of all three types of cultural traits. The main difference between groups will be in the importance each group ascribes to each of them. Nomadic hunter–gatherer groups have far fewer material culture products as they need to be able to carry all their belongings with them when they move between campsites, while modern city-dwellers are unable to move all their material belongings without the help of huge trucks. The main result of ethnographic work on humans has been to show the vast cultural diversity that prevails in our species and this has allowed us to uncover the dramatic diversity in human solutions. Moreover, only an ethnographic approach can allow us to start to understand the different factors that affect human cultural diversity in terms of climate, rainfall, altitude, environmental productivity, environmental change, and so on. A similarly complete knowledge of culture in chimpanzees or other animals will only emerge with detailed ethnographies like those that have been done in humans.

Ethnography is not only about listing behavioral differences; it is also, and equally importantly, about precisely documenting the ecological and social conditions that prevail in each population in order to understand how such cultural differences have evolved. Culture is embedded in the social and environmental life of the individuals and only an ethnographic approach can provide us with the contextual information that is so important to understanding it. Some have suggested that such an approach cannot help us to answer the question of culture in animals[5] because we cannot exclude possible fine-grained or past ecological influences on behavior or prove a social influence on the acquisition of the potential cultural behavior. However, if we were to accept this position, we could also not accept ethnographic information as evidence for culture in humans, a position I imagine no one would want to adopt. Furthermore, this position strongly underestimates some essential contributions of the ethnographic approach. First, only the ethnographic approach permits the identification of candidates of cultural behavior by centering on comparisons between populations. Second, only the ethnographic approach can

[5] See Galef (1992, 2004), Laland and Janik (2006) and Tomasello (1990).

possibly identify the natural ecological or social factors affecting the presence of the different cultural candidates. Third, the ethnographic approach alone will allow us to document how these candidate cultural behavior patterns are acquired.[6]

Imo, the cultural innovator

Imo, a juvenile female Japanese macaque, was a member of the Koshima troop, which lived on a small island in the Japanese archipelago. This troop was provisioned daily with sweet potatoes, which the animal keepers threw on a beach for them. The macaques quickly learned to come to the beach and eat the potatoes, and at the same time, to tolerate the close presence of the human observers. One day, Imo carried a sweet potato from the sandy beach into the water, where she washed the sand away from the potato and then ate it. The next morning, she continued to use this new behavior, which resulted in sand-free, saltier tasting potatoes. Interestingly, until this point, the macaques had avoided the water, but this new innovation required them to at least come into contact with it. Imo's playmates were the first to copy her behavior and the younger infants followed soon after. The mothers of the young potato-washers also learned the behavior, but at a later time. In the end, everyone except for some of the big adult males of the troop acquired Imo's innovation. Interestingly, while washing the potatoes, the macaques started to go into deeper water, and eventually, new babies lost any kind of aversion to the water and started to play and swim in it. A few years later, Imo invented another new behavior; she collected a handful of wheat grains that the caretakers had thrown on the sand and carried them to the water. Once there, she threw everything into the water, but as only the wheat grains and not the sand floated on the surface, she was able to eat them much more rapidly than if she had to sort them from the sand on land. This new invention also spread to the majority of the group. So thanks to Imo's inventions, a new culture began in the troop that included not only potato-washing and separating wheat from sand, but also bathing, playing, and swimming in the water.

Japanese people have long adopted a more holistic view of living things than we have in the western world, so when the Japanese observers saw this in the 1950s, they immediately called it "culture."[7] They were not bothered by the fact that it resulted from human intervention, as these macaques had been living in close contact with humans for generations and had long been part of Japanese culture. The primatological community immediately welcomed their observations, while some psychologists adopted a more critical attitude and claimed it could not be culture

[6] Boesch (1996, 2003), Whiten *et al.* (2003), Whiten and van Schaik (2007), van Schaik *et al.* (2003).

[7] Imanishi (1952) was the first to start using the word "culture" without limiting its application to humans, and he was followed by a number of observers who reported observations of the invention and propagation of new behaviors in different populations of Japanese macaques (Kawai 1965, Kawamura 1959, 1965). Frans de Waal, in his book *The ape and the sushi master* (2001), nicely describes this important difference between the Japanese and western approaches to culture.

because these traits spread too slowly within the group and not everyone picked them up.[8] This led to seemingly endless discussions about how quickly a behavioral trait should spread within a social group for it to be considered cultural transmission and how we can differentiate a social transmission mechanism from an individual learning process. Furthermore, some psychologists have argued that it cannot be a cultural process as imitation has not been proven in this case.

Nevertheless, for the first time, Imo and her playmates' behavior opened the door to discussions of culture in animal species, a topic that until then had been restricted to humans in the fields of anthropology and psychology. Not everybody shares our enthusiasm, though. Some recent reviews of the concept of culture in the field of anthropology do not even mention these observations, and, consequently, there is no hint of a discussion about the possibility that cultures might be observed in other animal species.[9] But it remains impressive that Imo invented two new techniques that spread throughout the group and that these techniques resulted in a new series of behaviors related to water. To this day, Imo remains the individual that has had the strongest impact on discussions about nonhuman cultural abilities.

Social transmission of cultural traits

Culture rests on the fact that each of us learns most of ours from our social group. However, this requires a specific form of intelligence based on social learning. Wolfgang Koehler, the great German psychologist who pioneered animal observations by building a research center in Tenerife in the Canary Islands in 1913 and discovered many facets of chimpanzee intelligence, said, "Actually, humans cannot immediately and simply copy when they do not understand enough of the tasks and the steps needed."[10] When we see a chimpanzee crack nuts, we, like the young chimpanzees in the middle of Figure 2.3, see a lot of things happening all at once: there are two chimpanzees, sitting on two different roots from one of the many nearby trees, holding wooden clubs, with many other wooden branches, nuts, and dead leaves nearby on the forest floor. Furthermore, *Coula* nut-cracking is a very social activity and our youngster will often be surrounded by nut-crackers sitting in a variety of positions, holding differently shaped hammers, some with their right hand, some with their left, and hitting the nuts at different angles and from different heights. As Wolfgang Koehler was saying, you must put some order to all these aspects and how you do so is an essential part of culture. Koehler insisted that copying implies that "one immediately understands, or gains insight into, what another's actions mean, and by doing so, they constitute a 'solution' to the present situation." "What" and "how" cultural transmission is learned are still passionately discussed questions between those on different sides of the culture debate.

[8] See Galef (1992, 2004), Tomasello (1990).
[9] Kuper (1999), Barnard (2000).
[10] See Koehler (1959).

Figure 2.3 **What can youngsters learn from social models** when cracking *Coula* nuts in the Taï forest? Only a few of the many aspects that are relevant to this typical nut-cracking situation are highlighted. The aspects related to objects in nut-cracking are in light gray, the aspect related to the goal of the nut-cracking – eating the nuts – is in black, and those related to the behavior of the models that can be copied are in dark gray.

How do humans acquire their cultural practices? Some psychologists insist that imitation and teaching are key to human cultural transmission, but surprisingly few observations have been done with humans in daily life to confirm which social learning mechanisms are used to acquire culture traits. When available, however, cultural traits are described to be acquired by a mixture of individual trial-and-error learning with some copying of successful individuals. Superb observations of arrowhead diversity in the prehistoric Great Basin in North America have shown that in regions where life was easier for humans, like California, a higher diversity of forms have been found, while in regions where harder living conditions prevailed, like Nevada, the diversity of arrowhead varieties was much lower, which points to more systematic selective copying of successful individuals. The result of this is that cultural traits in human societies are copied from group members with limited fidelity. For example, the correlation of the habits of parents with the habits of their children is 13% for sports and 16% for entertainment, and reaches an average of 34% for attitude towards feminism, but 80–94% for political party affiliation.[11] Similarly, new studies reveal that young children balance prior knowledge of the causal relationship of the task to acquire with new evidence presented to them by adults and with the knowledge of the adult's intention, resulting in them

[11] Boyd and Richerson (1985), Cavalli-Sforza *et al.* (1982).

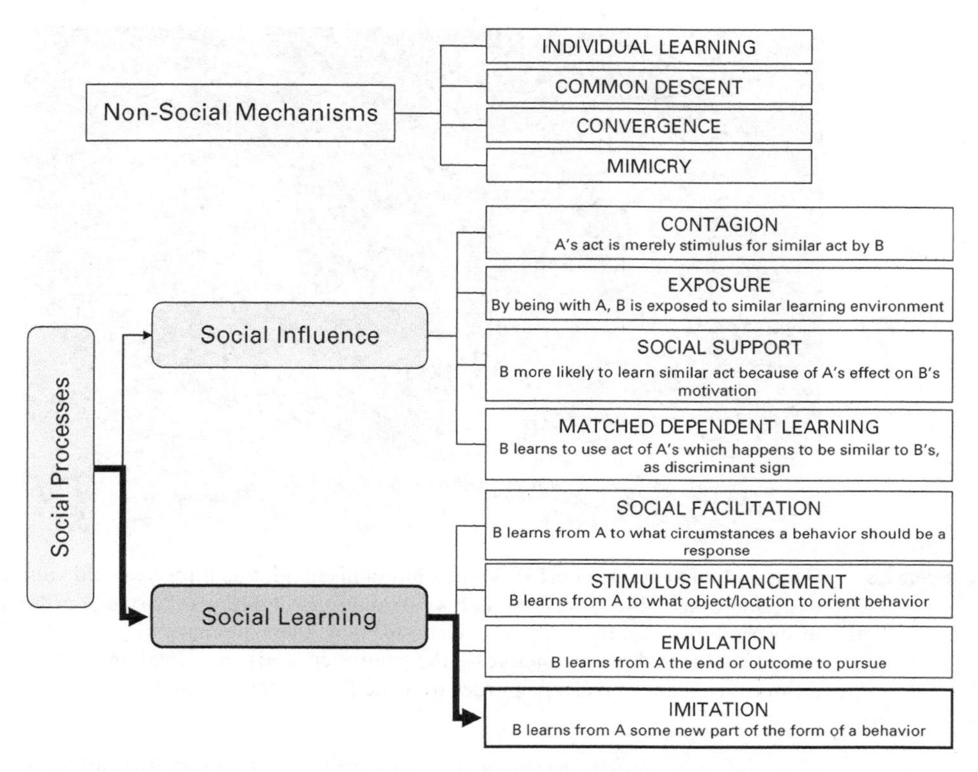

Figure 2.4 Different learning mechanisms that might be involved in cultural learning (adapted from Whiten and Ham 1992 and Whiten *et al.* 2009). For some, any *social processes* make culture possible (e.g. Kummer 1971, Wilson 1975); to others, specific *social learning* mechanisms are required (e.g. Heyes 1993, Boesch 1996, Boyd and Richerson 1996, Caldwell and Whiten 2002, Byrne 2007, Whiten *et al.* 2009); while for some others, only *imitation* will permit culture to develop and persist (e.g. Galef 1988, 1992, Tomasello 1990, Tomasello and Call 1997).[12]

copying different aspects of the tasks and different models to varying degree.[13] Hence, social learning in humans is often based on social influence, whereby the performance of the naïve individual is simply improved due to the mere presence of other individuals, combined with a lot of individual variation in participants' use

[12] There is a flurry of publications proposing different mechanisms by which one individual might "copy" or "ape" another social group member (Galef 1988, 1992, Visalberghi and Fragaszy 1990, Tomasello 1990, Whiten and Ham 1992, Tomasello *et al.* 1993, Heyes 1993, Tomasello and Call 1997, Whiten and Custance 1996, Zentall 1996, Byrne and Russon 1998, Caldwell and Whiten 2002, Gergely *et al.* 2002, Whiten *et al.* 2004, 2009, Huber *et al.* 2009, Bates and Byrne 2010, Rendell *et al.* 2011). The central objective has been to propose some classifications of social learning based on possible different levels of social intelligence in the naïve individual learning the technique, while at the same time, some have suggested that some mechanisms, namely imitation, would lead to more exact copying of the behavior or the intention of the other one resulting in more faithful cultural transmission.

[13] Buchsbaum *et al.* (2011), Corriveau and Harris (2009), Over and Carpenter (2009), Harris and Corriveau (2011).

Figure 2.5 **What do I do to crack a nut?** Holding a nut in his hand, this three-year-old youngster needs to decide which objects to select in his surroundings and how to correctly order his actions to successfully open the nuts. As you can see here, there are many objects in the Taï forest that could be used to crack nuts and the youngster needs a minimal understanding of the technical challenge involved in order to make the correct decisions.

of social information, so that a sizeable number of participants fail to conform with what is seen in the majority of the individuals within the group. This observation makes sense as without individual learning, how can changes in the environment be tracked to keep adopting the better solutions? Furthermore, the need to be able to adapt to environmental changes explains why social learning should be selective, as it makes no sense to copy a bad model – for example, one who uses a bad or inefficient method to crack nuts – and one should be able to distinguish successful models within the group and selectively copy them – selectively because while a chimpanzee mother might be a good model for how to correctly crack nuts, she will not be a good model for learning to hunt in groups, for example.

To clarify the debate about the different social learning mechanisms, a simple example is depicted in Figure 2.3. Sartre in the middle of the picture is only 6 years old, an age at which chimpanzees are not yet able to master the technical challenges of nut-cracking and still make many errors (see next chapter), and he sits in the middle of a group of nut-cracking chimpanzees that includes his mother, Salomé, with her newborn baby on her shoulder. What should Sartre copy to become an efficient nut-cracker? First, as illustrated in light gray, he could be influenced by the objects his mother Salomé is working with, such as the root, hammer, nuts, or dead logs found around her, and Sartre could start to manipulate the objects that are present (this is called *social facilitation* or *social enhancement*, Figure 2.4). More precisely, Sartre could be influenced by the fact that Salomé is bringing the hammer into contact with the nuts, and, thus, he could concentrate his manipulation specifically on these two objects (called *stimulus enhancement*, Figure 2.4). Alternatively, as shown in black in Figure 2.3, Sartre could focus his attention on the fact that Salomé is eating the nuts, and he would therefore try whatever comes to mind in order to break the nuts and eat what is inside (called *emulation*, Figure 2.4). Finally, as illustrated in dark gray, Sartre could be influenced specifically by the movement Salomé uses to actually hit the nut with the hammer and he would copy that precise up-and-down movement (called *imitation*, Figure 2.4). Yet, beyond just watching, Sartre could "see through" the movement and realize that his mother intends to crack the nuts with hammers and independently of the difficulties she may encounter if no good hammers are around, he would copy the parts of the movement he sees as the relevant ones to reach the goal (called *rational imitation*). As readers can imagine, with such finely graded distinctions, scientists have plenty of points to argue about. Further, more than one mechanism could be used at the same time, so that stimulus enhancement about the hammer and nuts could be used to limit the ways used to open the nuts through emulation.

Box 2.2: When is social learning expected?

Social learning includes all learning that has been influenced in some way by social group members (see Figure 2.4; those interested in such distinctions should refer to the references below). It is important to realize that many aspects in a model can be copied in different ways at the same time and, therefore, it can be very hard to tease apart the contribution of each mechanism to the learning of complex behavioral traits, like nut-cracking, termite-fishing, or piano playing.

Despite the attention given to social learning, there is still much uncertainty about which social learning mechanisms should be expected for what kinds of behaviors. Is imitation more frequent for tool-use behavior or for symbolic words? Would emulation be more efficient for technically complex behavior rather than social displays? Similarly, what mixture of different learning mechanisms should we expect for the learning of different types of behavior and what level of previous experience is needed to allow the full development of learning abilities? Studies with children and chimpanzees have shown that different learning mechanisms will be used for different tasks depending on the complexity of the task itself as well as the familiarity and the accuracy of the model proposed to the child (e.g. Horner and Whiten 2007, McGuigan *et al.* 2007, Corriveau and Harris 2009, Horner *et al.* 2010, Harris and Corriveau 2011).

Imitation has generally been proposed to be the most important copying mechanisms allowing for precise reproduction of the model's behavior (Piaget 1935). However, reviews of studies on imitation have shown that different groups of researchers and different experimental designs have found conflicting results or different interpretations of the same results, with some finding that chimpanzees are able and others finding that chimpanzees are unable to imitate (e.g. Caldwell and Whiten 2002). Nevertheless, five out of nine reviews agree that chimpanzees can imitate. In a disturbing comparison, all studies done by one group (Tomasello's) concluded that apes are strongly limited in their ability to imitate, while all studies in another group (Whiten's) concluded the exact opposite. This highlights the challenges of interpreting the work by research teams that have adopted different definitions and work with captive individuals with very different backgrounds and experience (Rumbaugh *et al.* 2008, Lyn *et al.* 2010, Leavens *et al.* 2010, Leavens and Bard 2011).

References: Tomasello 1990, Visalberghi and Fragaszy 1990, Whiten and Ham 1992, Zentall 1996, Tomasello and Call 1997, Caldwell and Whiten 2002, Whiten *et al.* 2003, 2004, 2009, Gergely *et al.* 2002, Horowitz 2003, Rumbaugh *et al.* 2008, Rendell *et al.* 2011.

Opinions diverge decisively at this level, with some arguing that pictures like Figure 2.6 are totally misleading and chimpanzees and other animal species are not able to copy other group members and individuals are mainly reinventing the behavioral traits each time they acquire them. Others, however, argue that there is plenty of evidence for imitation in animal species (see Box 2.2). However, the culture debate also develops on the other side of the scene, namely on what mothers can do to help their offspring acquire a task more efficiently.

Figure 2.6 **Kummer looking intently at the nut his mother Kiri is cracking**. Young chimpanzees demonstrate a strong interest in their mother's actions with nuts and spend a great deal of time over a period of many years watching them. In this picture, Kummer is less than a year old and has not yet tried nut-cracking, but he is already watching what his mother is doing with great interest.

In other words, are animals capable of teaching? Here again, we can distinguish between different levels, as Salomé, noticing that Sartre is starting to occasionally crack nuts, may provide him with simple situations where he could attempt to crack nuts, as when she leaves intact nuts and places the hammer on the anvil for him to use – scientists call this *stimulation*. Alternatively, noticing that Sartre is using a bad hammer, Salomé could help by lending him her good hammer while she collects some more nuts – scientists call this *facilitation*. Also, Salomé could use Sartre's tool to show him the correct way to crack nuts – called *demonstration* or *scaffolding*. Or Salomé could "see" Sartre persisting in making a mistake and Salomé could herself correct her son's error – we would call this *active teaching*. Finally, Salomé could manipulate Sartre's own body to show him the correct way to perform the task – called *molding*. These last two forms of teaching have been proposed to be more demanding as Salomé must realize that Sartre has not reached the same level of understanding or expertise that she has and she must put herself in his place to see what could help him. This process requires Salomé to be able to differentiate between what she knows and can do from what Sartre knows and can do.

The paradox of studying "culture outside of culture"

Originally, the idea of culture in animals came from the ethnographic approach by which the behavioral repertoire of different wild populations of chimpanzees, whales, and birds were compared. From such comparisons, propositions for potential cultural differences were made. However, a new criterion was quickly imposed, namely the need to exclude genetic and environmental influences as an explanation for those differences. The justification for this additional criterion came from the many examples in different animal species of individuals in different populations that readily adapt to different environmental conditions, and we needed to make sure that we were not simply looking at an ecological adaptive response. However, by adding such a criterion, we imposed more stringent requirements to accept the presence of culture in animals than has ever been required for humans.

Box 2.3: Can genes explain cultural differences?

By comparing different populations of an animal species, the ethnographic approach tends to assume that possible genetic differences are controlled for. Nevertheless, it would be preferable to directly test the contribution of genetic differences on behavioral differences. However, testing whether cultural differences could be explained by genetic differences has proven to be difficult for three main reasons. First, genes have not yet been found that code for any proposed cultural behavior. This could be seen as proof that genes do not code for such behaviors, but since so few genes' functions are known, such a conclusion would be premature. Second, tests of genetic influence are based on non-coding genes and therefore track only population history. As individuals with successful cultural traits might reproduce better and disperse between social groups, they will also spread their genes. So we might detect a correlation that is a by-product of successful individual dissemination and not one between genes and culture. Third, genetic characterizations of individual populations are often incomplete or absent. Studies of human cultural evolution have shown that, for some well-studied cases, genetic influences were proposed to account for 70% of the similarity among cultural patterns observed in different groups (Tehrani and Collard 2002).

Therefore, one way to test for the role of genes would be to adopt a cladistic approach that compares cultural differences between populations by assuming that, if genes were important, we should find more cultural similarities between geographically closed populations. As one study found this to not be the case, it was suggested that the genetic explanation was not supported (Lycett *et al.* 2007, 2010). However, this analysis did not include any genetic data and therefore had to rely on unproven assumptions about genetic structures between chimpanzee populations.

Thanks to the development of non-invasive genetic techniques, a recent study showed that genetic and proposed cultural differences between nine chimpanzee populations were strongly correlated (Langergraber *et al.* 2011). However, these authors recognized that they were not able to exclude the possibility that culturally successful groups could fare better than others and therefore successful cultural traits might disperse at the same time as genes and produce the observed correlation. Therefore, this result should not be a surprise. More to the point, however, for genetically similar chimpanzee populations living in the same stretch of forest, as in the Mahale Mountains or in Taï forest, about 52% of potential cultural differences could *not* be explained by genetic differences, thus providing strong support for the cultural explanation (Langergraber *et al.* 2011). As expected, this study showed that an interaction between cultural and genetic effects must be seriously considered.

This new criterion has led to a debate about the possible influences of past and present environmental factors on potential cultural traits in animals. Leaving aside the fact that it is difficult to detail past environmental conditions, it would not explain the maintenance of the present-day observed behavioral differences. As a consequence and to control for possible ecological influences, psychologists have started to favor studying cultural learning mechanisms with captive animals because it would be possible to concentrate on cultural transmission mechanisms in the absence of any of the potentially confounding environmental effects. Thus, these psychologists have argued that captive studies allow them to tease apart the social from the ecological influences. Ironically, what they were really proposing was to study "culture outside of culture." Since captive chimpanzees mainly come from a few West African countries, such as Guinea and Liberia, and many of the proposed cultural behaviors are observed in East African chimpanzees, such captive studies would be akin to studying the culture of the Aka Pygmies of Central Africa with Nigerian prisoners in German prisons!

Captive studies certainly allow one to exclude many of the factors that influence the learning of a task in the real world. However, the risk with such experiments is that the captive contexts might be so far removed from real-world conditions that it becomes impossible to say how much their performances can be generalized to the whole species. By imposing too much control over what may be relevant factors, captivity might result in such impoverished environments that it prevents animals from using the combination of stimuli that enable them to make assessments of others' behavior.[14] As expert psychologists, Dorothy Cheney and Robert Seyfarth, who have been working for decades with baboons in Kenya and Botswana, said, "laboratory experiments can illuminate a species' abilities only if their results can be placed within the context of an animal's natural social behavior. In the absence of such grounding, they remain difficult, if not impossible, to interpret."[15]

Favoring captive experiments over other approaches to studying animal culture leads to some contradictions. First, proposed cultural traits should be studied in the social group that perform it naturally. However, unless you assume that all behaviors in chimpanzees are cultural, it does not make much sense to study cultural transmission with non-cultural behaviors. Some examples of behaviors used to study the social learning mechanisms in captive chimpanzees include throwing sand, raking bananas towards a cage, lifting or sliding doors on different kinds of boxes, pulling ropes, using keys to open doors, and so on. None of these tasks are related to the natural tasks that chimpanzees encounter in the wild and, therefore, none of them have any affinity with behaviors that have been proposed to be cultural in primates. Such studies are relevant to understanding the ability of captive individuals in learning artificial tasks, but they have no direct relevance in helping us to understand cultural transmission in animals.

Furthermore, the "culture outside of culture" approach suffers from a total underestimation of the social dimension of the culture process. A review of the published studies on social learning in chimpanzees that have been done in captivity shows that 14 out of 22 used a human and not a chimpanzee as a model to copy, and in 17 of these studies, only one model was shown, contrary to the many existing in any social group in the wild.[16] Moreover, Figure 2.7 illustrates the classical experimental procedure used to compare chimpanzees with humans. The human child typically undergoes testing while in the same room as his mother/father and the experimenters, and he often remains in close physical contact with his parent,

[14] Philosophers like Allen (2004), Allen and Bekoff (2007) and Bekoff (2004) have stressed this aspect and required that care be taken when one considers the value of such experiments. Colin Allen stresses that "greater faith in experiments than in free observation is practically a hallmark of the difference between psychologists and ethologists." Ethologists are the same individuals as the field workers I mention in the text.

[15] Cheney and Seyfarth (2007, p. 26).

[16] Bard *et al.* (2011).

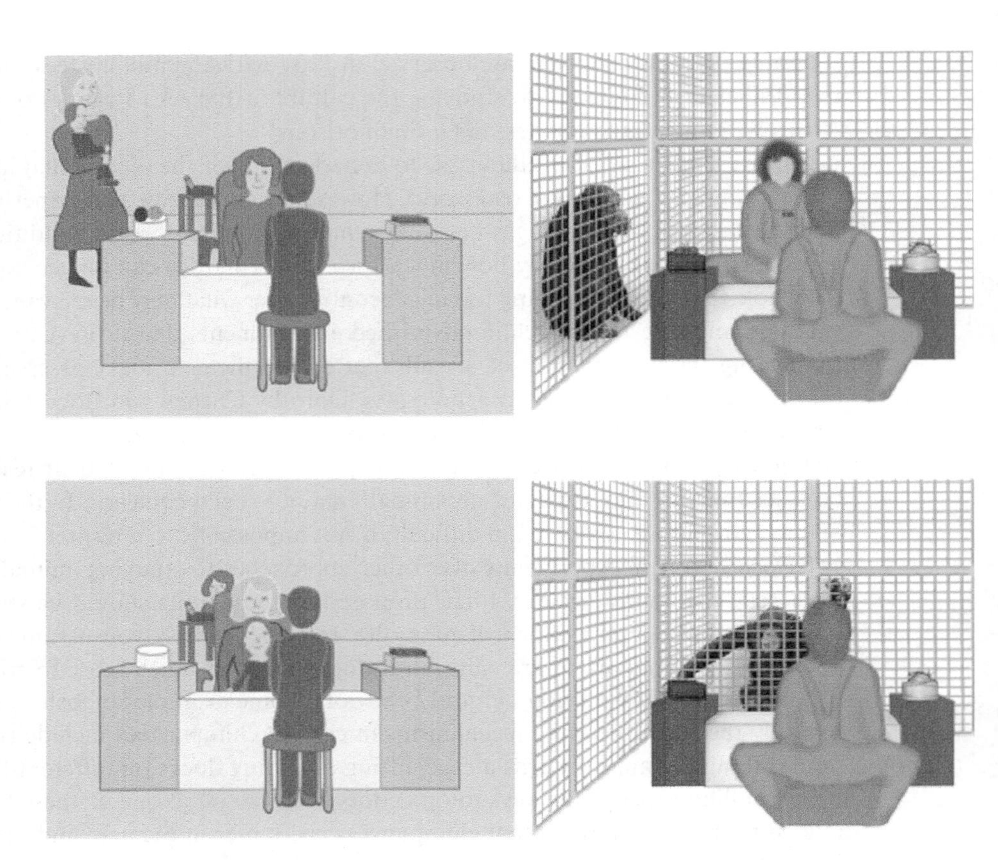

Figure 2.7 **Experimental settings imposed on captive chimpanzees** when compared with free-living humans (taken directly from Liszkowski *et al.* 2009). The fact that isolated captive chimpanzees failed to point to absent objects but human children sitting on the laps of their parents succeeded was suggested to be a species difference. However, it could just as easily have been the result of the differences in the testing procedure used for the two species, of the different living conditions (free versus captive), or of the different motivators used (a present parent plus vocal stimulation for the child versus two grapes for the chimpanzee). Sadly, these authors neither mention nor test for these alternative explanations in this publication.

is able to verbally communicate with all of them, and has direct contact with the object used in the test. The ape subject, on the other hand, is isolated during testing from all other group members, is separated from the experimenter and the test objects by a barrier (either glass or bars), and cannot communicate directly with anyone. So much for equivalent conditions! It is difficult to understand what we can conclude from experiments performed with so many differences on two different species.

Finally, a social group is more than just some animals living together. For example, all wild chimpanzees live in communities made up of 30–120 individuals that share large territories between 15 and 30 km^2. They usually remain in small subgroups, also called parties, with ever-changing compositions, where some individuals can meet less than once per month. Adult female chimpanzees are regularly absent from the main activities for up to six months and males can be absent for two to three months at a time. This flexible social grouping pattern, which is called fission–fusion grouping, requires special flexibility from its members as one's position is always changing as a function of which individuals are present within the party and the social relationships must be constantly updated as one meets others and has to monitor changes that might have happened in his or her absence. This is in stark contrast to captive social conditions where a few individuals, rarely with more than one adult male, are kept constantly together in a comparatively tiny space and many individuals have experienced different captive condition history. In other words, the social complexities and daily challenges faced in the wild provide young chimpanzees with very different social and environmental stimulations than most captive settings.

For humans as well, the social dimension of life is central to culture. If we do not experience the reassuring, loving, and stimulating contact and presence of a parent early in life, we do not develop as we would have otherwise. Furthermore, later in life, we profit from the additional presence of siblings and long-term friends and colleagues. Although this might sound so obvious that it should not even need to be said, this sentiment is all too easily forgotten. For many humans, this is a luxury that they can only dream about and this has dramatic consequences on their development. The scandalous conditions of some orphanages in eastern Europe that were uncovered when the Berlin Wall fell has been abundantly reported in the media, but the scandal concerned not only the suffering of the infants in those institutions but also the lasting negative impacts the neglect had on their cognitive development and their abilities to form intense social bonds. Orphans "reared in institutions showed greatly diminished intellectual performance (borderline mental retardation) relative to children reared in their families of origin." After being placed in foster families, they presented significant gains in cognitive functioning and the earlier this happened, the greater the gains.[17] In a naturalistic study, social influences on the development of cognitive abilities have also been shown when comparing parenting styles and subsequent cognitive development in three human cultures.[18] African women typically always carry their young babies with them as they go to the market, cook, dance, and sleep, and this permanent bodily contact results in quicker developing and earlier walking infants, and results in them being

[17] This study suggested a sensitive phase after which cognitive recovery is limited (Nelson *et al.* 2007). In other cases where there were limitations in their ability to form normal social bonds, normal attachment in children was still lacking years later in school (McCall 2011, Bakermans-Kranenburg *et al.* 2008).

[18] Heidi Keller and her team worked intensively on this issue and have produced new and very convincing data on the cultural effect of parenting and future cognitive specializations (Keller *et al.* 2004a, 2004b, 2005). Recently, a study investigating the rate of development of mirror recognition in seven different human societies – USA, Canada, Kenya, Fiji, Saint Luca, Granada, and Peru (Broesch *et al.* 2011) – has confirmed the dramatic difference parenting style can make on the development of such a cognitive ability.

better able to self-regulate their behavior than western babies, who are separated from their mothers at a very early age and often for very long periods of time. Researchers have been clearly able to show that such divergent parenting styles lead to infants developing in different ways; due to this greater physical distance between mothers and babies in the western world, both exchange mutual glances more often and for longer periods of time, and, as a result, children learn much earlier to recognize themselves in a mirror than children who were constantly in bodily contact with their mothers when they were babies.

Chimpanzees are similarly influenced by social deprivation in captive settings. While young female chimpanzees in the wild have plenty of time to learn baby handling and nursing with their younger siblings under the watchful and tolerant eyes of their mothers, most captive female chimpanzees have no such experience and too often reject or neglect their infants when they give birth (for example, 21 out of 30 female chimpanzees who gave birth in the Yerkes National Primate Research Center in Atlanta, USA, rejected their babies). Caretakers normally place these rejected babies in nursery groups, which you can see in many zoos, so that young infants can be hand-fed with bottles by caretakers and at the same time have some social interaction. These are indeed luxurious conditions compared to some of the worst human orphanages, but they are still orphanages and very abnormal social situations. Detailed studies revealed that being carried by a human for 4 hours a day decreased the problem of over-attachment with a caretaker by 30% and the need to be attached to an object by 45% while at the same time significantly increasing their results in general cognitive tests.[19] This is especially relevant as such peer groups of young chimpanzees have often been used by scientists to study different aspects of intelligence, including those directly related to cultural phenomena like imitation, social learning, and social transmission.

Recognizing the importance of the social dimension, some researchers have built on the fact that some primate facilities keep chimpanzees in larger groups to see if it would be possible to mimic a cultural spread.[20] When confronted with an artificial box in which food could be attained using two different techniques, captive chimpanzees rapidly imitated the one technique introduced by an expert group mate. Cultural transmission from one group to a neighboring group was readily observed under such social conditions, as the technique used by one group was passed on to a neighboring group that could see the first one performing it through a window.

[19] The best study to date of this effect has been done by Kim Bard and colleagues (van Ijzendoorn *et al.* 2009, Bard *et al.* 2011). Although much more detailed, this study nevertheless only confirmed the negative effects of poor captive living conditions that have already been demonstrated decades ago in many different studies (Davenport and Rogers 1970, Menzel *et al.* 1970, Gardner and Gardner 1969, Bard *et al.* 2005).

[20] Andrew Whiten, Frans de Waal, and colleagues very successfully pioneered this approach of replacing the social dimension at the center of an experimental approach with captive chimpanzees. They also recognized that the social dimension is not limited to group size but also includes the social status of the individual introducing a novel cultural trait within the group and therefore used dominant females as models. Further, they used two chimpanzee groups from the Yerkes Primate Research Center in Georgia and six chimpanzee groups in Bastrop, Texas, as subjects in these ingenious tests (Horner *et al.* 2006, Whiten *et al.* 2005, 2007).

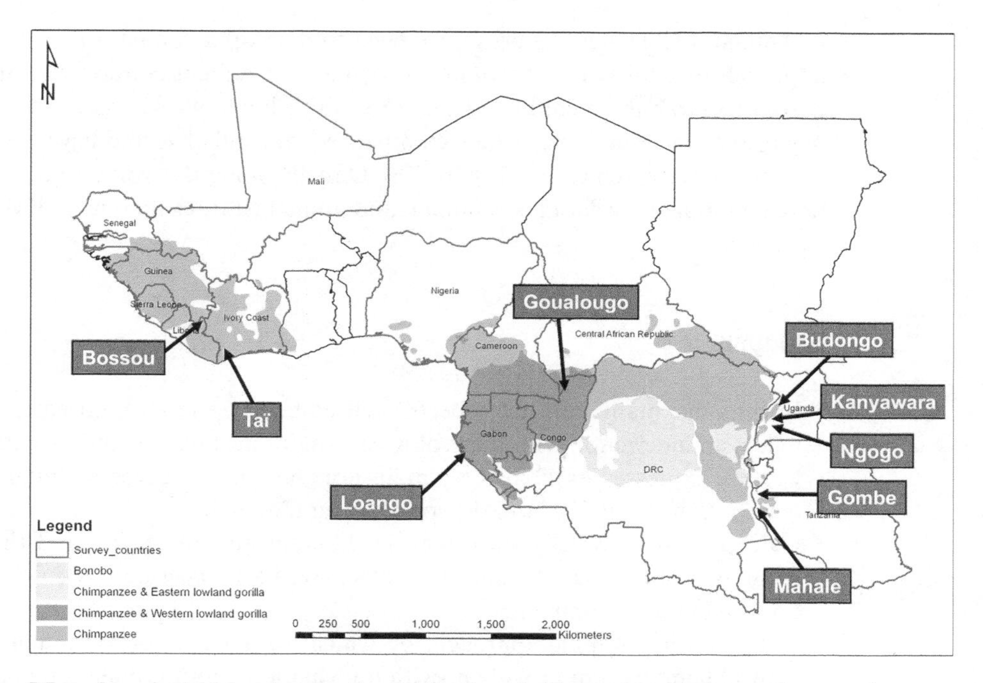

Figure 2.8 **Map of tropical Africa showing the distribution of chimpanzees** on the continent and highlighting some of the chimpanzee populations with detailed behavioral observations that will be regularly mentioned throughout this book. This map also illustrates how limited our knowledge of chimpanzees presently is and how many more populations are still totally unknown.

These remarkable studies have shown that the cultural spread of a totally arbitrary behavior can occur in captive chimpanzees as long as the social dimension is preserved. As Montaigne might have said, treating others as barbarians has been too readily done in the past and just reflects one's own "*esprit de clocher*" or shortsightedness. The success of those teams in overcoming some of the shortcomings of captive experiments – including using an untrustworthy model of another species or using barriers between the model and subject – shows that progress can be achieved in understanding more about other species by being ready to consider the natural socio-ecological conditions they normally face in the wild.

Animal cultures to learn about human cultures

It is time to follow in the footsteps of Margaret Mead and Claude Levi-Strauss and make a detailed ethnography of chimpanzee culture if we want to obtain a better understanding of what is unique about and what is shared by human and chimpanzee cultures. Experiments studying culture outside of culture with captive animals will have only limited relevance to wild cultures and in helping us to understand human cultural intelligence. What we need is to confront humans with examples of animal intelligence. To do so, we need to develop a comparative ethnographic approach that takes into account the richness of animal cultures and then compares them to human cultures. Therefore, in this book, we will make the long journey to the places in Africa where chimpanzees are still alive and have been studied by different research teams (Figure 2.8). Only by doing this will we progress in our understanding of what makes human and animal cultures unique and what makes them similar.

Synopsis

- In humans, many cultural aspects such as beliefs, taboos, and categorizations reflect an interaction with the ecological conditions faced by the individuals.
- Culture expresses itself in three main domains that must be distinguished: the material, the symbolic, and the social cultural domain.
- Cultural behaviors have been seen in different animal species and the potato-washing of the Japanese macaques has played a pioneering role in opening the door to culture in animals.
- Only an ethnographic approach, by which we would observe animals in their natural habitats, can provide us with information about population and cultural differences.
- Cultural traits can be transmitted by individuals copying different aspects of the actions of group members.
- The psychologists' tendency to favor experiments with captive animals over wild observations is equivalent to studying "culture outside of culture" and this has limited value in helping us to understand the culture phenomenon.

3 Shaping nature into home
About material culture

The most that the animal can achieve is to collect; man produces, he prepares the means of life, which without him nature would not have produced … makes the environment serve his ends, masters it.

Frederick Engels, *The dialectics of nature* (1934).

… the monkey is not a variety of human, not only because he is unable to speak, but more so because we are sure that his kind is unable to improve, which is the specific character of the human species …

Jean-Jacques Rousseau,
Discours sur l'origine de l'inégalité parmi les hommes (1755).

Taï Forest, December 1990

Salomé, Falstaff's closest female friend, is a strong, assertive mother of two sons. Her eldest, Snoopy, is Falstaff's main hunting partner. Her youngest son, 4-year-old Sartre, is a very active, spoiled toddler, keen to make the most of his mother's generosity while at the same time learning to become independent. At this time of year, some of the most abundant fruits are becoming rare and the hunting season is nearing its end. Searching for food is getting more difficult. The abundant Coula nuts are starting to ripen, but their hard shells prevent the monkeys from accessing them and the bushpigs and elephants, which have strong enough teeth to break them open, have to wait another month for the nuts to fall to the ground. However, Salomé, like the other chimpanzees, has a trick that allows her to access this very rich food source: she is able to use tools. When Salomé enters the ridge of the valley south of our camp, which I know is rich in Coula trees, she collects a hard piece of dead wood she found under a Coula tree where it had previously been used as a hammer. She puts it in her mouth and climbs up this large Coula tree. Shifting the wooden hammer into the groin of her left leg, she enters the highest, sunniest part of the tree and uses her free hand to collect eight ripe Coula nuts, which she places in her mouth. She moves to another branch and collects six more nuts, holding them in her free right foot. Then, followed by Sartre, she moves to a thicker branch higher up in the tree and begins to crack the nuts with the hammer using her left hand. She hits each seven or eight times and then, "crack," the nut opens. Sartre extends his hand to beg for the nut and Salomé often lets him take the entire nut she has just cracked. Six, seven hits and "crack," another nut is open.

Salomé shifts the spare nuts from her mouth to her left hand, and while holding the hammer with her left foot, she eats the nuts held in her right hand. When she finishes eating a nut, she stores the nuts back in her mouth and uses the hammer with her left hand while stabilizing the next nut between hits with the thumb and index finger of her right hand. Just before she strikes, she quickly removes her fingers from the nut so that she does not hit herself and as soon as she lifts the hammer from the nut, she puts her fingers back to prevent the nut from falling. Five, six, seven hits accompanied by these rapid movements with both of her hands, then "crack" ... and shifting the nuts from her mouth to her hand, she again eats the nut while holding the hammer with her foot. Her dexterity amazes me!

That day, Salomé cracked nuts in three neighboring trees for 65 minutes without coming down to the ground and shared an impressive 45% of the nuts she opened with Sartre. At one point, while I was carefully recording Salomé's nut-cracking and -sharing in my small notebook, I heard rustling sounds behind me. Turning around, I saw two gray mangabey monkeys looking alternatively between her and me. Mangabeys are often attracted by the sounds of chimpanzees cracking nuts, but even though they also love these nuts, I have never seen them using hammers. So they waited for the chimpanzees to finish and leave the cracking site so they could then eat the scraps left behind. Although it requires quite a bit of effort, full-grown male mangabeys can use their large canine teeth to crack some nuts, but the smaller monkeys seem to specialize in scavenging on the remains that the chimpanzees leave behind. Somehow these monkeys do not "see" that it is possible to use a hammer to crack open the nuts. They do not understand that an inanimate hard object forcefully hit against a nut can break it open. This reveals an interesting limitation in their understanding of the effect objects can have on each another.

Nut-cracking is a very important activity for the Taï chimpanzees as the *Coula* trees are very abundant and produce nuts every year between December and March. During the nut season, chimpanzees crack an average of 2 nuts per minute, for an average period of 2 hours and 15 minutes per day. After following them for the entire *Coula* nut season, we estimated that chimpanzees gain over 3000 kcal per day from cracking nuts.[1] So the use of tools allows these chimpanzees to exploit a very rich food source, one that they would not otherwise have access to, for 4 months of every year (Figures 3.1 and 3.2). Except for Tina, a young female chimpanzee with two badly deformed hands, all of the chimpanzees in the Taï forest learn to crack nuts by the time they are five years of age. Chimpanzees "see" that a hammer will allow them to achieve what they cannot achieve with their bare hands alone and that they can use an external inanimate object to do so. In other words, to be a tool user, you need to "see" that cause and effect of external objects can be controlled for your own benefit.

[1] We have detailed the nut-cracking behavior of the chimpanzees in the Taï forest in many publications (Boesch and Boesch 1981, 1983, 1984a, b, Boesch 1991a, 1993a, 1996, Boesch and Boesch-Achermann 2000, Boesch *et al.* 1994).

Figure 3.1 **An adult female, Malibu, cracks *Coula* nuts in a tree** early in the season. She stands against a vertical branch while her infant Fidji, who can be seen here holding onto her mother's left thigh, seems to watch her actions closely. Malibu stabilizes the nut with her right thumb and forefinger, quickly removing them when she hit the nuts with a small stone hammer she is holding in her left hand. Malibu holds some spare nuts in her right foot. Such vertical branches are regularly used as anvils because they are the largest and thus the most stable branches high up in trees. From this position, Malibu can collect more nuts almost without moving (as can be seen from the one nut hanging in the branches below her left foot). Thus, cracking nuts in trees allows individuals to efficiently nut-crack before the nuts fall to the ground later in the season.

Traces of nut-cracking – cracked shells, hammers with pounding traces, and roots with pitted holes – are found in all parts of the Taï National Park as well as in the forests situated west of the large Sassandra River that runs north to south in western Côte d'Ivoire. During the nut season, chimpanzee social groups are relatively large so that about 15–20 chimpanzees come together in a *Coula* region and the whole forest resonates with pounding sounds so loud that it sounds as if a group of carpenters were hammering nails into the wooden roof of a house. These sounds impressed the colonial French army so much that they wondered if people were working iron ore in the park and they mounted an expedition to cross the entire park during World War II to search for them!

Of the five different species of nuts eaten by the Taï chimpanzees, *Coula edulis* is by far the most abundant and the most often consumed. *Panda oleosa* trees produce an especially hard nut that is rich in protein but these trees are much rarer and more dispersed. Female chimpanzees crack *Panda* nuts for hours on end using heavy stones that can easily weigh over 10 kg. Not only are they much harder, but each

Figure 3.2 **Nut-cracking in Taï chimpanzees.** An adult female holding a hard wooden branch with her two hands pounds the *Coula* nuts she placed in a small hole on an apparent root of a large tree. On the ground, adult females open and eat on average 2.26 nuts per minute and use 6.26 hits to open each of them.

nut contains three to four different nuts that must be extracted one at a time. Young chimpanzees start learning to crack them later and need more time to acquire the technique than for the *Coula* nuts. Another nut they eat is produced by the very large but also quite rare *Parinari excelsa* trees. These trees produce vast quantities of nuts though, so they are quite popular. The two last species, *Detarium senegalense* and *Sacoglottis gabonensis*, are cracked less systematically because the kernels of the former germinate and thus can be eaten without tools later in the season and the kernels of the latter are very small.[2]

However, while these nuts are found from Liberia to Congo, nut-cracking is not observed everywhere where these nuts occur. On the contrary, this behavior is strictly limited to a zone in the west of Côte d'Ivoire, in the southeast corner of Guinea, and in eastern Liberia. While I was walking through the forest in Loango National Park in Gabon, I was amazed to see the abundance of *Coula* trees in many places. Bushpigs and elephants in this area eat the rich nuts – we were regularly guided to large groups of bushpigs by the sounds of the nuts they were cracking with their teeth – but there is no evidence that the chimpanzees there eat these nuts. The chimpanzees west of the Sassandra River in Côte d'Ivoire use tools to gain access to a bonanza of rich nuts, while the chimpanzees east of the River – which are sometimes only separated by a few kilometers – do not use tools or eat these

[2] Intriguingly, the chimpanzees in the Taï forest some 60 km south near the village of Guiroutou crack the *Sacoglottis* nuts systematically and do so at each larger tree within their territory. This suggests another intriguing cultural difference that we are presently studying.

Figure 3.3 **An adult mangabey eating nut remains from a chimpanzee anvil.** He pushes away the hammer used by the chimpanzee to access the small nut scraps that the chimpanzee left behind. Despite much exposure to nut-cracking chimpanzees, the mangabeys have not learned to use tools to access these nuts.

nuts. Recently, chimpanzees in the Ebo forest in the south of Cameroon, more than 1700 km away from the Taï forest, were seen cracking *Coula* nuts.[3] This shows that this tool technique was invented independently more than once, but did not disseminate to all of the areas where *Coula* nuts are found.

Using tools has opened a whole new rich food source to chimpanzees in a small restricted part of West Africa. In areas without this invention, such as in so many forests of Central Africa, chimpanzees simply ignore this food source and have to rely on other resources. Thus, as is often seen in human populations, tool use allows chimpanzees to shape the food availability to their needs. During the nut season in the Taï forest, monkeys, which, like the mangabeys, have not learned to use tools, scavenge on the nut remains left behind by the chimpanzees (Figure 3.3). Only those mammals with highly specialized morphological traits are able to access nuts in the forest – for example, squirrels with their razor-sharp incisors, and bushpigs and elephants with their powerful crushing molars. Other animals, like the forest antelope, the buffalos, and the 10 species of monkeys present, have no access to them. Thus, tool use is a major way to extend the food spectrum and create a unique cultural niche.

[3] Morgan and Abwe (2006) reported this observation in a small area within this forest. It is not clear yet how widely distributed this behavior is and if the behavior has been present for a long time or is a new invention that might be spreading. Alternatively, it has been suggested that nut-cracking could have once existed in all regions between Cameroon and Côte d'Ivoire and has now become extinct (Wrangham 2006). This, however, would require quite a lot of independent losses within the numerous chimpanzee groups living in Gabon, Nigeria, Ghana, and eastern Côte d'Ivoire.

Tool makers in evolution

In 1960, Louis Leakey, the famous anthropologist, discovered the first fossil remains of what he called *Homo habilis*, the oldest representative of our genus, *Homo*. He decided to name him *habilis* because in the Olduvai Gorge in Tanzania where he found the remains, he also discovered stone tools. To him, such stone tools were a hallmark of humanity in the sense that those stones showed evidence of pounding and flaking, which were presumed to be only within the abilities of a human ancestor. *Homo habilis*, the first tool maker, became the founder of the *Homo* lineage. Accordingly, we see in the *Homo* lineage only species that became increasingly and successively more skilled at making tools, including *Homo erectus*, the inventor of the famous hand-axes – those wonderfully symmetrical, flat-shaped stone axes that were produced in almost the same shape for over a million years – followed by the Neanderthals, and eventually ourselves, *Homo sapiens*. In the bushy tree leading to modern humans, we find different species, most of which have been discovered within the last 50 years or so; these belong to either the genus *Australopithecus/ Paranthropus*, those southern apes that were ancestral to the *Homo* line but did not use tools and survived quite late, even co-existing with representatives of *Homo* for a brief time, or to the genus *Ardipithecus*, which included the robust forms of southern ape that, at some point, co-existed with the two first genera in Africa. There are undoubtedly many morphological characteristics that distinguish the representatives of these genera from each other, but tool use has generally been restricted to *Homo*.

Louis Leakey was very interested in understanding how much the environment could explain the evolution of human traits like tool use or hunting, which were still considered to be uniquely human behaviors at the time. To test for this, he suggested that Jane Goodall initiate a study on a population of chimpanzees that lived not in the dense rainforests of Africa where most chimpanzee populations live but, instead, in a much more open habitat to see if this would favor any of those human-like traits. This was what led Jane Goodall to begin her pioneering work on the wild chimpanzees in Gombe National Park in Tanzania, which was not too far away from Olduvai Gorge. Jane exceeded all of her mentor's wildest expectations and within two years had made direct observations of chimpanzees hunting small baboons and making and using tools – in Gombe, the chimpanzees use sticks, herbs, and stems to fish for termites and driver ants, and use leaves to clean their bodies and as sponges to drink water out of holes in tree trunks.[4] In other words, it looked as if Leakey was right and the Gombe chimpanzees' skills in using tools and hunting confirmed how important the environment is in favoring the development of such skills. This supported the prevailing theory of human evolution that proposes that it was our ancestors leaving the forest that made them develop their uniquely

[4] These observations had a profound effect on our perception of what humans are and set the stage for a long series of new observations of chimpanzees in the wild (Goodall 1963, 1964).

human traits.[5] Not long after Jane Goodall began her work in Gombe, a Japanese team led by Toshisada Nishida started a project on the chimpanzees of Mahale Mountains National Park some 200 km south of Gombe on Lake Tanganyika. Mahale is a mosaic habitat that includes savanna, woodlands, and gallery forests and they also found that the chimpanzees there hunt and use tools.

Does the forest really prevent tool use by chimpanzees? It was with this question in mind that I decided to go to the Taï forest, where some previous reports had suggested that the chimpanzees were using tools. As we know by now, the chimpanzees in the Taï forest not only use tools but use them almost daily for a good part of the year. During a typical day, Salomé, Héra, Brutus, Falstaff, and the other Taï chimpanzees also use tools to access hard-to-reach insects. Even during the nut-cracking season, they use sponges to drink water out of holes in tree trunks, prepare wands to dip for ants, fashion short, stout sticks to extract grubs from the nests of wood-boring bees, prepare different twigs to uncover grubs inserted under the bark in the wood of large dead tree trunks or to inspect holes and different cavities in the forest, make small sticks to get at bone marrow in their colobus monkey prey, and make long sticks to dip for honey in large beehives found in tree hollows. However, tools are not only used to feed; they are also regularly used in aggressive contexts when males either pull long branches along the ground to increase the noise they make when they display or throw branches directly at opponents. Furthermore, in extreme cases, they have been seen to use larger branches as weapons against leopards that had been threatening them. So, indeed, forest chimpanzees use tools in many different contexts. In fact, recent comparisons show that the forest chimpanzees from Taï National Park and Goualougo Triangle National Park in Congo use more tools than the chimpanzees living in more open habitats like Gombe or Mahale National Park. Thus, what led to the acquisition of tool use in chimpanzees and possibly our ancestors is more complex than a simple environmental difference.[6]

Material culture shapes one's own world

Culture has been proposed to allow humans to shape the environment in which they live. The argument goes that thanks to our advanced technology, humans have become much less constrained by the limits set by the environments in which we live. For example, building houses and using fire have allowed us to survive in extremely cold climates and the use of weapons have allowed us to hunt for animals that are

[5] The "Out of the Savanna" and the "West Side Story" models proposed by Dart (1925), Leakey (1961), and Coppens (1983) suggest a complex process whereby our ancestors left the forest as tectonic changes in East Africa moved the plates of Kenya and Tanzania upwards while the Congolese forests east of the Nile Valley remained lower. The simple scenario is that the ancestors to the great apes trapped in the forest became chimpanzees, while those trapped in the East had to adapt to a higher altitude and, therefore, drier habitat that was only partly forested. This paved the way for the australopithecines and then for the appearance of *Homo* – all in eastern or southern Africa.

[6] See Boesch (2009) for a more complete discussion of the important chimpanzee population differences in tool use and in many social domains.

much larger than we are. Thanks to an especially developed sense of causality, we became able to understand the benefits of using tools to exploit otherwise inaccessible resources. The most important would have been hard-to-process foods, like hard nuts, honey protected in beehives, and underground resources, including different kinds of tubers, water, and animals deep in cavities. As experts are likely to share rewards, the benefits gained from such resources would be profitable to the entire social group. Furthermore, the unlikelihood of discovering some of the more complex techniques by chance makes them obvious candidates for social transmission. These have been the general arguments for the evolution of tool use in humans.[7] While it is obvious that humans have built cities and become incredibly technologically advanced, the question remains whether the basic skills needed to do so are really unique to humans, or if we already see signs of the presence of similar skills in other species.

As we have seen, chimpanzees similarly shape their own world as they are able to access many forest resources that are simply not accessible to other species or only for much more limited amounts of time. The rich nuts of the forest became easily accessible once they had invented the hammer, while large mammals such as elephants and bushpigs can only eat them after they have fallen to the ground. Other tools facilitate their access to honey, driver ants, numerous insect grubs, and eggs embedded well inside of tree trunks, all of which are precious nutritional supplements. Many monkeys in the Taï forest are insect-eaters and spend a large amount of time each day chasing them in the trees but honey, driver ants, and many grubs are not accessible to them. In other regions of Africa, chimpanzees also use technology to open new habitats for them. For example, chimpanzees have conquered water-poor regions with well-digging cultures, as seen in the chimpanzees of the dry regions of the Niokoloba National Park in Senegal or on the other side of the continent in the Semliki Game Reserve in Uganda.[8] The chimpanzees from Semliki have also been observed to dig wells during more water favorable periods, either near running water or algae-choked water, probably to filter parasites or dirt from the nearby accessible source of water. During very dry periods, these chimpanzees use leaf sponges to extract water from the deep wells they had dug during exceptionally dry years. Furthermore, chimpanzees from Fongoli, which is close to Niokoloba National Park in Senegal, face an extremely hot climate and during the hottest days (when temperatures can reach 42°C), they have been seen to seek shelter in caves, where maximum temperatures are only 29°C.[9] The Fongoli chimpanzees were also seen to immerse themselves in cool water during the very hot hours of the day, a rare sight indeed in a species famous for avoiding water.

Both humans and chimpanzees have been seen to greatly lengthen their termite consumption time window through tool use. In general, the flying termites that

[7] See, among others, Leakey (1980), Johanson and Edey (1981), Mithen (1996).

[8] For Uganda: Hunt *et al.* (1999), for Senegal: Galat-Luong and Galat (2000).

[9] Jill Pruetz studies chimpanzees in one of the driest and hottest habitats known to be used by this species in Africa and made these astounding discoveries of new chimpanzee behaviors to regulate body temperatures (Pruetz 2007).

Figure 3.4 **Masudi, a male chimpanzee from Mahale, fishing for carpenter ants** with a thin stick. These ants make nests in holes in live tree trunks and often the nest entrance is just a tiny opening that can only be penetrated with fine tools. Here, Masudi successfully removes the stick with one ant biting on it, which he will immediately put in his mouth and chew. Masudi will fish for these ants for over 35 minutes, switching to new tools as the one he is using breaks.

swarm in the millions during their rare yearly flight are very attractive to many species. When this begins, chimpanzees and baboons keep guard near the flying spot of the termite mound and try to catch as many as they can during the few hours that the swarming lasts. Humans also appreciate their buttery almond taste; in many places throughout Africa, they place buckets of water under lit areas to collect the masses of swarming insects attracted by the light. Since Jane Goodall's first report of termite fishing in the early 1960s, this behavior has been the hallmark of tool skills for the Gombe chimpanzees. These chimpanzees somehow discovered that before swarming, the upper part of the mound is filled with active termites. Termite fishing is based on this insight and by inserting a soft piece of bark or herb about 25–30 cm down into the curved tunnels of the mound, they reach deep enough to elicit the soldiers protecting the nest to bite the tool in defense. By carefully removing it, the chimpanzee can then gather the soldiers attached to it and eat them one by one – skilled individuals can obtain between one and six soldiers. Thanks to tools, the Gombe chimpanzees have been seen to eat termites for about two months out of every year, significantly increasing the amount of termites they consume compared to the baboons that can eat them only during the few swarming days. So, similar to the case of nut-cracking in the Taï forest, using tools allows these chimpanzees to reach a new and rich food source that would otherwise be unavailable. Similarly, human societies such as the Hadza, a group of foragers from Tanzania, take advantage of the fact that the swarming always happens after a heavy rainfall and have discovered that by tapping on the termite mounds with sticks in a way that mimics the rainfall, they can lure the termites to swarm and thereby catch them before the rain starts. We see here two species extending the duration of the consumption period of a rich food source.

Tables 3.1 and 3.2 illustrate some of the ethnographic observations of material culture in chimpanzees with some of the populations for which we have enough observations. Table 3.1 is a preliminary list of traits and trait distribution for 11 chimpanzee communities belonging to 9 different populations. It is "preliminary" in the sense that new observations are regularly being described, even as I am making this list. For each cultural trait, the *presence* in each community is very different and sometimes the *form* of the behavior varies as well. In Table 3.2, a more detailed analysis compares the distribution of some termite and bee species in relation to the consumption behavior of the chimpanzees. This illustrates the care needed when discussing cultural differences or abilities. In his 1996 book, *The prehistory of the mind*, Steve Mithen says, "The failure of Taï chimpanzees to use termite sticks is most likely to arise simply from the fact that no individual ... has ever thought of doing such a thing ... or managed to learn it from another chimp before that chimp forgot how to do it, or passed away with his great tool-use secret. This is not cultural behaviour; it is simply not being very good at thinking about making and using physical objects" (p. 77). In reality, the termites that the Gombe chimpanzees are fishing for simply do not exist in the Taï forest! The ecology of the prey, including its presence, influences a cultural behavior, but as we see in Table 3.2, presence alone does not always explain the form of the behavior.

Technology boosts chimpanzee cultural ethnography

Until recently, we have had barely any information about the behavior of the chimpanzees living in the huge Congo Basin forest that extends for over 4 million km^2 through Cameroon, Gabon, Congo, Central African Republic, and the Democratic Republic of Congo. Many field workers have started projects on apes in those forests where chimpanzees cohabit with gorillas, but chimpanzees have proven to be simply too hard to habituate to the presence of human observers and it is a sad reality but without new results, it quickly becomes impossible to finance a research project. However, as I have already mentioned, we run the risk of badly underestimating our knowledge about chimpanzees if all long-term studies on chimpanzees are geographically restricted to West and East Africa. Without data on the Central African chimpanzee populations, how much about tool use and cultural behavior in chimpanzees are we missing? I knew of a new project in the Goualougo Triangle in Congo, where the chimpanzees had very little fear of humans and where new behaviors had already been uncovered,[10] but more studies were needed. This was

[10] Crickette Sanz and David Morgan have been studying the chimpanzees of the Goualougo Triangle for many years and have made some very important new observations on the use of tools by this population (Sanz and Morgan 2007, 2009, Sanz *et al.* 2004, 2009).

Table 3.1 Material culture diversity in chimpanzees. Preliminary list of material culture elements found when comparing a sample of 11 different chimpanzee communities ordered from West to East Africa. First listed are potential cultural traits, including tool use, followed by notations for each community explaining its status: "+", behavior is present; "–", behavior has not been observed; "(–)", behavior has not been observed but an ecological explanation can be found to explain this absence; "blank", no data found about this behavior

	Chimpanzee population										
Pattern	Bossou	Fongoli	Taï N	Taï S	Loango	Goualougo	Gombe	Mahale	Kanyawara	Ngogo	Budongo
With tool											
Driver ant-dip	short/long	long	short	short		C2	long	–	–	–	–
Wood ant-fish	+		(–)	(–)		–	–	+	–	–	–
Termite fish	–	S	(–)	(–)	(–)	C2	S	–/+[1]	(–)	(–)	(–)
Grub extraction	–		+	+	+	–				+	–
Honey dip	–		+	+	C5	C3	+	–	+	+	+
Bee probe	–	+	+	+	+	–	+	–	+	–	–
Ground puncture	–		–	–	B	T	–	–		(–)?	–
Nut crack	+[1]	(–)	+[2]	+[2]	–	–	–[1]	–[1]	(–)	(–)	(–)
Nut extract	–	(–)	+	+	–	–	–	–	(–)	(–)	(–)
Anvil prop	+		–	–			–	–	(–)	(–)	(–)
Inspect stick			+	+	+	+	–	–		+	–
Leaf sponge	+	+[1]	+	+	+	+	+	+	+	+	+
Leaf wadge	+		+	+		+	+	+	+	+	+
Leaf wipe	–		+[1,3]	+[1]		+[1,2]	+[3]	+		+[1,2]	+[3]
Leaf napkin	–	+	–	–		+	+	–		+	+
Leaf dab	–		–	–			–	–	+		–
Wound inspect	–		+	+			+	–		+	+
Marrow pick	(–)		+	+		+	+	–	–	–	–
Brain eat	(–)		+	+		–	–	–		+	–
Eye eat	(–)		+	+		–	–	–		–	–

Table 3.1 *(cont.)*

Pattern	Chimpanzee population										
	Bossou	Fongoli	Taï N	Taï S	Loango	Goualougo	Gombe	Mahale	Kanyawara	Ngogo	Budongo
Missile throw	+	+	+	+		+	+	+		+	+
Stick club	+	+	+	+		–	+	+		+	–
Flail twig	+	+	+	+		+	+	+		+	+
Fly whisk	+		+	+		+	+	–		(–)	
Branch drag	+		+	+	+	+	+	+	+	+	+
Lever	–		–	–	+	+	+	–		+	–
Pestle pound	+		–	–		(–)	–	–		(–)	(–)
Algae scoop	+		(–)	(–)		–	–	–		(–)	(–)
Hook stick	+		–	–		–	–	–		–	–
Self tickle	–		–	–			+	–	–		–
Leaf rain cover	–		–	–	–	+	–	–		–	–
Without tool											
Termite mound-pound	(–)		+	–	(–)	–	–	–		(–)	–
Driver ant-hand			+[1]	+[2]		–				(–)?	
Herbal pith	–		–	+						+	+
Strychnos pound	–		+[1]	+[2]	+	–	+[2]	–		(–)	–
Fruit pound			+	+		+	+			(–)	+
Skull pound	(–)		+	–		–	–	–		–	–
Muzzle rub			–	–			–	+		–	–
Ground day-nest	–		+	–		+	–			–	+
Day cushion	–		+	+		+		+		+	
Day nest			+	+			+	+		+	+

For driver ant-dip: Short = short wand (mean = 37 cm), Long = long wand (mean = 70 cm), S = single tool use, C = complex or sequential tool use with indication of the maximum numbers of tools used in the sequence.

For termite fish: 1 = tool use not reported for M Group, but for B and K groups.

For ground puncture: B = done for bee nest, T = done for termite nest.

For nut crack: 1 = nuts of *Elaeis guinensis* only, 2 = nuts of *Coula edulis*, *Panda oleaso*, *Parinari excelsa*, *Detarium senegalense*, and *Sacoglottis gabonensis.*

For leaf sponge: 1 = water collected with a twig and not with leaves.

For leaf-wipe: 1 = wipe off feces, 2 = wipe off semen, 3 = wipe off sticky fruit juice.

For driver ant-hand: 1 = hand inserted deeper than elbow in the ant nest, 2 = hand inserted to the wrist.

For *Strychnos* pound: 1 = pound fruits when decayed and black, 2 = pound fruits when fresh and green.

References: **Bossou:** Sugiyama and Koman (1979), Sugiyama (1994), Matsuzawa *et al.* (2001), Nakamura and Nishida (2006). **Fongoli:** Pruetz and Bertolani (2007), Bogart and Pruetz (2008, 2011), Pruetz (pers. com). **Taï North and South:** Boesch and Boesch (1990, pers. obs.). **Loango:** Boesch *et al.* (2009, pers. obs.), Head (pers. com). **Goualougo:** Sanz *et al.* (2004, 2009), Sanz and Morgan (2007, pers. com). **Gombe:** Goodall (1986). **Mahale**; Nishida and Hiraiwa (1982), Nishida and Uehara (1980), Uehara (1982), Collins and McGrew (1987), Corp *et al.* (2009). **Kanyawara:** Whiten *et al.* (1999). **Ngogo:** Watts (2008), Langergraber *et al.* (2011, pers. com). **Budongo:** Whiten *et al.* (1999), Zuberbuehler, Wittig and Crockford (pers. com).

Table 3.2 Ethnographic approach to material culture diversity in chimpanzees when consuming termite and bee species at different sites across Africa (adapted from Sanz *et al*. 2004, Sanz and Morgan 2007, Bogart and Pruetz 2008).

		Chimpanzee population																				
		Fongoli			Taï			Loango			Goualougo			Gombe			Mahale			Bili		
Termite species		P	C	T	P	C	T	P	C	T	P	C	T	P	C	T	P	C	T	P	C	T
Macrotermes	bellicosus	+	+	S	–			–			+	+	C2	+	+	S	–					
	subhyalinus	+	+	S	–			–			+	+	C2	–			–					
	ivorensis	–			+	+	–	–			–			–			–					
	muelleri	–			–						+	+	?	–						+	–	
Thoracotermes		–			+	+	B	+	–	–	+	–?		–			–			+	+	B
Pseudocanthotermes	spp. ?	–			–									+	–		+	+	S			
Cubitermes	sp.				+	+	–													+	+	B

		Chimpanzee population														
		Taï			Loango			Goualougo			Gombe			Mahale		
Bee species		P	C	T	P	C	T	P	C	T	P	C	T	P	C	T
Apis	mellifera	+	+	S	+	+	C3	+	+	C3	+	+	S	+	+	S
Hypotrigona	gribodoi				–			+	+	C3						
	ruspolii				–			+	–							
Meliponula	bocandei	+	+	–	+	+	C	+	–							
	nebulata				+	+	C5	+	+	C3						
	lendliana				+	+	C3	+	–							
Xylocopa	spp.	+	+	S	+	?		–			+	–		+	+	S

P = Species present, C = Species consumed by chimpanzees. T = Species consumed with the help of tools, either S = Single tool used, or C = complex or sequential tool use with indication of the maximum numbers of tools used in the sequence. B = Termite mound banged against hard surface (this behavior is not tool use but sometimes named "proto-tool use").

References: **Fongoli:** Bogart and Pruetz (2008, 2011). **Taï:** Boesch and Boesch (1990). **Loango:** (Boesch et al. 2009, pers. obs.). **Goualougo:** Sanz *et al.* (2004), Sanz and Morgan (2007). **Gombe:** McGrew *et al.* (1979), McGrew (1992). **Mahale:** Collins and McGrew (1987), Nishida and Uehara (1980). **Bili:** Hicks (2010).

my main motivation when in 2005 I began a field project to study the chimpanzees and gorillas in Loango National Park in Gabon, one of the most beautiful, pristine places in Africa. We quickly discovered that the forest in this park was very impressive and very wet; positioned between the Atlantic Ocean and a very large lagoon, where large sections of the forest were inundated during the rainy season and intercepted by large, permanent swamps that both chimpanzees and gorillas were happy to cross. We had to quickly learn how to crawl through the water and remove the leeches that live there and were only too happy to feast on us like their other warm-blooded prey. We were regularly trapped in impossibly dark water full of spiny raffia leaves, with each of us struggling to overcome our irrational fear of being attacked by crocodiles, electric fish, or buffaloes. We then had to paddle back to find better ways to pass through swamps that were new to us. However, GPS technology[11] had opened up a whole new world and we rapidly mapped all the swamps so that we could avoid those crossings that we estimated to be too treacherous. Following the apes through this forest remains an adventure; there are breathtakingly beautiful places and extremely muddy crossings, but in the end, our efforts paid off.

One day, after having spent hours searching in vain for the elusive chimpanzees, I was in the camp with Josephine Head, the student leading the project, when she mentioned that two months ago they had followed the calls of chimpanzees to an old bee nest surrounded by many sticks covered with honey. I jumped off my chair at the thought of what sounded like tools! She was not sure, but luckily she had taken a perfect GPS bearing and we decided to go and have a look the next morning. We reached the tree after walking for just two hours and I could not believe my eyes. Under a fallen tree in which a large hole revealed an old bee nest, there was a large pile of sticks and leaves on the ground (see Figure 3.5), and Josephine assured me that they were all covered with honey when she first found them. A closer look revealed that many of the sticks had been cut at both ends to have a specific length and the leafy side branches had been removed. In addition, we could clearly see from the wear presented at either end that they must have been used in different ways. Four medium-sized branches up to 1.60 m long seemed to have been used to pound the nest entrance that had been closed by the bees with a very hard resin and wood. The 22 thinner sticks could have been used to break open chambers within the nest and enlarge the holes in order to access to the honey. Sixty-five sticks seemed to have been used as collectors to extract the honey; those were easily recognizable as they all had brushed ends which would have allowed more of the honey to be removed from the up to 1 m deep nest. Finally, some 15 sticks had

[11] Global Positioning System (GPS) devices are small, hand-held devices that, once connected with at least three satellites, are able to provide precise coordinates of our position wherever we are in the world. That is the theory anyway. However, out in the African jungle, it was at first very frustrating as you needed to be in a relatively open space within the forest and had to wait for at least 30 minutes to get a bearing, which was not very useful when following shy and quick-moving animals. However, after a few years, the technology has improved and more satellites are available so that now, you can track in real time the exact position of yourself or your target animal with a precision of less than 10 m. A wonderful and tremendously useful progress for fieldwork!

Figure 3.5 The first *Melipone* bee nest that Josephine Head showed me in April 2005, where we collected 106 tools after the chimpanzees had eaten all the honey. The insert on the right provides details of some of the types of tools found at the site, including a thick pounder, several brushed collectors, and a medium-sized enlarger.

both brushed and blunt ends, suggesting that those were used for both enlarging and breaking open chambers in the nest and for collecting the honey. I was totally thrilled; after 30 years of following the chimpanzees in West and East Africa, for the first time I was seeing evidence of obligate sequential tool use, in which chimpanzees used different types of tools one after the other to reach a goal. In this case, they had to have used a minimum of three tools – a "pounder" followed by an "enlarger" and finally a "collector" – to get at the honey.

Seeing sequential use of tools in chimpanzees was a first for me. Was there something special about honey that would require such a sophisticated technique? Surveying the forest around camp, we realized that the forest of Loango is full of bees of different species, some making small nests, some large ones, some high up in trees, and some underground. Loango chimpanzees were eating honey from three different types of nests.[12] In addition, chimpanzees in Congo, Cameroon, Central African Republic, and Nigeria had also been suggested to use similar sequential tools and in all cases, it was to acquire honey. Honey is a very energy-rich food as it is full of sugar. Unlike in Europe, where the honey we eat is

[12] To better compare the honey extraction of the Loango and Goualougo chimpanzees, we systematically collected specimens of some of the bee species present and had a specialist identify them (see Sanz and Morgan 2009). For observations of potential tool use to extract honey in other Central African chimpanzee populations, see Fay and Carroll (1994), Hicks *et al.* (2005), Deblauwe *et al.* (2006).

produced only by honeybees that often cause painful stings, in Africa, the honey is also produced by stingless bees, called *Melipone*, which are much easier to raid. However, *Melipone* bees carefully choose tree holes that are high up in trees and have small, hard-to-access entrances that they seal with resin. Consequently, different tools are required to access this well-protected food source. Once again, a material culture was born out of necessity – hammers for hard nuts, sequential tools for honey.

Although it was now known that chimpanzees use sequential tools, it was unknown exactly how they did so as there had not yet been any direct observations. More technology was required to uncover this chimpanzee inventiveness. As we were uncovering sequential tools in Loango, Crickette Sanz and David Morgan had the brilliant idea of using remote camera traps to resolve one of the mysteries they encountered in Goualougo Triangle forest. When I visited David Morgan in the Goualougo Triangle in early 2003, he showed me one of these places in the forest where they had found sticks inserted in the ground around a subterranean termite nest. At that time, he did not know exactly what the chimpanzees were doing with them, although he assumed it had something to do with the almost invisible termite nest. Taking one of the sticks left on the ground by a chimpanzee, I tried to insert it 30 cm into the ground vertically around the area where the chimpanzee had left it. When I first perforated the ground, nothing happened. The second time, I tried perforating the ground more to the left but again nothing happened. As I pushed it into the ground a third time, I suddenly felt less resistance in the soil; I had reached something (see Figure 3.6). Removing the stick and taking a thinner twig, which David said he sometimes finds associated with the thicker one, I inserted it into the tunnel left by the thicker stick, and copying the movement of the Gombe chimpanzees when fishing for termites, I shook the twig gently inside the tunnel to stimulate soldiers to bite and then pulled it out. Nothing … Were we totally wrong about what we thought these chimpanzees were doing? I tried three more times and the last time, in front of David's incredulous eyes, three large termite soldiers were attached to it! I tried three more times and each time caught three to four soldiers which I shared with David; so this looked to be an especially astute way to get at termites situated well underground. The video camera traps gave us a definitive answer.[13] Over the years, Crickette and David caught on film hours of chimpanzees using the two types of tools and very efficiently fishing termite soldiers from their underground nests. This was revolutionary in that, for the first time, we could see chimpanzees feeding on hard-to-access underground resources, an ability that had previously

[13] Video camera traps are normal, small video cameras combined with motion detectors, similar to what you use to light your garden at night to find your way home. Both are packed in a waterproof box to protect them from the high humidity of the forest and are placed so that they systematically film all large animals that move in front of the camera and continue to film as long as any movement is detected. As the video cameras are totally silent and hidden behind leaves, they are rapidly accepted by even very shy animals. The huge benefit in this case is that no human needs to be present and very detailed sequences of the activity of the animals are filmed. All that is needed is to regularly download the captured video sequences and change the cameras' batteries. This has opened the way to unobtrusively observing animals in remote locations.

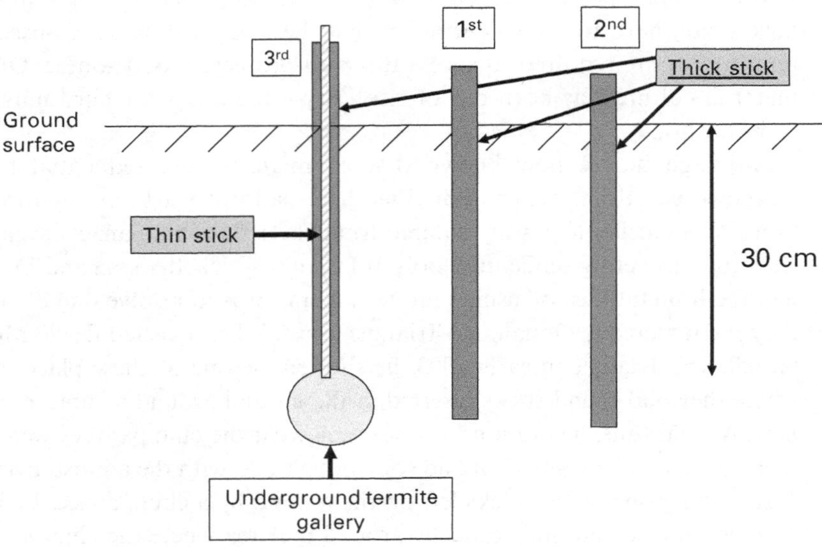

Figure 3.6 **Goualougo chimpanzees' technique to access underground termite nests** in the Nouabele Ndoki National Park of Congo (adapted from Sanz *et al.* 2004). The chimpanzee, 1st and 2nd, punctuates the ground with a thick stick to search for the invisible nest structure. Once located, they insert a thinner stick, 3rd, into the tunnel made by the thick stick that reaches the nest and fish out the soldier termites that bite the intruding thinner stick.

been considered to be unique to humans among primates.[14] Thus, we have another example of human uniqueness crumbling in the face of evidence from the field!

There were also mysteries awaiting us in Loango and camera traps proved equally helpful in uncovering them. We regularly found 30–40 cm deep holes in the forest with sticks nearby that had been modified and were now covered with earth. We searched many of these spots carefully but remained unable to determine how or why such holes had been dug. Then, after two years, we found such a hole that had two sticks vertically inserted into the ground around it. I tried to see if I could reach something by perforating the ground around the hole like I had done in Goualougo. I repeatedly inserted the stick 50 cm into the ground at different distances around the hole. Suddenly, while I was perforating close to the edge of

[14] We already had some hints that chimpanzees might extract underground tubers in the dry Ugalla Game Reserve in western Tanzania (Hernandez-Aguilar *et al.* 2007) and in the Tongo region in eastern DRC (Lanjouw 2002). Exploitation of underground resources was often seen as a major way for a group of individuals to sustain itself in poor and very dry habitats (Hatley and Kappelman 1980). The case of the Bushmen in Botswana that heavily rely on underground tubers for food and water during many months of the year clearly illustrates how important this can be and it suggests how such an innovation can open a new habitat type to humans that was previously considered to be hostile (Marlowe 2010).

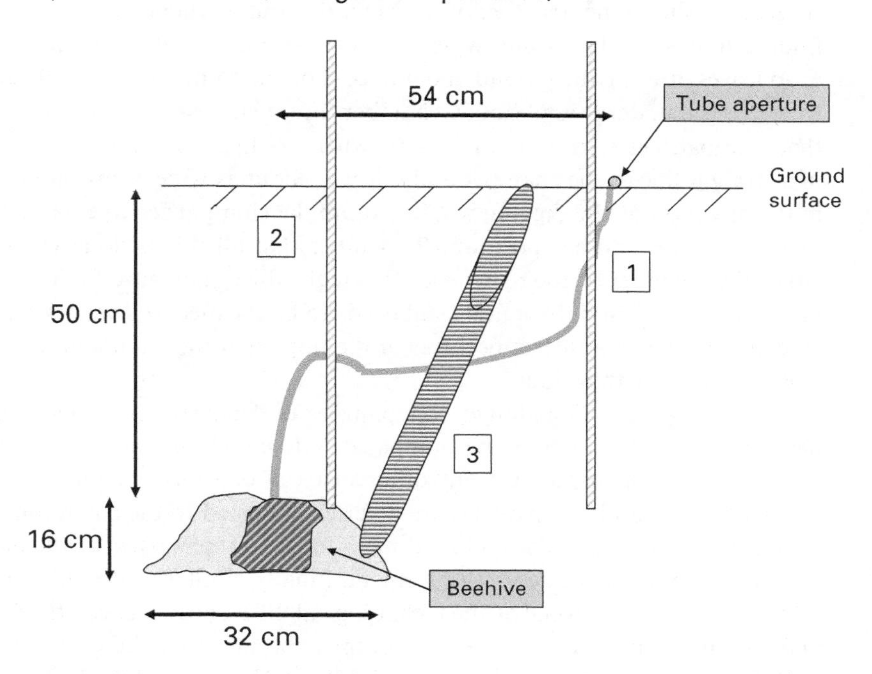

Figure 3.7 **Underground honey extraction by chimpanzees** in Loango National Park, Gabon. The tiny stingless bees enter the small tube aperture at the surface and then follow an irregular tube until they reach the hive, where honey cells surround the comb with all the eggs and the queen. The hive (in striped dark gray) is located within an open chamber in the ground (in plain gray) that seems to allow the hive to grow in size with varying honey production. The sophisticated technique used by these chimpanzees combines the use of perforating tools to locate the honey (1 and 2), followed by the digging of a tunnel to reach it, which is often done by hand with the assistance of a 2nd tool (3). It remains unclear how the chimpanzees decide where to dig the tunnel to reach the honey.

the hole, I felt a change in the resistance of the soil and when I took the stick out, the tip was covered in honey. There was a beehive down there and I had struck a chamber full of honey! I asked our local pygmy trackers if they knew about this variety of bees. They told me, yes, they knew about them, but they did not eat them because the village elders said they are too deep and therefore too difficult to reach. Following this discovery, I inspected many places where the chimpanzees had dug into the ground in different places in the forest and realized that all of them were near underground beehives. Whenever successful, they seemed to dig an oblique tunnel so as to reach the underground honey chamber horizontally, possibly to avoid too much soil mixing with the honey (see Figure 3.7).

Honey chambers can be as deep as 1 m underground. How do the chimpanzees know where these chambers are? In addition, the bees build a small wax tunnel to reach the chamber, which can be positioned up to 70 cm laterally from the tunnel opening and the wall of this tunnel is very soft and follows a very irregular course (Figure 3.7 shows the structure of a nest we dug out to understand how challenging finding honey is). The 4-mm wide wax tube opening on the surface is hidden by dead leaves littering the ground and it once took me 45 minutes and 30 perforations to locate the underground honey chamber. From successful excavations, we know that chimpanzees seem to know exactly where to dig as the tunnel they make goes in a straight line to the chamber and it is just about as wide as my hand. Any error in the direction of the digging would require the chimpanzee to enlarge the tunnel or restart from another position. Remember that all this is done without seeing either the chamber or the many roots through which they have to dig to reach the honey. How they find these nests still needs to be clarified and we have now placed 30 camera traps near intact beehives in the hope that the chimpanzees themselves will show us how they do it.

The Loango and Goualougo chimpanzees perfectly fit the general pattern we have seen in the Taï and Gombe chimpanzees; to enrich their natural diet in significant ways, they incorporate many different types of tools in their repertoire. The tool techniques and complexities are precisely adjusted to the environmental challenges they face and this has allowed them to exploit new parts of the habitat that would otherwise not be accessible to them. Finally, similar to the cultural distribution of nut-cracking, some of the technological differences between the Goualougo and Loango chimpanzees developed despite the fact that these two populations exploit exactly the same species of stingless bees. However, while the Loango chimpanzees use up to five types of tools to access the honey from the *Meliponula nebulata* beehives, the Goualougo chimpanzees only use three types of tools to raid the hives of this same bee species (Table 3.2). These technical cultural processes permit access to new resources and enlarge the niche in which animals live and reproduce.

Cumulative cultural evolution in chimpanzees

How did sequential tool use evolve? Reconstructing past behaviors is always challenging because behavior does not leave fossil remains behind. Here, I suggest a scenario based on all our present-day observations of chimpanzee behavior (see Figure 3.8). All chimpanzee populations have been seen to extract different types of food from holes with their hands, but only a limited number of populations have added a level of complexity to this simple behavior by using a tool to access food, mostly in difficult-to-reach holes. This is the case for the Taï, Loango, and Goualougo chimpanzees, who all extract honey from beehives with one type of stick (see the second box from the top in Figure 3.8).[15] From observations of the Taï

[15] For the honey extraction technique, I mention only the Loango and Goualougo chimpanzees here as we have directly observed this behavior in these groups. We should not forget that tool remains found in association with beehives have suggested sequential tool use in different populations of chimpanzees throughout the Central African range of this species (see footnote 12).

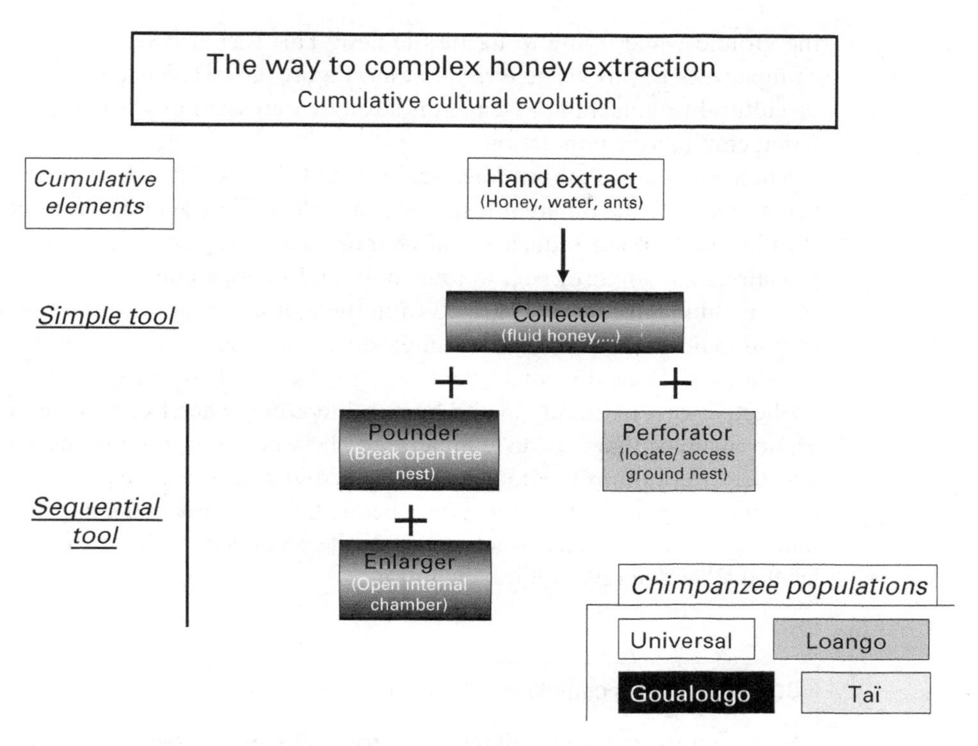

Figure 3.8 **Cumulative cultural evolution leading to the complex honey extraction technique**, as suggested by the observations of different techniques seen in wild chimpanzee populations. Each of the steps presented in this diagram has been observed to be used by three different chimpanzee populations (each shade of gray in the boxes represents one population where the level has been observed). Taï, Loango, and Goualougo chimpanzees have been seen to use one simple collector to extract honey from large open beehives (second box from the top), while only the Loango and Goualougo chimpanzees have been seen to first use a pounder to access protected beehives (third box from the top on the left). Enlargers have been further added to the technique to access the internal chambers of especially large and deep hives (lowest left box). In contrast, for underground beehives, the Loango chimpanzees have been seen to add a perforator to the simpler collector technique (third box from the top on the right). Thus, a progressive accumulation has led to more complex tool use.

chimpanzees, we know this is most successful when trees with beehives inside fall down, thus making the honey more easily accessible to the chimpanzees. However, if one wants to access honey from other bee species, then a thick branch used to pound open the protection surrounding intact nest entrances must be incorporated into the behavioral sequence. This is what we have observed in Goualougo and Loango, where the chimpanzees use pounders to break open nests before they begin using extraction tools. Finally, for the deeper and more structured nests of the large stingless bees commonly found in Central African forests, chimpanzees would need to add a third tool type in their tool set to open chambers within the nests. Alternatively, for underground stingless beehives, the Loango chimpanzees did not include a pounder in the sequence, but instead incorporated a stick to perforate the ground while trying to locate the nest. This was also seen in the Goualougo chimpanzees but to locate termite nests (Figure 3.6). Hence, a cumulative process of cultural technical improvement is strongly suggested by the comparison of different chimpanzee populations.

These cumulative innovations were most likely not due to one single individual using them, for the different techniques are all still in use in different populations. Within one tool-use sequence, Taï chimpanzees have been seen to use two tools, Goualougo chimpanzees up to four tools, and Loango chimpanzees up to five. As the Taï chimpanzees are separated from these other groups by about 3000 km and several major rivers, it would be unreasonable to assume one inventor for all these innovations. Thus, this might be a very good example of what has been labeled a "collective cast of mind" in the human literature, whereby the invention of one group member that has disseminated to the whole group could be improved by another invention from another individual, and that collective product would then disseminate again to the whole group before being improved once again. This accumulation of innovations, which improves the product at each step, is what has been labeled a "cumulative cultural evolution."

Box 3.1: What is cumulative cultural evolution?

Many human cultural artifacts seem too complex to have been invented at one time by one person, and therefore, many cultural products are thought to result from a cumulative evolutionary process. Cumulative cultural evolution, by which one individual improves on a socially acquired cultural trait and this new innovation is then transmitted to other group members who might, at a later time, improve on it further, has gained special status as it was proposed to be at the base of the cultural complexity we see in many human cultural products (Ford 1962, O'Brien *et al.* 2001). It is this process of successive accumulation of changes by different individuals that characterizes this cumulative process.

In humans, this process has been suggested to have led to many observed cultural changes, such as changes in the shape and size of pottery, carpet weaving, basketry and projectile points (e.g. Ford 1962, Shennan 2000, O'Brien *et al.* 2001, Tehrani and Collard 2002, Jordan and Shennan 2003, Beaune 2004). Psychologists, who sometimes call this process the "ratchet effect," have argued that only humans attain enough faithfulness in their copying of others' innovations to make such a process evolutionarily beneficial (e.g. Galef 1992, Boyd and Richerson 1996, Tomasello 1999, Tennie *et al.* 2009).

Although archaeologists have very nicely shown evidence for accumulation of cultural changes, how a cultural trait is defined remains mainly descriptive without any direct functional connection. Therefore, it is difficult to understand why a pot with a 2- versus a 3-cm neck should be considered as two distinct cultural traits (e.g. Ford 1962). To circumvent this issue, I suggest that we should consider only changes that result in a measurable improvement in efficiency or a change in function. To clarify, below are five suggested criteria that I believe need to be considered before cumulative cultural evolution can be claimed in any animal species, including humans.

- *Heritable continuity*: accumulation of changes within one technical chain or lineage.
- *Functional efficiency*: each level within the technical chain should represent a technical improvement in solving the task.
- *Temporal increment*: accumulation of changes should happen over a time period that excludes successive improvements by the same individual.

When no archaeological data are available:

- *Nested distribution*: the distribution range of the more complex technical level should be more restrictive than the one of simpler level.
- *Concomitant use*: different technical levels should be observed concurrently.

I am fully aware that, as seen in humans, accumulation of changes could also lead to simplification of a task or that successful levels could spread more widely and therefore contradict the nested distribution criteria. However, for comparative purposes, we need criteria that could be compared between species, including those for which no archaeological data are available. The best example of such a cumulative cultural process comes from New Caledonian crows, which, in a large area, use straight twigs to extract grubs from branches, but in a smaller, more restricted area, fashion more efficient hooked twigs to do so (Hunt and Gray 2003).

When taking into account the most recent knowledge on tool use in chimpanzees (Boesch *et al.* 2009, Sanz *et al.* 2004, Sanz and Morgan 2007), a cumulative cultural evolutionary process can be observed in some technological chains as illustrated in Figures 3.8, 3.9, and 3.10 (see also Boesch 1993a, 2003, Boesch and Tomasello 1998, Matsuzawa 2001).

For a discussion about cumulative evolution, see: Ford 1962, Galef 1992, Boyd and Richerson 1996, Boesch and Tomasello 1998, Tomasello 1999, Ambrose 2001, 2010, Shennan 2000, O'Brien *et al.* 2001, 2010, Boesch 2003, Whiten *et al.* 2003, Tennie *et al.* 2009.

A similar cumulative cultural evolution process can be suggested in the case of hammer use by chimpanzees (see Figure 3.9). These improvements center around the increased complexity in objects included in the hitting movements. First, flailing of insects, conspecifics, or snakes with attached leafy sapling branches has been reported in all chimpanzee populations, as has clubbing playmates or social competitors with wooden sticks (see the boxes in white in Figure 3.9). One innovation was added to this universal by the Gombe and Taï chimpanzees, who pound hard fruit directly against tree trunks or roots to break them open. This is similar to clubbing, but in this case, the food is held in the hand and it serves a different purpose, namely to access food (see Figure 3.9). A second innovation was seen in West Africa, where it is now a hammer that is used to pound the hard food that is first placed on a

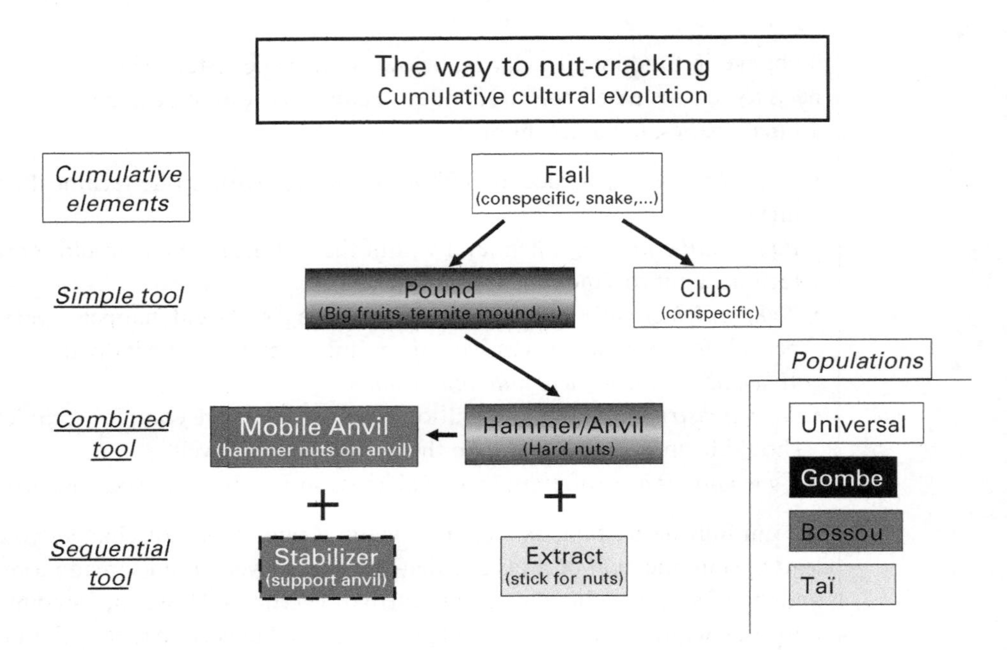

Figure 3.9 **Cumulative cultural evolution leading to nut-cracking techniques.** Here again, a behavior seen in all chimpanzee populations is elaborated through the successive addition of new elements. First, a simple tool is integrated into the pounding movements, as seen in the Gombe, Bossou, and Taï chimpanzees (second box on the top left). Then the introduction of a second tool led to the invention of combined tool use as seen in the Taï and Bossou chimpanzees (third box on the right). Combined tool use occurs when at least two tools are used at the same time (e.g. an anvil is selected and used at the same time as a hammer). For each step of the accumulation process seen in a given population, I have listed the different resources gained or involved. The stabilizer box is dashed as it has only been observed twice so far in the Bossou chimpanzees.

hard surface. From this last innovation, two parallel additions can be distinguished, both examples of sequential tool use: the Taï chimpanzees added a stick in order to extract nut remains embedded in cracked shells and the Bossou chimpanzees added mobile anvils on which they placed nuts. A final innovation that has been observed a few times in the Bossou chimpanzees was the placement of a stone under the mobile anvil to stabilize it before hitting the nut with the hammer (see Figure 3.9). As for the honey extraction example, for each step in the cumulative process, the tool technique improved the access to a given food resource and is used effectively in at least one chimpanzee population, but it was later elaborated for a new food source. Thus, for both honey extraction and nut-cracking, we observe a parallel and similar cumulative cultural process developing from a universal, simple chimpanzee technique.

Cumulative cultural evolution has recently been proposed by some psychologists to be a central element in distinguishing humans from other animal species.[16] Their

[16] Galef (1992), Heyes (1993), Tomasello *et al.* (1993), Tennie *et al.* (2009).

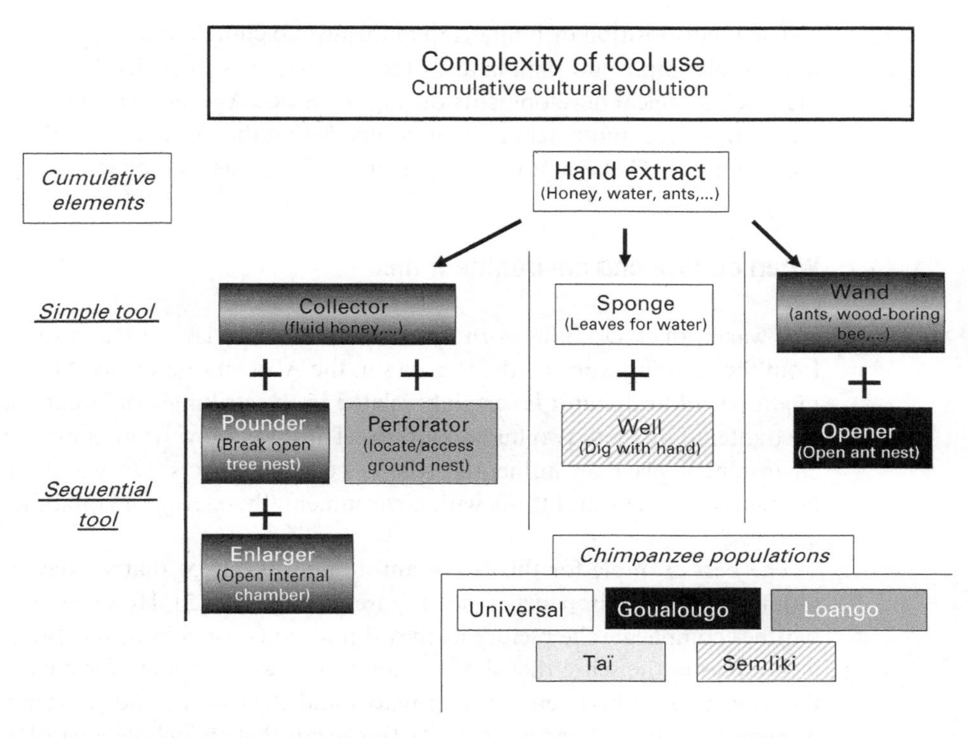

Figure 3.10 **Cumulative cultural evolution in chimpanzees** starting with extracting objects from holes with the hand. All chimpanzee groups have been seen to extract different objects/food from holes with their hands. From this, a cumulative cultural process is developing and leading to more complex techniques in three different directions in wild chimpanzees, each of which is seen in one or more different chimpanzee populations.

argument is that only humans are able to faithfully copy the actions of others, while chimpanzees and some other animals are only able to copy the objects or the effects of objects (see Figure 2.3); therefore, according to them, cultural evolution is rare in animals. As a result, they will probably greet the scenarios I present in Figures 3.8 and 3.9 with some skepticism. However, the ethnographic comparison of the techniques used in different chimpanzee populations leaves no doubt that a recurrent increase in the complexity of cultural traits based on similar behavioral elements exists. Obviously the alternative explanation would be that each step has been invented totally independently from one another; however, that would not explain the striking geographic clustering of changes within a specific technical lineage. Furthermore, each of those steps has been seen to increase the performance of the previous steps in the conditions it is performed (see Figure 3.10); a perforator improves the technique of using only a collector when searching for underground beehives, while a well allows for cleaner drinking water than would the use of a sponge alone, and a stick allows one to extract more of a nut kernel than hammering the nut alone would have. Finally, within each technical lineage, the simpler form has a much wider distribution range than the more complex form, suggesting that the latter originated from the former.

The predisposition of humans to innovate and cumulate change might indeed be much more important than in any other species, as is so evidently seen in the complex technological developments during the Stone Age, accelerating into the more recent Iron Age and reaching high points during the industrial revolution and the computer age. This is a point I will return to in the last two chapters of this book.

When culture and environment mix

The warm, thick clothing worn by the Inuit in Greenland is easily distinguishable from the clothing worn by the Pygmies in the African rainforests. Much about this might be cultural, but it is certainly related to the strikingly different temperatures encountered by these two human groups. Thus, we know from human culture that environment plays an influential role in cultural aspects. Do we observe a similar blending of cultural traits with environmentally relevant adaptations in animal cultures?

The best example for this is the ant-dipping behavior that we have seen in the Gombe and Taï chimpanzees (see Figures 3.11 and 3.12). However, Bossou chimpanzees complicate the picture as they dip for ants using both the Taï and Gombe techniques in the sense that they use long tools when dipping for ants on trails, as the insects move between foraging places and their nests, and shorter tools when dipping for ants in their nests. Does this mean that the whole idea of two culturally different solutions is nonsense or should we look for another explanation? As expected, those on the "human superiority side" glow with pleasure at this example and immediately suggest that ant-dipping is mainly affected by ecological conditions.[17] Actually, this has turned out to be an illuminating example of how *cultural solutions and environmental factors interact* to shape material culture. Bossou is in Guinea, on the border of Liberia and Côte d'Ivoire, only some 200 km away from Taï. To understand what led to this difference in ant-dipping behavior between the Bossou and Taï chimpanzees, we looked at how much of it could be explained by the behavior of the ants, especially the speed the ants climb up the stick, how many ants attack intruders and how persistently they do so. Taï chimpanzees use shorter sticks than Bossou chimpanzees when dipping for ants at the nest and this difference could not be explained by a difference in ant behavior. Tool length seems to be a response to how quickly the ants climb the stick in Bossou chimpanzees, as they use longer sticks to fish ants at their nest, where they are so much more numerous,

[17] The long-term research project on the Bossou chimpanzees was initiated by Yukimaru Sugiyama in 1976 and, together with Taï, has provided the main source of information on the chimpanzees of West Africa. The chimpanzees in this research site are confronted with a special problem: they have a village in the middle of their territory, which has led to their total isolation from contact with other wild chimpanzee groups (Sugiyama and Koman 1979, Sugiyama 1981, 1984). This study of ant-dipping (Humle and Matsuzawa 2002) was solely intended to explain differences in ant-dipping techniques within the Bossou chimpanzees, and not to make any comparisons between different chimpanzee populations as some misrepresented (e.g. Laland and Hoppitt 2003, Laland and Janik 2006, Tennie *et al.* 2009). Only if it had done this might it have had an influence on the question of culture in chimpanzees.

Figure 3.11 **Ant-dipping culture in Gombe chimpanzees.** An adult male holds a long stick in the nest entrance (upper left) and once ants have swarmed about halfway up the length of the stick, he rapidly encircles the stick with his right hand and pulls the stick through with his left hand (central part), before swiftly bringing all of the ants on the stick into his mouth in one clump (upper right).

than on trails. However, it does not explain the difference in the length of the tools found between Bossou and Taï chimpanzees. In fact, the Taï chimpanzees should use longer tools if the speed of the ants was as important there as it seems to be in Bossou, but exactly the opposite is true. Finally, how aggressively the large black driver ants attack the stick explains why the Bossou chimpanzees do not extract these ants with their hands, but this does not prevent the Taï chimpanzees from extracting them with their hands alone. In Taï, it seems that the cultural traditions are stronger motivators than the pain and aggressiveness of the ants.[18]

Showing the presence of cultural traits is complex as some aspects within the same cultural behavior technique, as in ant-dipping, for example, can be explained by the presence of an insect species or some aspect of the insect's behavior, while

[18] In a nice example of collaboration between two research groups, we developed a joint protocol to measure the effect of ecological factors on the ant-dipping behavior of the chimpanzees in Taï and Bossou (Möbius *et al.* 2008). We precisely measured how quickly the ants climbed up a stick that had been placed in a nest entrance, like the chimpanzees so often do, as well as counted the number of ants that attacked it. Furthermore, we precisely identified the five (not two, as I had previously thought) different species of driver ants eaten by the chimpanzees in both populations. This illustrates the complexity of controlling for all possible environmental factors while at the same time showing that when we do, we can obtain a much more precise idea of how culture intermingles with environment.

Figure 3.12 **Ant-dipping culture in Taï chimpanzees.** In the center picture, an adult female in Taï arrives at a driver ant nest. In anticipation of the need for more than one tool, this female fashioned two sticks before arriving at the nest entrance. She then inserts the end of one stick into the ant nest (upper right), and once the ants have climbed about 10 cm up the stick, she will eat them directly with her mouth (lower right).

others are cultural. This produces a blend of environmental and cultural responses, which probably exist in many other examples.

Material culture in other species

Orangutans have also been seen to use tools and the best example is their use of sticks to extract seeds from the very spiny *Neesia* fruit in northern Sumatra. Once the fruit are partly open during the dehiscence process, the visible seeds are extracted. The orangutans on the northern side of the Alas River have been seen to use sticks that they hold in their mouths to extract the seeds, while the orangutans on the eastern side of the river forcefully split the fruit open with their hands and eat the seeds without using tools.[19] So here we see a very similar scenario to the one we have seen for the nut-cracking technique in chimpanzees, where a river acts as a boundary between the two material cultural traits.

Capuchin monkeys have surprised the scientific world for how extensively some populations use tools. Seeing a bipedally standing, 3 kg capuchin monkey lifting a

[19] van Schaik and Knott (2001).

2.5 kg hammer to crack nuts is very impressive! For example, the capuchin monkeys from Boa Vista in the province of Piaui, Brazil, select and carry stone hammers and nuts to the large stone anvils that they crack the nuts on.[20] On average, they use stones that weigh one-quarter to one-half as much as they themselves do. Populations of bearded capuchin monkeys in the dry forests of Caatinga in Brazil have been seen to not only use stones to crack nuts but also to dig for underground resources and use twigs to inspect holes in trees and on the ground. At the same time, other populations of bearded capuchins have not been observed to use tools and at least some of them live in richer forests. Therefore, we might see in this species of capuchins again a mix between ecological influences and cultural solutions.

Many other animal species, including elephants, California sea otters, dolphins, some fish, and many monkey and bird species, use tools to get at food or to protect themselves from predators or intruders. Clearly the use of tools has emerged in many species depending on the specific needs they have encountered. For some birds and fish, it could be argued that a genetic component might be responsible for their relatively rigid forms of tool use – rigid in the sense that there are one or two single forms of tool use observed in a taxonomically homogeneous unit, like the California sea otters – but this does not apply to the more flexible types of tool use observed in other species. The New Caledonian crow provides a very nice example of complex tool use in birds. They manufacture and use hook-tools to aid in capturing prey hidden in tree holes.[21] These crows live in forests in family groups and they have been observed to use tools to forage in live and dead trees for different types of insects, including millipedes and cockroaches. They even carry tools around as they explore different trees for sometimes more than 30 minutes at a time. Using their beak skillfully, they cut one end of the leaf in such a way as to produce a clear hook-shaped end with which it is possible to extract insects from their holes. An experiment done with captive hand-reared juvenile New Caledonian crows showed that, without any previous experience, they spontaneously made and used tools, which suggests that at least part of this skill could have a genetic component.[22] However, in the wild, New Caledonian crows produce two types of tools in a rather standardized way and it was shown that different populations produced tools with different levels of complexity. As the more complex tool type was only found in a limited area within the geographical distribution of the simpler tool use, it was suggested that one was an earlier form of the other. This pattern makes it clear that if there is a genetic component, it is not enough to explain the observations, which

[20] Visalberghi *et al.* (2007, 2009) have described this impressive behavior of capuchin monkeys and how inventive and skilled they are at solving this environmental challenge. Caatinga is a dry forest where a lower species diversity and lower net productivity has been found than in many other dry and wetter forests. The capuchins there might face severe energy bottlenecks and they would gain much from using probing twigs and stones to get at underground resources and crack nuts (Moura and Lee 2004, Lee and Moura 2005, Canale *et al.* 2009).

[21] Hunt (1996) and Hunt and Gray (2003) provide vivid descriptions of the tool-use behavior of the New Caledonian crows and the important flexibility within this species.

[22] Kenward *et al.* (2005) showed that part of the abilities required for tool use in New Caledonian crows are found in captive hand-reared juvenile birds.

points to a similar mixture of social and environmental influence like that we have seen in chimpanzees. Thus, this strongly suggests a social learning influence in a bird group famous for their intelligence.

As we have previously seen with the example of Imo, culture is observed in many macaque troops, including those living in environments where humans have always played an important role. In this species, culture includes the social propagation of new feeding habits as well as some non-adaptive behaviors, like stone handling, but it also allows for the expansion of their niche within a habitat by gaining, for example, access to water. This can have important consequences as it was shown that some males from Imo's troop swam across bodies of water to reach other islands and this has altered important aspects about the dispersal possibilities of the troop.

Material culture seems to be present in a number of different animal species, all of which increases their access to new food resources or increases the duration of access to a rich food source. This flexible way of improving access to food is generally advantageous and we should not be surprised if it can be found in a large variety of species. Flexible learning opens the door to material culture and this has been shown to directly increase with increases in brain size in different animal groups.[23]

History of material culture: chimpanzee Stone Age

Human culture has been shown to have a long history and the whole field of archaeology has been dedicated to the study of our past cultures. History is the best testimony of the transmission of cultural traits over generations, something some have denied to animals. Archaeology is the study of uncovered signs of an old civilization. But how do you know where to find old buried nut-cracking sites in the middle of a rainforest? Chimpanzee archaeology began when we excavated a well-known *Panda* nut-cracking site, Panda 100. This tree was special to Hedwige and me as it was at this tree that we had practiced our skills at silently approaching nut-cracking chimpanzees so that they would not notice our presence. It was a very large and productive *Panda* tree and during the early days of our work in Taï, we had tried approaching nut-cracking chimpanzees for several seasons without much success. We needed to improve if we wanted to observe them so either Hedwige or I would sit facing Panda 100, just like a chimpanzee would, while the other would move closer and closer until he or she was close enough to see the other one without being heard. In such a dense forest with the ground covered by dead leaves, advancing unnoticed is quite a challenge, but we progressively improved. In addition, Panda 100 was in the middle of a delta formed by three rivulets that were regularly inundated and therefore quite rich in *Panda* trees. We had marked all 20 *Panda* trees and all the stones available nearby and spent weeks there during the *Panda* nut-cracking season to precisely follow how the stones were transported by the chimpanzees in

[23] Flexible learning has been shown to correlate with brain size in birds as well as mammals (Lefebvre *et al.* 2002, 2004).

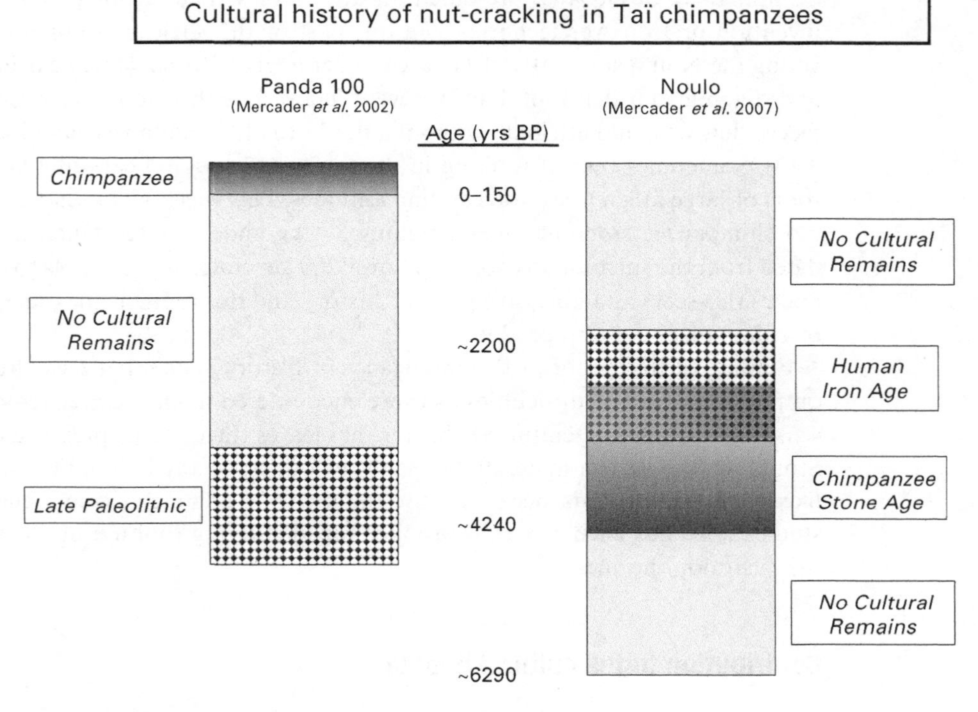

Figure 3.13 **Remains of chimpanzee nut-cracking activities over the past 6000 years**. We excavated two locations around Panda 100 and were lucky enough to find plenty of remains of chimpanzee as well as human activity. The primary old chimpanzee layer was found below a human Iron Age layer at the Noulo site. Confirmation that the stone tools were used by chimpanzees came from the starch remains attached to the stones, which proved that they were used mainly for nut species used exclusively by chimpanzees.

order to crack these nuts. But when Panda 100 stopped producing and fell over in 1984, we thought that would be the best place to initiate the chimpanzee archaeology project without depriving the chimpanzees. We were lucky.

Material culture in the context of nut-cracking has been present for a minimum of 4300 years in the Taï forest, which is equivalent to about 225 generations of chimpanzees.[24] To give you an idea of the time lapse that this represents, the

[24] Mercader *et al.* (2002, 2007) presented our results in detail. It is a minimal estimate of the age of nut-cracking in chimpanzees as it was not possible to attain older layers of soil in this very acidic and humid ground, where even the hardest stones degrade after 6000 years, which was the age of the oldest layer we could reach and date. As we also found human-produced artifacts in the same excavation sites, we tested the morphology of the stones we found with human stone technology specialists and analyzed the stones for the presence of chimpanzee-specific nut remains. This led us to say that out of the 206 stone pieces of all sizes that we excavated, a minimum of 21 medium-sized pieces were used by chimpanzees to crack nuts. To distinguish stone pieces used by humans from those used by chimpanzees, we used the structure of the starch grains that stick to the stones and remain well preserved in this very humid soil. Starch grain structure is very variable and can be very specific in some cases. Luckily, for three nut species known to be exclusively cracked by chimpanzees and not by humans,

"Chimpanzee Stone Age" pre-dated the advent of settled farming villages and the invention of Iron Age technology in this part of the African continent. By excavating the Noulo site, a 10 × 10 m area under a large *Parinari* tree near Panda 100, and digging to a depth of 1 m, we were lucky enough to uncover over 21 stone pieces that we could attribute without a doubt to old chimpanzee activities (Figure 3.13). Sometimes by just walking in the forest and looking carefully between the roots of large fallen trees, you can find antiques. This was how I found a 1250-year-old chimpanzee stone hammer weighing 3.4 kg under a large fallen tree that we dated from the surrounding soil. That over 200 generations used tools to crack nuts makes this seem an impressive piece of history and this remains the oldest example of chimpanzee cultural products.

Finding further proof of the importance of history in chimpanzee culture hinges on the same kind of difficulties as those encountered in classic archaeology in the sense that you need a cultural behavior that leaves traces of its presence over time. Stones are the perfect material, but sadly, wood, especially in humid environments like tropical rainforests, decays far too rapidly to be of much use in archaeological studies. This has been a central problem for identifying tool use in our ancestors, the australopithecines.

Contribution to the cultural debate

After reading this chapter, the hardcore experimental psychologist[25] will certainly remain skeptical and suggest that, however interesting the material presented here is, it should not distract from the fact that humans with their highly advanced technological culture must have gone through a very different scale of cumulative cultural evolution than that seen in chimpanzees, and, therefore, we cannot pretend it is based on the same cultural processes. Furthermore, the hardcore psychologist would argue that he still sees ecology as playing a decisive role in molding the behaviors adopted by a population, and thus, if ethnographic studies are essential, he still sees it as only a first step. Without experimental studies, he does not see how I could prove the social learning dimension of any of the behaviors presented in this chapter.

I would argue that the point of discussion is not whether human cultures are the same as animal cultures or whether human cultural products are more complex than any seen in animals. Instead, the point is to uncover the specific mechanisms that support the production of material culture in each species. On this point, ecology, mathematics, and physics have taught us that small quantitative differences can

Panda oleosa, *Parinari excelsa*, and *Detarium senegalense*, the starch grains are very specific and were found on 42.5% of the stone pieces (Mercader *et al.* 2007).

[25] I am going to use the imaginary character of the "hardcore experimental psychologist" at the end of each chapter to summarize the opinions of the skeptics towards the notion of animal culture in front of the material presented in each chapter and that will then summarize the main argument in favor and against culture in animals.

lead to large effects when given enough time, and thus, it could be expected that the cultural mechanisms seen in chimpanzees, when placed in a context where regular invention due to regular environmental changes would be necessary, could rapidly accumulate quite a number of complex cultural innovations. And we know that our ancestors, unlike the chimpanzees which have remained in the tropical African rainforests, invaded many different habitats and required different material cultures in their dispersals both within and out of Africa.

One aspect of material culture is that innovation and dissemination occur within a given environment. This means that the interaction of cultural and environmental factors that we have already seen led to chimpanzee ant-dipping behavior, for example, might be typical for many material cultural traits and this interaction will produce the material culture specific to each population. The classical criteria of culture in animals that requires the exclusion of any environmental causes will therefore by definition exclude many cultural skills in the material culture domain.[26] This could possibly explain why such a criterion has never been considered for human material culture – no one has ever suggested the exclusion of shelters, clothing, or particular culinary specialties from human cultural achievements just because they are influenced by climatic conditions. So then why should we do so for other species?

Material culture is the least arbitrary cultural domain and the one that is most oriented towards solving concrete environmental problems. As such, it might well be subject to different acquisition and dissemination mechanisms than cultural traits with more communicative or symbolic content. Material culture involves technology and technology is constrained by the physical dimension of the task. For example, cracking nuts requires one to exercise some amount of force on the nuts and it is energetically more efficient to strike them powerfully rather than to simply apply firm pressure. Therefore, the social learning contribution could be restricted to the type of tools and to the effect pursued. Similarly, the learning of stone flaking by human ancestors could be limited to some aspects of the tools and the function of the flakes, rather than to the hitting movements that are technically confined by the physics of stone flaking. Imitation need not be so important as long as the right material is being used. For material culture, we should expect less imitation but more social canalization or emulation that considers the objects being used.

Cumulative culture, by which an invention is added to an already existing cultural trait, is strongly suggested for some sophisticated material culture products in chimpanzees, including tools sets for extracting nuts from broken shells and sequential tools used to access honey in beehives. Granted, this remains very simple in comparison to the innumerable improvements that have been made to computers,

[26] There were some very good historical reasons for imposing this criterion during a certain period of time in the sense that most animal species are known to adapt to the different environmental conditions they face. Ecology is the science discipline studying the adaptation of species to different ecological conditions. Now that culture has been shown in many animal species, it would be time to accept that culture and ecology mix together both in human and nonhuman animal species.

for example, but the basic ability to improve on cultural traits that were invented by and learned from others is present.

In the next chapter, we will see how important social culture is to chimpanzees and how it has to be based on different but complementary social learning processes.

Synopsis

- Material culture in chimpanzees opens access to new food resources and directly contributes to some independence from nature's limitations.
- Material culture is very different in different populations of chimpanzees and sometimes presents complex interactions with the environment.
- Cumulative cultural evolution is suggested in some cases of tool technology, in which successive accumulations of innovations have led to complex tool use.
- Material culture is also seen in other animal species, especially other primates, although the high propensity of chimpanzees to use tools makes this domain of culture more developed in them than in other nonhuman species.
- Many similarities are found between human and chimpanzee material cultures; the main distinction is the diversity of environments faced by humans, which selects for diverse material cultures, especially when confronted with more hostile and cold habitats.

4 One for all and all for one
About social culture

Le singe est capable tout au plus de combattre avec les grues, tandis que l'Homme fait dompter l'éléphant et vaincre le lion

George de Buffon, *Histoire naturelle des quadrupèdes* (1762, p. 38)

Taï Forest, September 1976

As a total newcomer during my first stay in the Taï forest, I am not sure where I should start my study and so I have been exploring different locations that could be of interest. This brought me so far into the middle of the park today that I was a two-day walk from any human settlement. While searching for chimpanzees in the pouring rain, I hear loud chimpanzee screams not too far away. I immediately move in the direction I think the screams are coming from, although I am not confident because of the noisy downpour. Luckily, the chimpanzees are constantly calling and eventually I locate them behind an area that had been cleared by a fallen tree. Hiding behind a tree trunk, I cannot believe my eyes: high up in a tree, an adult male chimpanzee is holding a large, still alive red colobus monkey by the leg and around him, two screaming adults are moving to get access to the meat. I move forward to get a better view but I am immediately spotted by the chimpanzees, who flee, dragging their prey along with them. I remain alone in the rain, upset that I had been stupid enough to expose myself and trying to make sense of the completely unexpected scene I had just witnessed!

This was the first observation by a scientist of hunting in West Africa chimpanzees, contradicting a published report suggesting that chimpanzees in that region did not hunt! This was just one in a series of observations that showed me how hasty conclusions can be in the scientific literature. It was another seven years before I saw an actual hunt, but with time I was able to gain a better understanding of the complexity of the hunts in Taï chimpanzees.

Taï Forest, October 1992

Falstaff and Snoopy are resting on the ground when they hear a short call from a nearby red colobus group. They get up immediately, exchange glances, and move silently in the direction of the call. Macho, Kendo, and Brutus catch up with them shortly afterwards and they all slow down to peer cautiously at a group of monkeys in the trees. As usual, Snoopy makes the first move and climbs 5 m up the tree under the monkeys, which are still unaware of the presence of the hunters, while Brutus and Falstaff continue farther ahead of the monkeys on the ground for some 80 m. Snoopy slowly climbs a nearby tree while Kendo follows behind. When Snoopy reaches about 15 m, the monkeys give their first alarm calls and all start to move away, right in the direction where Brutus and Falstaff sat looking at the monkeys' reaction to Snoopy's presence. Despite the fact that the monkeys have already left his tree, Snoopy climbs right to the top and slowly follows them into the canopy. The monkeys weigh about 10 kg at most, while an adult chimpanzee weighs at least 40 kg. Therefore, while it is natural for the monkeys to jump over tree gaps, the chimpanzees must balance the tree with their weight until they can grasp the branches of the opposite tree and progressively shift their weight from one tree to the other to bridge the gap. Snoopy follows the monkeys slowly, obviously not trying to catch up but instead "herding" them in a given direction. He could not possibly see Falstaff and Brutus moving into position on the ground so Snoopy may have remembered from past hunts what they would do while he was herding the monkeys. Falstaff and Brutus start to run on the ground in order to remain in front of the first fleeing monkeys. Macho, trying to catch the monkeys, climbs at full speed underneath them. Falstaff and Brutus race a good 100 m ahead on the monkeys' escape route and cross a river, where both of them swiftly climb different trees to wait silently for the monkeys to arrive. Brutus dashes higher up to exactly the right place as the first monkeys, now herded by Macho and Snoopy, enter his tree. Brutus forcefully shakes the branches and barks aggressively. Surprised, some of the monkeys turn back, while others leap to the highest of the small branches. Falstaff, approaching from the side, snatches an adult male monkey who is ferociously threatening him and drags it to the ground to a screaming group of six females who appeared after hearing the monkeys' alarm calls.

Having more experience since the first hunt I witnessed, I immediately see more than just a wild scrum. I see Brutus joining Falstaff and seizing the legs of the monkeys, while Ondine and Salomé, his female friends, sit screaming with excitement at his side. Macho and Kendo, who are empty-handed, are displaying around the scrum and are greeted by screams and threats from the females. Ondine and Salomé are opening the monkey's belly and Loukoum and Malibu are helping them pull out the bowels while Snoopy sneaks in and takes the liver and another piece. Falstaff is watching Brutus, who is bent forward and tries to cut the monkey in half. This lasts seven minutes, during which time Kendo and Macho continue to display and the females scream and bark at them. With a final joint effort, Falstaff and Brutus pull with all their strength to rip the fur holding the two halves of the monkey together. Falstaff, along with Macho, Kendo, and another female, is left with the upper half, and Brutus and the remaining females are left with the lower half. Twenty minutes after the capture, the excitement ceases and all that can be heard is the sound of bones cracking while the chimpanzees calmly eat the meat along with some fresh leaves.

Figure 4.1 **Hunters and adult females rush to eat meat after a successful hunt** in the Taï forest. The capture of an adult prey, as seen in the center of the picture, always attracts a large number of group members, including those who were not active during the hunt but who were alerted to join by the loud calls of the excited hunters. The high level of competition we see does not preclude some clear meat-sharing rules that ensure that successful hunters receive a larger amount of meat than others.

I have seen hundreds of hunts by the Taï chimpanzees and I have carefully analyzed all of them to determine the amount of meat each individual is given and eats and from whom he or she receives meat. From this, I have determined that what appears to be a wild scrum is, in fact, clearly organized and that meat is not distributed randomly at all. In the Taï forest, groups of highly coordinated and organized hunters are more successful at catching prey. Lone hunters in this forest are successful at capturing prey in 17% of the hunts, while groups with more than two hunters are successful in 64% of the hunts. This system, however, should only persist if the hunters are rewarded with meat, which means that the prey must be shared in such a way as to be profitable to them. Meat-sharing in Taï chimpanzees seems to follow such an expectation.

When Kendo was the dominant male, I often saw him trying to get meat by displaying towards those who had it and positioning himself close to it, but he was repeatedly turned away by the others because he had not taken part in the hunt. He would continue to insist for some time but did not succeed in getting any meat. Furthermore, only 0.5% of the meat transfers resulted from theft (when one individual takes the meat away from the protesting meat owner), which underlines the prevailing consensus about meat-sharing in the group. The value assigned to the role of hunters was so precise that capturers obtained, on average, 85 g of meat per hunt, which is not surprising as they were the ones controlling the meat; somewhat more surprising was that ambushers, who had undoubtedly contributed to the hunt but had not captured and did not control the prey, also received an average of 85 g of meat. Individuals who were present during the meat-sharing but who did not hunt received about 32 g of meat per hunt. The group somehow remembered each male's contribution to the hunt and let them have more meat than non-hunters. Thus, the Taï chimpanzees' meat-sharing rules enforce the support of cooperating individuals by allowing hunters and those performing the most important actions more access to meat than the other chimpanzees.[1]

Gombe National Park, Tanzania, April 1992

I went to Gombe to compare the hunting behavior of different groups of chimpanzees, and it was there that I started to appreciate how diverse hunting cultures are in chimpanzees. Frodo was the largest male in the community and the grandson of the legendary female, Flo, who Jane Goodall had portrayed so vividly in her famous book, *In the shadow of man*. When I visited Gombe, he was a real troublemaker. His older brother, Freud, was the second highest ranking male in the group and naturally Frodo wanted to take advantage of this and push forward his own personal agenda. He often displayed aggressively towards the other chimpanzees in the group, even towards the alpha male, Wilkie. But what I loved about Frodo and the reason I always followed him on the steep slopes of Gombe was that Frodo was the best hunter in the history of Gombe and a great fighter against red colobus monkeys.

16 April 16:30: I am high up in the hills of Gombe with a perfect view over the entire valley and Lake Tanganyika, which is glowing below us in the sun. I am with 4 chimpanzees when I hear Frodo screaming in the direction of a group of about 20 red colobus monkeys. The group of monkeys is isolated in a large tree in the grassland with only two small nearby trees to use as an escape route back into the forest. Frodo climbs 2 m up the tree the monkeys are feeding in and he looks at them without approaching. He gives the impression that he wants to approach but does not dare to and does not really know what else to do. The other chimpanzees sit waiting and watching Frodo. Two large male monkeys immediately begin screaming aggressively and charge in Frodo's direction. Ten minutes into this close standoff, during which time Frodo continuously screams and looks towards the other male chimpanzees, one of the male monkeys jumps onto Frodo, who then rushes to the ground while simultaneously being attacked by this and three other male monkeys. The four pursue the screaming Frodo through the grassland towards the forest, while some of the other chimpanzees watch without doing anything and others actually seek refuge in the small trees. Eventually

[1] The hunting behavior of the Taï chimpanzees has been described in many publications (Boesch and Boesch 1989, Boesch 1994a,b, 2002, 2009, Boesch and Boesch-Achermann 2000, Boesch *et al.* 2002).

Figure 4.2 **Gombe males during a grooming session on a savanna ridge**. This shows the open grassland intermingled with wooded areas that allows the hunters to successfully corner monkeys when they hunt. Evered, the oldest male of the community, is in the middle grooming the alpha male, Wilkie. I saw Evered perform collaborative hunting roles on many occasions, including completely anticipating the escape movements of the red colobus prey on the steep slopes of this park.

all return to the same tree, with Frodo still wanting to hunt but not daring to do so. He is attacked two more times by a male monkey and has to twice shake him off his back but he never retreats. In the end, the colobus group, including all the females, start to move back into the forest. This was what Frodo had been waiting for, as they are no longer in a position where they can easily defend themselves and their females! He swiftly maneuvers himself through the monkey males and grabs a baby from its mother's belly. As he does so, two other chimpanzees take advantage of the rising panic in the monkeys and grab two more babies from their mothers.

Even in my wildest thoughts I could not have imagined more dramatic differences between chimpanzee groups! In all my years at Taï, I had never seen colobus monkeys chase adult hunters away like I had just seen at Gombe. The monkeys would have all been eaten. All over Africa, adult male colobus are known to threaten and attack without hesitation chimpanzees who come too close and if the intruder does not retreat, up to four of them have been reported to jump on a chimpanzee and easily pierce his tough skin with their small but extremely sharp canines. Except in Taï so far, wherever chimpanzees live with red colobus monkeys, they have been described to be fearful when faced with male colobus. One day in Gombe, I was alone with Wilkie, who was passing under a group of red colobus and moving towards a fruiting tree farther ahead. Without any warning and right in front of my bewildered eyes, a brown ball of fur called aggressively and fell directly onto Wilkie's back. Wilkie ran away trying to get the large colobus male off his back. Typically, Frodo, being the bravest, would face the monkeys; he would scream with fear but still hold his ground and then when some disorganization in the monkeys became visible, he would zigzag between them to reach the mothers the males had been protecting. To his credit, he was extremely successful in doing this and caught, by far, more baby monkeys than any other male in the group.

Hunting cultures in chimpanzees

How did such different hunting cultures evolve in chimpanzees? The continuous forest canopy in Taï provides the monkeys with plenty of escape possibilities; because of this, lone hunters in Taï are rarely successful – only 17% of lone hunts are successful after hunting for 39 minutes on average – and the chimpanzees are forced to organize themselves in order to increase hunting success. In contrast, the woodlands and forests of Gombe are regularly interrupted and by cornering monkeys against such interruptions, lone hunters can be quite successful – 50% of lone hunts are successful after only 7.2 minutes on average.[2] Thus, a single ecological difference seems to force chimpanzees to hunt in groups, which in turn leads to a suite of social differences no longer directly related to ecology. For example, Gombe chimpanzees are mainly solitary hunters and when they do hunt in groups, they do not coordinate their actions, they retreat when faced with threatening monkeys, they avoid adult male monkeys to the point that they may not eat even those they have killed, they specialize in killing baby monkeys, and they often reject mothers when they come into contact with them. Thus, the fact remains that Gombe chimpanzees are capable of cooperating during a hunt, even if it is rarely seen, but hunting in a forest where prey can often be cornered by lone hunters does not favor cooperation.[3] Taï hunters, in contrast, hunt mostly in groups and coordinate their actions as a team, always hold their ground when faced with threatening monkeys, look for adult prey, and rarely take babies without also capturing the mothers (see Table 4.1). If hunting party size seems directly influenced by the forest structure where the hunt takes place, all other differences constitute aspects of a hunting culture based on a population-specific social dynamic towards the prey. Thus, here we could be facing another example of how ecological and cultural influences affect the behav-

[2] In Gombe, 64% of the observed hunts were done by solitary hunters, compared to 16% in Taï. In addition, when we compared only group hunts, we found that coordination of different complementary hunting movements between the hunters was seen in 19% of the group hunts in Gombe but in 77% of the group hunts in Taï. Numerous studies have analyzed the hunting behavior of the Gombe chimpanzees (Goodall 1968, 1986, Teleki 1973a, Busse 1977, 1978, Boesch 1994a,b, Stanford *et al.* 1994a,b, Stanford 1998, Gilby 2006).

[3] In my view, this difference is not a question of ability but a response to the conditions prevailing during hunts. I have seen some very illustrative examples of coordinated group hunting in Gombe, all of which included Frodo as one of the hunters.

ioral differences we see in different chimpanzee populations, like we have already seen in the cases of ant-dipping and honey extraction (see Chapter 3).

Box 4.1: How do we detect cooperation in group hunting?

All chimpanzee populations hunt mainly small-bodied, arboreal monkeys that are able to run quickly over high branches and jump between relatively large tree gaps. For fieldworkers who are following a group activity, it can quickly become overwhelming if there are many individuals involved. Therefore, studying the hunting behavior of chimpanzees in trees has always been challenging as observers need to be able to answer two different questions as the hunt unfolds.

- Who is present during a hunt?

Chimpanzees hunt on a voluntary basis and, thus, the number of individuals actively participating in a hunt is constantly changing, with some individuals participating for long periods of time and others participating only very briefly, then retreating to the ground and possibly becoming active again at a later time. In addition, some group members, mostly females but also males, may join the hunt just to follow and watch from the ground. Basically, the observer has to scan all group members present and determine how long and during which phase of the hunt each of them is active for.

- What is each participant doing during a hunt?

Ideally, each hunter should then be followed to see what he is doing and where he is placing himself in relation to the prey and the other hunters. As hunts happen high up in trees, this requires the observer on the ground to constantly move below the hunters. For my first six years studying hunting, I was constantly drawing diagrams in my field book with the positions of the hunters, the prey, and the main trees so I could better visualize how the hunt unfolded. These rapid sketches allowed me to keep track of the position of each hunter and know where I needed to look to be sure of their ever-changing positions.

It is obviously more difficult to determine who is active and what they are doing when there are more chimpanzees participating. Luckily in Taï, hunting group size was an average of 4 hunters with a maximum of 8 hunters, and this happened in a primary rainforest where observers can freely run on the ground below the hunters if necessary. In contrast, in Gombe and Mahale, the woodland undergrowth is sometimes so thick that it completely blocks the visibility into the trees where the hunt takes place and prevents observers from moving rapidly. Alternatively, the forest at Ngogo is similarly open like it is in Taï, but the much larger number of chimpanzees that are present during a hunt – an average of 24 individuals – makes it very hard to keep track of all individuals. Thus, lower visibility and higher number of chimpanzees could both make it difficult for observers to record hunts with enough detail to follow each hunter and see their specific contribution during the hunt.

Thus, while group hunting and hunting success has been reliably recorded in all studied chimpanzee populations (e.g. Goodall 1986, Nishida *et al.* 1983, Mitani *et al.* 2002, Gilby *et al.* 2010, Mitani 2009), it has proven more difficult to obtain detailed descriptions of the hunters from all of them. This has proven so difficult that in Ngogo and sometimes Gombe, a distinction between hunters and bystanders has been considered to be unreliable (e.g. Teleki 1973a, Mitani and Watts 1999, 2001, Watts and Mitani 2000, 2002, Gilby 2006) and this has prevented an evaluation of the individual benefits of hunting and limited our ability to discuss the evolution of hunting in chimpanzees. Nevertheless, much effort has been invested to document the hunt precisely and, whenever possible, detailed descriptions are provided (e.g. Boesch 1994b, 2002, Gilby and Wrangham 2007, Gilby *et al.* 2010)

For the Taï chimpanzees, the sharing of meat ensures that hunting pays off. Even during the wildest and largest scrum, where dozens of individuals are charging around trying to get meat and tearing off and eating pieces from the prey, those hunters who contributed the most are assured of receiving the largest share of the meat. To ensure that this happens, females, who rarely, if ever, hunt, actively support the prey owner and willingly initiate fights if non-hunter males try to access too much meat. In addition, it was not uncommon to see Salomé and Ondine holding onto a full carcass until Brutus, who was off chasing a male or interrupting a fight between two individuals, came back; in both cases, when Brutus returned, they would then give the prey back to him. Furthermore, about 40% of the meat-sharing resulted from the meat owner actively allowing others to access the meat, either by holding the meat towards the other individual or by directly cutting off a piece of meat and handing it to them. When the kill was under the control of one of the big males, whose authority over the prey was recognized by all, meat-sharing went relatively smoothly and everyone who had hunted obtained his share. Furthermore, meat is regularly shared between males and females, presumably because females in Taï copulate more frequently with males who share meat with them.[4]

In contrast, chimpanzees in Gombe who hunt are not guaranteed more meat than non-hunters, who regularly receive more meat. High-ranking males are regularly seen forcefully taking meat or even entire carcasses away from hunters and then sharing a small amount with the females and other high-ranking males. The lower respect for meat owners is also accompanied by harassing forms of begging. Beggars in Gombe are reported to regularly place their hands in front of the meat-eater's mouth and actively prevent him from eating and to push the meat away from

[4] In a detailed study including over 22 months of observations, Cristina Gomes showed that males who share more meat with females also had more sex with them. Furthermore, she found that they do so on a long-term basis as this effect does not depend on whether females were in estrus at the time that the meat was shared with them (Gomes and Boesch 2009). It was not possible to determine if this effect was due to the males sharing more or the females soliciting the males more.

the owner's mouth with their hand. Observers believe that Gombe chimpanzees are forced to share meat in order to get rid of pressure from beggars.[5] Interestingly, such begging gestures have never been seen in Taï, where sharing appears to be more voluntary. Thus, more meat is shared with harassers in Gombe, whereas more meat is shared with hunters, especially the best hunters, and sexual partners in Taï.

So, the two distinct hunting cultures that have developed in Taï and Gombe are also visible in the meat-sharing rules that predominate in those groups. If forest structure might have played a role in determining the hunting success of lone hunters, the many subsequent differences seen in hunting strategies and meat-sharing have only a very indirect connection with the environment and seem more directly affected by the specific social dynamic prevailing within each community.

The social hunting behavior of the Ngogo chimpanzees living in the Kibale National Park, Uganda, provides a nice illustration of the role of social factors. This community represents one extreme of the social scale as it is famous for being the largest community ever studied – this community includes over 170 chimpanzees with more than 20 adult males – and since male chimpanzees are very gregarious, it is not uncommon to meet more than 20 males together. This provides a very different social dimension to hunting; in contrast to both Gombe and Taï, finding more than four males to hunt at Ngogo is not an issue and, thus, group hunting is the rule (see Table 4.1). These chimpanzees can then overrun the defense of the red colobus monkeys by sheer number and in the ensuing panic, many different males try their luck. The result is not only that more hunters hunt at the same time, but that many more prey are captured during successful hunts.[6] This can have a dramatic effect on the red colobus population and after several years of such razzia hunting, most groups of monkeys within the territory were almost exterminated, which resulted in less frequent hunting by these chimpanzees. Meat-sharing is often reciprocated between social partner males at Ngogo and 20% of the meat is transferred as a result of stealing, which is relatively high compared to Taï. Thus, at Ngogo, hunting is strictly done by large groups of males who create panic in their prey and regularly capture large numbers of monkeys in one hunt.

Ngogo hunting culture is not only the result of the larger number of potential hunters, but, as in the case of Gombe, it is also influenced by the forest structure. A precise analysis showed that, in Ngogo, the hunting success rate for red colobus monkeys is much lower in primary forest than in areas with broken canopy. Thus, only 15% of their encounters with monkeys take place in primary forest compared to 64% in broken forest. Strikingly, the much larger hunting groups in Ngogo were

[5] Sharing as a result of harassment by beggars has been suggested by many who have studied the Gombe chimpanzees (Wrangham 1975, Gilby 2006).

[6] John Mitani and David Watts (1999, 2001, Watts and Mitani, 2002) have published extensively on the hunting behavior of this large community of chimpanzees. Average hunting group size is 24 individuals and they are successful in 85% of hunts and capture an average of 4 monkeys per hunt. They were kind enough to invite me to visit their site, where I followed some hunts in their company, and I agree with them about the key importance of demography in combination with the environmental conditions to understand the dynamic between hunter and prey. As it is relatively more forested than Gombe, Ngogo represents an intermediary situation between Gombe and Taï.

Table 4.1 Social cultural traits with some eco-sociological background data among four of the best studied chimpanzee populations

	Taï	Gombe	Mahale	Ngogo
Hunting				
Group hunt	84%	36%	72%	100%[1]
Collaboration	77%	19%	0%	rare
Adult prey captured	44%	11%	24%	26%
Meat-sharing rate	11.9			2.8
Active sharing of meat	40%	20%		
Stealing of prey	2%	28%		20%
Fear of red colobus	No	Yes	Yes	Yes
Meat-sharing favors	best hunter	harasser	α's allies	social partner
Territoriality				
Intergroup encounter	0.84/mth	present	present	1.97/mth
Coalition	common	common	common	common
Patrol	3.0/week	3.0/week	common	7.2/week
Support	45.2%	0%	0%	12.5%
Inter-community killing[2]	0.08 kill/y	0.83 kill/y	0.3 kill/y	1.83 kill/y
Predation defense				
Leopard presence	common	extinct	rare	extinct
Predator attacks	yes	–	rare[3]	–
Chimpanzee killed	11[4]	–	0	–
Chimpanzee injured	9	–	0	–
Tending injured	all	close kin		
Tending frequency	8.1'/h			
Chimpanzee support	100%	–	No	–
Cooperative support	yes	–	–	–
Affiliative behavior				
Adoption of orphans	50%	50%		
Male adopters	50%	–		

[1] Watts and Mitani (2002) suggest that solitary hunts on some monkey species might be underestimated.

[2] Inter-community killings are based on different observation time for each population (see Wrangham *et al.* 2006, Boesch *et al.* 2008).

[3] No direct observations was made, but chimpanzee hairs have been found in few lion scats (Tsukahara 1993).

[4] The number of chimpanzees killed and injured by leopards in Taï is listed as of February 2011 and takes into account the victims in both the North and South Group. It is therefore higher than the figures presented in Boesch and Boesch-Achermann (2000).

References: **Taï**: Boesch and Boesch-Achermann 2000, Gomes and Boesch 2009, Boesch *et al.* 2008, 2010, **Gombe**: Goodall 1986, Stanford *et al.* 1994a, Gilby 2006, Williams *et al.* 2004, **Mahale**: Nishida *et al.* 1983, 1985, 1992, Kutsukake and Matsusaka 2002, **Ngogo**: Watts and Mitani 2002, Mitani and Watts 2001, 2005, Watts *et al.* 2006.

successful in only 55% of hunts in primary forest, a success rate that is very similar to the Taï chimpanzees. In contrast, in broken forest, the Ngogo chimpanzees are successful in 92% of hunts.

Chimpanzees in all known populations hunt for meat but each seems to have its own specific hunting culture that integrates the specifics of both the environment and the social group. Mahale chimpanzees live not far from Gombe – about 200 km to the south – and they also hunt for the same red colobus species. However, when I arrived there with my Gombe and Taï glasses on, I was struck by how fearful the Mahale chimpanzees were of the red colobus males and by how they had developed their own specific way of dealing with them.

September 1999: a group of 10 chimpanzees, including 4 adult males, arrive under a group of red colobus. Clearly interested in the monkeys, the males on the ground look up at them as if they are waiting for an opportunity. After about an hour, some of the adult monkeys jump onto lower branches and threaten the chimpanzees by screaming and shaking branches towards them. Three male chimpanzees react by silently displaying under the monkeys. Two adolescent males climb towards the monkeys but they are immediately threatened by a male monkey, who jumps towards them onto lower branches and aggressively shakes a branch at the chimpanzees, who retreat to the ground. Almost two hours later, the chimpanzees move on without making any serious attempt to catch the monkeys.

On other occasions, I saw mob attacks by adult male colobuses that would completely block a potential hunting movement. The Mahale red colobus monkeys are bolder than their counterparts in Gombe and they resist the chimpanzee attacks even when a group of colobus has started to panic.[7] The fear of adult colobus could also explain why Mahale chimpanzees occasionally do not touch adult colobuses that they have successfully isolated, even when they are on the ground and surrounded by 10 chimpanzees. If I had not seen this with my own eyes, I would have simply not believed it! In the end, compared to the Gombe chimpanzees, those in Mahale are more indecisive about when to start a hunt and show more respect towards their adult prey, which, in the case of red colobus monkeys, are admittedly more aggressive than in Gombe.[8]

Cooperation: acting at the same time or acting together?

Alexander Dumas' heroes, "The Three Musketeers," represent the western ideal of cooperation: unfailing support to others, perfect reciprocity, abiding readiness

[7] I saw an extreme example in Mahale when an exceedingly pushy Japanese television crew was there to film a successful hunt. Each time the chimpanzees tried to make a capture, the filming team would cheer loudly, which increased the panic levels of the female monkeys. However, despite this, the male monkeys kept cool and faced the chimpanzees whenever they needed to protect their females and they successfully chased away the chimpanzee hunters.

[8] The aggressive nature of red colobus males makes them regularly violently threaten human observers, who suddenly appear next to them and they shake branches and scream at us. I never experienced such aggression levels from the Gombe red colobus monkeys (Boesch *et al.* 2002).

to make the supreme sacrifice, and unlimited sharing of resources. As their motto went, "All for one and one for all." When I was younger, I was a die-hard rugby player. In rugby, as each player fulfills a special role, cooperation between players is essential and a team's success only results from excellence in collective action. My youthful enthusiasm drove me to play early on Sunday mornings on playgrounds covered with snow and mud, but what I found increasingly difficult to tolerate was the human inability to cooperate. The demon of selfishness lurked around every corner and whenever one player had the impression that he could succeed alone, he would invariably forget the team and sadly be knocked to the ground. So we had to invest huge amounts of energy just to keep the ball on our side. How often did we forget that we were members of a team? How often did personal ambition overshadow the common goal? How often did this make our team lose? "I" is not a word heard in a rugby team and everyone who has played a group sport will know exactly what I am talking about. It is not enough to just have 11 or 15 players on the field; it is as essential that those players act as a team.[9] However, it is very difficult to put this theory into practice.

Psychologists have recently developed an increased interest in cooperation, building on the idea that an understanding of others as independent individuals is important for cooperation (see Box 4.2). Cooperation, in the sense of the synchronized action of two or more individuals as they pursue a common goal, is a complex social behavior as it implies, first, that the cooperators agree on pursuing a common goal, and second, that they act "together." In other words, they must not only *synchronize* their actions but also *coordinate* them in order to pursue that goal. So it is not only important that cooperation is a group activity; participants must also share a common goal. Individuals must somehow forgo their personal interest in favor of the common goal. Recent experiments comparing captive chimpanzees with free-living humans have shown that the former have difficulty understanding humans as independent partners with their own intentions and knowledge. From this, it was concluded that captive chimpanzees are unable to cooperate, and, therefore, that cooperation is a uniquely human ability.

Box 4.2: How to identify cooperation in a group activity

When more than one chimpanzee arrives under a group of red colobus monkeys, more than one individual might show an interest in the monkeys and start a hunt together. Do they hunt together because they arrived there and are hungry at the same time? Or do they hunt together because they want to profit from the panic that will ensue when the monkeys see one chimpanzee climbing the tree so that

[9] I am aware of the impressive cooperative actions seen in humans, like our supreme sacrifices in wars, our heroic behaviors in modern armies, and our construction of huge dams and pyramids. Most of them, however, are sustained by strong financial incentives and also strict enforcement rules, such as when army deserters are imprisoned or killed. Spontaneous cooperation as seen in chimpanzees obviously exists in humans, but there is a constant conflict between group and personal interest.

each might increase their individual success? Or do they hunt together because they want to coordinate their actions so that they have more success as a team?

These three alternatives have been at the core of a long debate about whether cooperation requires an understanding of others in order to be effective. When can acting together be seen as joint team work (see Boesch and Boesch 1989, Bratman 1992, Butterfill 2012)? Different explanations for group activity can be proposed.

- *Similar goal*: this corresponds to the first alternative in the sense that the two individuals happen to act at the same time towards the same goal but for an external, independent reason. To biologists, the essential point is whether the two individuals increase their likelihood of success when they act together compared to when they act alone, and if so, they call this cooperation (this is an economic definition).
- *Common goal*: the acting together can be influenced by each participant synchronizing their actions in time or space as this permits for some synergetic effect that could potentially increase success. The simultaneous occurrence of actions during a hunt could benefit each participant as the prey become disorganized when they are faced with many attackers and this could allow each hunter to capture prey more easily. Such simultaneous hunts where each hunter chases different prey has been seen in all group hunts in chimpanzees with differing frequencies (Boesch and Boesch 1989, Boesch and Boesch-Achermann 2000). Some spatial coordination also exists between the participants and such hunts have also been seen in all chimpanzee populations, albeit less frequently (Boesch and Boesch 1989). In a sense, those hunts could still be described as "simultaneous solitary hunts."
- *Shared goal*: here, the acting together is directed to one and the same goal and each individual acts so as to make the combined actions complementary to reach the shared goal. The hunters now collaborate to capture the same prey and they often perform very different actions, each of which is aimed at capturing the prey based on the expectations that the other will fulfill his part, even when he is out of sight (Boesch and Boesch 1989). Shared goals coordinate multiple agents' goal-directed activities around an outcome to be achieved as a common effect of their efforts (Butterfill 2012).

Some have suggested that shared goals imply detailed mind-reading of the other participants with shared intentions and shared planning and that this would be a specific ability of humans that is not seen in chimpanzees (see Tomasello *et al.* 2005, Moll and Tomasello 2007, Tomasello 2009). The discussion is still open as to what is really required for an individual to be able to pursue a common, shared goal (see Bratman 1992, Butterfill 2012).

References: Boesch and Boesch 1989, Bratman 1992, 2009, Tomasello *et al.* 2005, Moll and Tomasello 2007, Carpenter 2009, Butterfill 2012.

Therefore, the descriptions by eighteenth-century naturalist explorers about cooperative behavior in animals are extraordinary. These include Charles-George Le Roy's 1761 description of wolves hunting in packs:

> If they want to attack a herd, the she-wolf would present herself to the herding dog, while the male would infiltrate the pack and carry away a sheep that the dog was too far away from to protect. If they are attacking a wild animal, the roles would be distributed according the forces: the male wolf would find the prey, chase and attack it until out of breath, and at that time, the she-wolf, who had positioned herself at the right place, would replace him with fresh forces and render the fight too uneven.[10]

Today, in the savannas of East Africa, tourists can admire lions, hyenas, and wild dogs hunting successfully in groups and bringing down large prey like zebra, buffaloes, and baby elephants. At the same time, many tourists have seen lions and wild dogs start a hunt but fail after chasing their prey for several hundreds of meters. After this, the carnivores were seen trying again against another prey or quitting for the day. Group hunting is risky and we know of no species that is always successful – average group hunting success is only 3.9% in wolves hunting moose, 30% in lions, 7.6–63% for hyenas hunting calves or adult gemsbok, respectively, and ranges from 50% to 84% in Taï and Gombe compared to Ngogo chimpanzees, respectively.[11] This is a major hurdle, as the investment in the present hunt may pay off only at quite a later time, and no one is very good at waiting. Thus, a prerequisite to cooperation is the ability to forego an immediate advantage for future potential rewards.

Meat access is a second persistent uncertainty of group hunting. Only one individual succeeds in making the capture and all other hunters are left without meat. As when humans play rugby, the demon of selfishness makes regular appearances in chimpanzees. On 7 July 1992, at the end of a long, hard hunt in Taï forest in which the monkeys became very dispersed, Brutus made an incredibly long anticipation of a possible prey's escape movement. I ran behind him because I now fully trusted him in his hunting judgments and after a mad climb after something I could not see, he caught a juvenile red colobus monkey. By now he was already alone in the forest, and noticing this, he continued on for 60 more meters to silently eat his quarry alone on the forest floor. Brutus, who was by far the biggest meat provider to the group and easily controlled very large sharing clusters, did not hesitate to keep it all for himself on that occasion. The disappearance of the captor with his prey has been reported so regularly in Gombe that it is often difficult to even know for sure if a hunt was successful and how many prey were

[10] « S'il est question d'attaquer un troupeau, la louve va se présenter au chien, pendant que le mâle insulte le parc et emporte un mouton, que le chien n'est plus à portée de défendre. S'il faut attaquer quelque bête fauve, les rôles se partagent en raison des forces : le loup se met en quête, attaque l'animal, le poursuit et le met hors d'haleine, lorsque la louve, qui d'avance s'était placée à quelque détroit, le reprend avec des forces fraîches, et rend en peu de temps le combat trop inégal. » Charles-George Le Roy, *Lettres sur les animaux* (1761–1781).

[11] For more on the lion studies, see Schaller (1972) or Packer *et al.* (1990); for the hyena study, see Mills (1990); for the wolf study, see Mech (1970).

Figure 4.3 **What is the best way for me to obtain meat?** Brutus, an adult male, looks up towards the monkeys and other hunters, possibly evaluating what he should do. His perspective might miss some of the important aspects of the joint group actions and, if possible, he should evaluate the whole from the group perspective.

captured. Capturing small prey, like baby or juvenile red colobus monkeys, which weigh less than 1 kg and are easily killed with one bite to the head, allows a hunter to disappear quietly. In Taï, this is the very rare exception and captors do usually not run away from other hunters (see Table 4.1). On the contrary, they usually pull the prey down to the ground, where a large number of chimpanzees can more easily access it.

Hunting in groups does not imply hunting together. Humans might be very good at cooperating, but in the heat of the action, the conflict between self- and group-interest is very evident. This is also complicated by perspective; what seems to be the best course of action to player A might look hopeless to player B. If A succeeds, we say he is a great player, but if he fails, we say he is selfish. We might have the ability to look at things from another person's perspective, but during such a game, it becomes clear that this is very difficult and, thus, very few of us are great football or rugby players. The challenges of group hunting are not only about coordinating with others' actions, but also about evaluating what is the best position to take in order to enhance the chances of success from the perspective of the group. As a hunt unfolds, a hunter might think he has an opportunity and forget about the group actions. This might be partly why learning to hunt in a group is so difficult and is really only fully developed in experienced individuals: Brutus and Falstaff were the only hunters in the Taï community who systematically anticipated the effects of other hunters on the escape movements of prey – they fully anticipated movements 24% and 20% of the time, respectively, compared to 2–15% for younger males, and both of them were over 30 years old. This is reminiscent of the Ache in Paraguay, where young hunters acquire the skills only progressively and the most accomplished hunters are over 40 years of age, with success rates slowly decreasing later.[12]

Cooperation in high-risk situations

Why would we find different cooperative cultures in chimpanzees? All animal species are driven by the need to survive and reproduce and the pressures they face in order to meet these needs are very important selective factors. The first and possibly most important pressure is the presence of predators – the higher the predation risk, the greater the need to respond to it. In Taï forest, leopards are abundant and regularly attack the chimpanzees.[13] Because leopards are equipped with deadly canines and claws, chimpanzees have had to adopt solutions when confronted with the big cats.

23 February 1989: I am with four mothers, including Salomé and Ondine, who are carrying their babies on their bellies, when we hear a very loud anguished chimpanzee call from about 100 m ahead. The females immediately rush to the spot while barking aggressively and within seconds, they reach Ella and her two sons, who are 5 m up a tree and barking weakly. Her face, shoulders, and rump are covered with fresh cuts (Figure 4.4). Within seconds, Brutus and five other males rush past Ella in the direction she is looking, and they bark aggressively and drum on tree trunks as they chase the leopard away. For several minutes, the whole forest resounds with aggressive barks as the chimpanzees chase the leopard. Kendo, Ella's eldest son, comes back quickly and carefully approaches her in the tree. After looking at her injuries, he immediately begins licking the blood from her wounds. Ella's wounds, which are typical of a leopard attack, include cuts on the neck and shoulders, probably because she spontaneously bent forward during the attack to avoid the leopard's dangerous claws, and slashes on her side and back, possibly from defending herself during the fight on the ground. On the ground near her tree, I see fresh leopard tracks as well as clear signs of a fight. Fitz, Ella's second son, also joins her and she immediately presents the wounds on her head for him to lick. Her youngest son, Gérald, remained motionless but unharmed above her.

Twenty minutes after the attack, Ella slowly comes down to the ground and lays on her uninjured side as more of the males return from the pursuit. The boss, Brutus, Kendo, and the dominant female, Ondine, sit close to her. More adult females arrive and sit around quietly. It rapidly becomes clear that Brutus and Kendo are preventing young chimpanzees and some low-ranking females from coming close to Ella. Pokou and Ricci, two of Ella's close associates, and Brutus and Falstaff spend lots of time

[12] A wonderfully instructive monograph of the Ache hunter–gatherers of Paraguay following many years of very detailed and precise observations has shown this very convincingly (Hill and Hurtado 1996).

[13] Boesch (1991c, 2009) details the biological consequences that leopard predation has had on the social lives of the Taï chimpanzees.

Figure 4.4 **Ella with her son Gérald, just seconds after the leopard attack.** She is covered with injuries from the attack. Visible are a cut on her left eyebrows, which nearly blinded her, and two cuts on her left arm. She also has some serious cuts under her right arm that are bleeding profusely.

cleaning her wounds. After almost 2 hours, Ella moves on for 150 m and is followed by about 30 group members. She then rests again, while the others gather around her and her sons and the two females lick her wounds continuously. Brutus and Falstaff do so as well. For the rest of the afternoon, all 30 chimpanzees remain with Ella, adapting to her pace, caring for her, and tending to her wounds. Her two sons, Kendo and Fitz, tend to her wounds for 60 and 86 minutes, respectively, for the first 4 and a half hours after the attack, while Brutus does so for over 2 hours, Falstaff for over 1 hour, Pokou and Gala for over 45 minutes each, and some other females for 30 minutes.

The instantaneous support by first the females and then the males most likely scared the leopard away before he could inflict more serious injuries to Ella or even kill her. Leopards, which are ambush killers, cannot risk injuries as they may starve if they cannot hunt. Therefore, when faced with a group reaction like I had just witnessed, they will quickly flee. Such extensive group support was unfailingly provided whenever a leopard was detected and chimpanzees would rush to chase him away, or when an individual was injured and the others would tend to the wounds for weeks, if needed, which was especially welcome for injuries on hard to reach areas, like the back, head, and shoulders.[14] However, some leopard injuries

[14] During our time observing the Taï chimpanzees, one individual was attacked by leopards every three years, on average, and one individual was killed by leopards every 18 years (Boesch 1991c, 2009, Boesch and Boesch-Achermann 2000).

Figure 4.5 **Rousseau with his injured testis after surviving a leopard attack.** Leopards are ambush killers and Rousseau, an adult male, was surprised by a leopard on 27 November 1990. He was attacked from behind and suffered a long gash down the length of his back. The scrotum of his right testis was cut and retracted, leaving the naked flesh exposed. This injury never healed and became infected regularly. Rousseau eventually died from what we believed was an infection due to this injury.

were much worse than Ella's. For example, Rousseau was also injured by a leopard (Figure 4.5) and a leopard fatally punctured Falstaff's lung, an injury from which he never recovered (see below). Individuals in the Taï forest who alarm call are always rapidly supported by group members who rush to where the individual is, and in a few cases, I can confirm that the arrival of the group members deterred a leopard from attacking a solitary individual.

While Salomé and Ondine were able to help Ella without being injured, Falstaff was less lucky. On 18 September 1987, he received a total of 16 wounds, including one under his right arm that perforated his lung, from a leopard attack as he was seemingly trying to rescue a youngster. A whitish liquid ran from this puncture, and as time passed, we could detect Falstaff's presence from the terrible smell of this infection alone. Two weeks after the fight, he was too weak to follow the group and we last saw him on 4 November. This illustrates how much risk the chimpanzees in the Taï forest take to support group members.

Although leopards are present in Mahale, they have not been reported to be a threat to the chimpanzees. Mahale chimpanzees have been observed to react only very rarely to alarm-calling individuals, even if leopard calls were heard. However, in the recent past, chimpanzees have possibly been the victims of lion attacks (see

Table 4.1). In contrast, leopards are now absent in Gombe and they have never been reported to attack chimpanzees.[15] Nevertheless, whenever chimpanzees sustain injuries in Gombe, either through fights with other chimpanzees or accidents, observers are unanimous in saying that tending to the wounds is restricted to close kin relationships, such as between mothers and infants.[16]

There is, however, one danger that all chimpanzee groups face – male chimpanzees from neighboring communities. Chimpanzees are all highly territorial and defend their range against intruders with high levels of aggression (see Table 4.1).

---

3 October 1987: Brutus is still the uncontested leader of the community, which has now lost many members due to Ebola. While moving deep into the southern part of the territory, they suddenly hear chimpanzee calls farther to the south. Brutus, followed by three males and a young immigrant female, immediately move towards the calls, while the other females remain quietly in the back with their infants. At first, they move very slowly but once Kendo and Macho join them, they speed up and stay totally silent. After 80 or so meters, Brutus starts to lead the group towards the east, away from the calls we heard in the south. This is the first time I have seen him direct an attack to the side of the opponents. Avoiding a frontal confrontation seems to be a reaction to the diminishing number of males in his community. After 300 m, they suddenly begin running after someone. I try my best to keep up and when I catch them, all the males are surrounding an adult female with a baby from the neighbor community. The males are slapping her on her back as she crouches on the ground and protects her baby with her body. Some are biting her on her back while some hold her by her foot or hand to prevent her from moving. None of the males try to snatch the baby away. Unable to escape, she can only scream between hits. Not long after this begins, I hear the calls of a group of chimpanzees approaching noisily from the south. At the last minute, Brutus, Kendo, and the others open the circle around the female and turn to face the arriving southerners. The female escapes while her captors are busy chasing the southerners away. The immigrant female and another younger male remain behind and drum fiercely while barking loudly.

This territorial encounter highlights the main characteristics of such encounters in Taï chimpanzees: cooperative team actions, neighbor attacks only when a minimal number of participants are present, coordination in the approach, strong aggression towards strangers, and team support of subjugated individuals. In

[15] Leopards have not been seen in Gombe for a long time and there have been no reported cases of leopards attacking chimpanzees during the over 40 years of study there (Goodall 1986, pers. comm.). While leopards are present and regularly seen and heard in Mahale, the chimpanzees there rarely react to them and it remains unclear if they attack the chimpanzees (Nishida 1990, pers. comm.).

[16] Jane Goodall (1986) noted that group members seem to be disgusted by the sight of a wound and turn away from it. Group members were never seen to do more than look at the wounds and they never licked or cleaned them. However, offspring were seen to lick their mothers' wounds and vice versa. Saliva is a very good antiseptic and how rapidly wounds heal can often be related to the amount of licking that chimpanzees provide.

nearly half of all observed territorial encounters, support from non-involved group members is observed and they regularly succeed in helping individuals (Table 4.1). As can be seen in Table 4.1, different communities present in this context different levels of intergroup killing as well as support to group members. Such differences are puzzling but insert themselves in the differences already noticed in the level of group organization when hunting.

Cultural altruism in chimpanzees

Altruism has also been regularly seen in the Taï chimpanzees in the case of adult group members adopting young orphans. When Vanessa died of anthrax in December 2008, she left behind Victor, her 2.5-year-old son who was still completely dependent on her milk for his survival. I sadly realized that he would not survive and that we would have to watch him slowly die. Victor was not the first breast-feeding orphan I had seen and, like in Gombe, orphaned youngsters of that age do not survive for very long. From her experience, Jane Goodall said that an orphan younger than 5 years of age would not survive (she stressed this as evidence for the tremendously important bond between chimpanzee mothers and their infants). I saw exactly the same situation when a one-year-old infant, who had been adopted by her seven-year-old sister who could not yet produce milk, died after four agonizing months in the arms of her sister who tried to do all she could to help her younger sibling.

Much to my surprise, Fredy, the third-highest ranking male in the group, carried little Victor on his back! I have never seen an adult male do this before in another chimpanzee community, and it has never been reported in any other wild chimpanzee population, so this was a wonderful surprise! I spent a day with them during the *Coula* nut season in February 2009, and Fredy carried Victor on his back the entire day and held him tenderly against his body while they rested (Figure 4.6). Fredy cracked 290 nuts over almost 3 hours and, as is very typically seen in mother–infant pairs, Victor begged for nuts from Fredy, who shared 75% of all the nuts he cracked with him. This metamorphosis of a prime male into a caring foster father was most impressive![17] However, I still had little hope that Victor would survive as he was still very dependent on maternal milk and I did not see much hope in this adoption, despite Fredy's effort. Nevertheless, Victor survived for seven more months and Fredy remained a perfect foster mother the entire time. Sadly, Victor then disappeared, possibly during an aggressive intergroup encounter that resulted in some injuries to

[17] Fredy was not the only adult male to adopt such an orphan. Porthos adopted the young, unrelated orphan female, Gia, for over 18 months, and he was completely committed to her, taking her along in the riskiest of situations and sharing his nest and nuts with her (see more details about Porthos and Gia in Boesch 2009). Porthos and Fredy are simply examples of this prevalent tendency of adult male chimpanzees in Taï to behave very altruistically towards helpless young group members (Boesch *et al.* 2010).

Figure 4.6 **Fredy, an adult male, adopted the 2.5-year-old orphan, Victor**. Victor was still totally dependent upon the milk of his mother when she died. The important investment of Fredy, a male unrelated to him, saved his life. Fredy not only carried Victor on his back (top left), which adult male chimpanzees normally never do, he let him rest on him (top right), shared food with him (bottom left), groomed him regularly (bottom right), and shared his night nest with him (pictures from Tobias Deschner).

Fredy also.[18] To support the case for altruism, we know from genetic analyses that Fredy was not Victor's father; his biological father was in fact present in the group but did not care for him.[19]

Chimpanzees in the wild are regularly orphaned and adoption in Gombe and Taï, mostly by older siblings, has been observed regularly.[20] Most reports of adoption by adult male and female chimpanzees in the wild come from the Taï chimpanzees. In

[18] The Disneynature film "Chimpanzee," which is centered on the Taï chimpanzees and tells the story of Fredy and Victor, changed Victor's name and modified the story so that it would have a happy ending.

[19] Thanks to new, non-invasive methods, we can perform genetic analyses of wild chimpanzees from their feces without coming into contact with them. Here, I would like to thank Linda Vigilant and Grit Schubert, who did most of the genetic analyses of the Taï chimpanzees.

[20] In Taï, we observed 36 cases of infants being orphaned and an adoption was seen in 18 cases (Boesch *et al.* 2010). In Gombe, adoptions are also regularly seen to be performed by older siblings and some cases of support to youngsters by adults have been documented, but these cases do not reach the status of an adoption (Goodall 1986). No reports of adoption have yet been published in other chimpanzee populations, although I would expect that they occur in at least some.

some cases, we could confirm that the foster parent was not related to the orphan, and, thus, these represented pure cases of altruistic behavior. The ability to care for the welfare of others has been denied to chimpanzees by some authors on the basis of experimental results obtained from captive chimpanzees. This difference should not come as a surprise as we should not expect that individuals would care for others without any prior experience of group solidarity, and such solidarity will develop only if external pressures favoring it exist. In Taï, the daily combined pressures that individuals face from predators and neighboring chimpanzees will lead to within-group solidarity, while small captive groups totally isolated within a passive environment have no incentive to develop such a predisposition. In most East African chimpanzee populations, leopards are either absent or not known to attack chimpanzees, and so this predation pressure is not felt by the individuals (see Table 4.1). In that part of Africa, tending for injured individuals, systematic support for individuals in need, and adoptions by unrelated adult males have not yet been seen to the same extent that they are in Taï chimpanzees. These differences point to the importance of the specific social learning environment experienced by youngsters for the development of cooperation and altruism.

Social niche construction in animals

Material culture shapes the outside physical world and has a tremendous influence on the type and amount of food that chimpanzees can exploit in a given habitat. In this respect, the impact of social culture is even more complex as cooperative hunting has led to meat becoming a more important food source and within-group solidarity, in terms of altruism and helping of group members, does not directly affect feeding efficiency. However, they play an important role in promoting complex social interactions within a chimpanzee community. In Taï, the important role of meat has favored the emergence of a second dominance system beside the one based on aggressive interactions. This second dominance system is concerned with access to meat in that good hunters and some high-ranking females obtain the highest positions; in contrast, all males dominate all females in the main social dominance hierarchy.[21] In a similar way, some chimpanzee sign-codes enable low-ranking males to gain access to estrus females despite the proximity of higher-ranking males. However, this would not be possible in a more rigidly hierarchical society, where individuals are more limited by dominant individuals in the number of social strategies available to them, which in the end can strongly limit the

[21] In Taï, some females, like Ondine and Salomé, reached the highest positions in the meat-sharing hierarchy, even though they were dominated by all nine adult males in the social dominance hierarchy (Boesch and Boesch 1989). The alpha male, the highest-ranking male of the social dominance hierarchy, would regularly have difficulty gaining access to meat if he had not actively participated in the hunt and his access to meat would be prevented by individuals that he was clearly dominant over in other social circumstances.

reproductive success of lower-ranking individuals. This in itself produces a new selection field and imposes a different way of thinking and acting in the social domain.

Baboons have recently been proposed to possess social culture also.[22] Following the sudden death of most of the adult males in one troop, the remaining males perpetuated a much less autocratic and aggressive style of dominance, which was less often directed against low-ranking individuals, along with higher level of affiliate behaviors between group members, which permitted the increased participation of females. The fascinating aspect of this study is that the new social culture lasted despite the regular immigration of new males from neighboring groups who did not show this social style. In fact, the immigrant males – in baboons, it is the males that transfer between groups as young adults – adapted to the low-aggression and high-affiliation style of their new group and this culture persisted.

Social culture expresses itself in a large web of different social behaviors. Chimpanzee social structure includes many adult males and females, which leads to complex forms of feeding and sexual competition as well as making complex forms of cooperation possible. When facing high predation pressure from leopards and intense intergroup competitions with neighboring chimpanzee communities, the social setting seems to facilitate the emergence of social cultures. A complex web of interactions seems to have produced the Taï chimpanzees' social environment. Because of the high density of leopards in the Taï forest, the costs of leopard attacks can only be lessened in chimpanzees thanks to group members' help and support, and the Taï chimpanzees alone share a common tendency to care for the injured, to support one another in territorial fights, and to cooperate in hunts. In addition, because the nuts in the Taï forest are so difficult to open, all infants grow up profiting from the nuts that their mothers share with them, and as a result, sharing meat later in life seems to be much easier. In contrast, the Gombe chimpanzees, who are lucky enough to be spared from leopard attacks in a woodland landscape with smaller food patches, have developed another style of social culture. Theirs is characterized by less-social adult females who forage more in family units. In addition, due to reduced predation pressure, helping or tending to wounds seems to be limited to closely related group members. Finally, cooperation is not mandatory for hunting success, and meat and particularly other tool-acquired foods are shared less frequently and directed to harassers.

The cultural web of interactions has transformed social partners into social agents that can be alternatively "seen" as sexual competitors, food contestants, hunting partners, coalitionary friends, food providers, nurses, and grooming associates. Such a social environment selects for more flexibility and allows infants from very early on to experience group members in a more differentiated way and contribute to a more empathetic and compassionate social environment.

[22] See Sapolsky (2006).

Contribution to the cultural debate

Our hardcore psychologist would certainly challenge us to prove that Taï chimpanzee hunters really have a shared goal and plan their actions as a team rather than individually. One strong argument in his mind is that, in numerous experiments, nothing like this has been shown in captive chimpanzees, while it is so readily demonstrated in young children. Surely captive chimpanzees do not have the same levels and kinds of experiences and stimulations as their wild counterparts, but so many captive chimpanzees have been raised in social groups that, if this ability existed in chimpanzees, it should also express itself in captive individuals. Furthermore, he feels it is very important to distinguish how behaviors are performed from the underlying motivations and he believes that nothing like this can be shown about motivation without making use of controlled experiments. Finally, our hardcore psychologist would stress that while the learning environment is so important in humans, this does not imply that this has to be the case for other species, like chimpanzees, and as long as this has not been proven, there is no reason to refuse to give value to captive experimental studies.

This leads me to underline once again that population differences in chimpanzee groups have not been fully recognized by many, and some, in fact, suggest that if a behavior seen in one group has not been observed in another, something is wrong and therefore this observation is suspect. Because the Gombe chimpanzees behave one way does not mean that all chimpanzee populations do the same. An ethnography of chimpanzees is required if we want to progress in our understanding of this species. I realize that, at present, it remains hazardous to predict population differences. Some obvious predictions exist – for example, there can be no nut-cracking in areas where there are no nuts – but no one would have expected that begging gestures for meat would be so different between Gombe and Taï chimpanzees. So we have to be very careful and not assume that because we see something in one chimpanzee population, it will automatically be present elsewhere. As I said in the first chapter, Descartes has been proven wrong and it is time we get rid of this old-fashioned attitude. Differences between populations are one of the most important characteristics that were discovered in chimpanzees in the last 30 years.

Twenty years after we published evidence of cooperative hunts in Taï chimpanzees, some psychologists still refuse to believe that chimpanzees are capable of cooperation. Such a different opinion may long prevail, since some hardcore psychologists believe that only experiments can prove cooperation, but to that end, let me stress that they will need to improve the design of their experiments so that chimpanzees have to work with other chimpanzees in a way that would be meaningful to them.[23] Intriguingly, despite progress in making more social

[23] Captive experiments have long tested and found no evidence for cooperation in chimpanzees without taking into account the social dimension of cooperation; this means that they paired chimpanzees to work together without any consideration of their social relationships (Crawford 1937, Chalmeau 1994, Chalmeau and Gallo 1996, Povinelli and O'Neill 2000). When this was eventually taken into account and only individuals with good affiliative relationships were paired, cooperation was observed much more frequently (Melis *et al.* 2006).

experimental procedures and having more success at engineering cooperation with captive chimpanzees, some psychologists still resist admitting that Taï chimpanzees hunt cooperatively.[24] Contrary to captive conditions, Taï chimpanzees simply try to obtain meat under very challenging circumstances and teamwork has proven to be the best way to acquire it. Once under the colobus, the first individual climbing in a tree immediately looks at the others and waits for them to make intentional movements to correctly anticipate the future development of the hunt and disperse to place themselves in the proper positions for the common action. Only then will the first individual begin to herd the prey. Obviously a shared goal will be best adopted by individuals when an ecological or social goal cannot be reached alone, thereby providing an important incentive to work together (see Box 4.2). Thus, well-tuned captive experiments incorporating socio-ecological circumstances equivalent to those seen in nature might elicit cooperation more readily in animals. Sadly, though, captive conditions are not ideal for this and engineering experimental situations mimicking group hunting, predator attacks, or territory defense are probably not possible.

In my mind, the importance of ecology on the development of behavior and cognition has been amply documented in many animal species, including humans, and I would therefore argue that whatever deficits we find in captive populations are more likely to be artifacts of captivity than characteristics of the species. Furthermore, from an ethnographic point of view, we know that different ecologies will favor different behaviors and cognitions. If an experiment could address the point of social culture, I would certainly welcome it. However, we have seen how many aspects of an individual's social domain influence each another so this might be nearly impossible to implement when an individual is removed from his social environment. Motivations are surely extremely difficult to study in the wild, but they will also be very difficult to reproduce in captivity. We have to assume that captive animals will have different motivations than wild ones who are fighting for their survival.

In recent years, both psychologists and economists have emphasized the cooperative nature of humans and suggested that this is an essential and unique

[24] Nine years after we published our observation about cooperative hunting in Taï chimpanzees (Boesch and Boesch 1989), a review about primate intelligence declared that, in reality and from their reading of the hunts, "each individual is pursuing its own best interests with little understanding of others" (Tomasello and Call 1997). However, as the amount of meat gained by the individuals is a direct function of the type of contribution made during the hunt, individual interest should discourage an individual from driving the prey, which is the least rewarding action during a hunt. But that is not the case. These authors later wrote, "Although it is a complex social activity, as it develops each individual simply assesses the state of the chase at each moment and decides what is best for it to do. … If everything were really collaborative wouldn't everyone on the team get an equal winner's share (not the most going to the capturer)?" (Tomasello *et al.* 2005, p. 722). Simultaneously, they stated that, "This event clearly is a group action, because … the chimpanzees are *mutually responsive* as they coordinate their behaviors with one another in space and time. But what seems to be missing is the *togetherness* or jointness that distinguishes shared cooperative activities from other sorts of group actions" (p. 641). More recently still, Call (2009, p. 372) suggested that, "There is no question that the end result is a cooperative joint activity but a question that is still unresolved is whether chimpanzees have an intrinsic motivation to cooperate. Would chimpanzees cooperate if there was no need to do so? Probably not …".

ability of our species. However, this interesting proposition has been strongly clouded by the fact that only experiments done with captive chimpanzees or other animals have been considered to support such a claim. This Cartesian view has weakened this proposition; would anyone seriously argue that chimpanzees are uniquely able to form large cooperative groups based solely on a comparison of the Ngogo chimpanzees with a group of human prisoners kept alone or with only a few others? Who would seriously suggest that humans recognize themselves in the mirror later than chimpanzees based solely on a comparison between Nso children from Nigeria with Yerkes captive chimpanzees? I assume no one would, as we would rightly question the representativeness of the compared sample populations.

As for symbolic cultural traits, the learning of social cultural traits requires one to take others into account, in this case, as an agent to help an individual reach a goal and not as an auditor. Therefore, important similarities and differences exist between the two types of traits and in the invention and learning of them: social culture is based more on the behavioral reaction of others than symbolic culture and we should therefore expect that the learning of social cultural traits will be accompanied by more practicing while symbolic traits need to be understood more entirely before being used. However, in the social domains, much less trial-and-error learning will be tolerated by partners and therefore imitation will be expected to be more important than for material culture. So we should expect different ways of learning cultural traits, depending upon whether practice is possible, if causal effects can be observed or have to be inferred, and if costs are associated with mistakes.

Cumulative culture can again be presumed to occur in the social culture domain, as some social changes in one aspect of social life, like in cooperative predator defense, seem to be incorporated in many different social domains leading to a whole specific social type. So, for example, it seems that the higher level of support given to individuals attacked by leopards has led to more tending to injured individuals in general, independent of the cause of injury. Furthermore, it seems to have spilled over into other social situations, such as territorial conflicts with neighboring chimpanzees where a great deal of support is given to subjugated group members. Such a first level of generalization could be explained by a process based on long-term reciprocity, as every individual could be in a situation in which they could profit from such extended within-group solidarity. A second generalization would be when solidarity was extended to individuals who cannot reciprocate, like in adoptions of orphans and food-sharing with youngsters. I realize this is quite speculative, but it stresses the possibility that one innovation in the social domain might have been progressively generalized to quite different social domains.

It seems important now to turn to the third domain of culture, the symbolic domain, and see how such cultural traits have developed and how they are learned and transmitted in different animal species. Since the symbolic cultural domain will be about communication, we can expect to see differences with both material and social culture.

Synopsis

- Social culture impregnates many aspects of social life in chimpanzees.
- Cooperation in hunting, helping individuals in need, sharing food, and adopting orphans are domains in which important differences are documented between different chimpanzee populations.
- Like for material culture, ecological conditions have a certain influence on the existence of social culture traits.
- Social dynamic within each group plays a larger role in producing many of the different social cultural traits documented in chimpanzees.
- Social culture, like material culture, can play a decisive role in the survival of the individuals benefiting from it.
- Human social culture has produced cooperation in very large groups with clear cases of altruism that has not yet been observed in chimpanzees.

5 I want to have sex with you About symbolic culture

The origin of language was not due to the basic needs of humans; ... It was neither hunger nor thirst but love, hate, pity, and anger that made humans produce our first words. The fruits don't escape from our hands, one can take them without speaking. One stalks in silence the prey on which one would feast. But for moving a young heart or repelling an unjust aggressor, nature dictates accents, cries, lamentations. There we have the invention of the most ancient words; and that is why the first languages were singable and passionate before they became simple and methodical.

Rousseau Jean-Jacques, *Essay on the Origin of Language*, Chapter 2 (1781).

Mahale chimpanzees, Mahale National Park, Tanzania, September 1999[1]

Up-and-comer Alofu is with Fanana, the reigning boss, at the rear of a small group of chimpanzees walking on the path towards the northern part of the territory, when they sit down to rest during the midday heat. Carter, another young male, takes advantage of the prevailing quietness to move closer to Christina, an attractive adult female with a sexual swelling. He sits facing her with his erection on full display but gets no reaction. Carter then breaks a small leafy branch off of a nearby sapling and, taking a leaf blade between his teeth, rips it apart bit by bit, spitting the cut bits on the ground without eating any of the leaf. He continues to do this with four more leaves, eating nothing but producing this ripping sound, which is just loud enough for Christina to hear. At first she sits up and then she comes to present to Carter. He mates with her quickly as the big males are still sleeping not too far away.

Taï chimpanzees, November 1991

After they finish eating ants, Brutus, Falstaff, and Snoopy climb a big fallen tree that has formed a clearing in the forest and settle down to rest and groom. When young Saphir

[1] Toshisada Nishida initiated his pioneering project on the chimpanzees of the Mahale Mountains National Park in Tanzania in 1965 (Nishida 1968). The Mahale Chimpanzee Project constitutes the second pillar of the bulk of our knowledge about East African chimpanzees, which for decades was all that we knew about chimpanzees. I was honored to be invited by Toshisada Nishida to study the Mahale chimpanzees and share his life in camp for two incredibly instructive months in 1999. Seeing the fascinating behavioral differences between two populations of chimpanzees living only 200 km apart, Gombe and Mahale, was a real adventure into chimpanzee culture.

Figure 5.1 **An adult male chimpanzee from the Taï forest leaf clips** before starting to drum on one of the forest's large buttressed trees. He holds the base of the leaf with his right hand while biting off pieces of the leaf blade. The noise he produces as he tears pieces away signals his intention to listeners. He may repeat these actions on three to four leaves. None of the leaf pieces are eaten as he does so.

appears at the edge of the windfall opening with a small sexual swelling, young Snoopy moves towards her with an erection but gets no response. Persevering, Snoopy begins to knock against the trunk of a small sapling with his knuckle. Saphir looks at him then and as he continues his knuckle-knocking, she approaches and presents her sexual swelling to him. Snoopy then mates with her quickly while the big males are still busy grooming.

Gombe chimpanzees, June 1992

After they had rested for about 30 minutes, Frodo leads Prof to the feeding site down in the Kasekala Valley. Frodo and Prof approach the clearing very slowly. A group of females is already sitting there with their playful infants. As the first female notices the males' presence and begins to pant-grunt in submission, the two males charge down the slope straight through the group of females. The females immediately collect their babies on their bellies and hurry up some trees, pant-grunting all the while. Frodo stops downhill from where the females had sat and Fifi, his sister, runs to greet him. Prof moves decisively towards Gigi with an erection and after shaking some branches, he climbs towards her and mates with her without any further ceremony. Both then descend to join the others who are now resting again on the ground.

Similar to a "sign-code," leaf-clip by the Mahale chimpanzees mean, "I want to have sex with you" (see Figure 5.1). As Carter was demonstrating in my field notes, leaf-clipping in Mahale is used by young males to entice sexually active females in the presence of higher-ranking males; during a two-month period, I saw this performed repeatedly by different young males. In all chimpanzee populations, dominant males who want to mate with a female normally show their erect penis. They also sometimes stress their point by stamping the ground with their feet or shaking a sapling. Alternatively, to say "I want to have sex with you," a young male chimpanzee in the Taï North group will knock with his knuckles on a small tree trunk or log. Although the sound this produces is very different from the sounds produced by leaf-clipping, it expresses the same meaning as in Mahale and it is understood correctly by Taï chimpanzees. Intriguingly, the knuckle-knock was observed by all young males in the North group for more than 20 years of observations but never in the males of the South group (see Table 5.1 and Figure 5.6). In contrast, the young male Gombe chimpanzees have not developed any specific signal to attract females for mating. These signs are used by all group members in the group-specific context and are understood correctly by others without hesitation. Thus, the same meaning can be conveyed with different signs in different populations (Figure 5.2). In addition, the connection between a gesture and the meaning it conveys is totally arbitrary, as nothing about the sounds produced by leaf-clipping allows it to be more easily associated with "I want to have sex with you" than the sounds produced by knocking one's

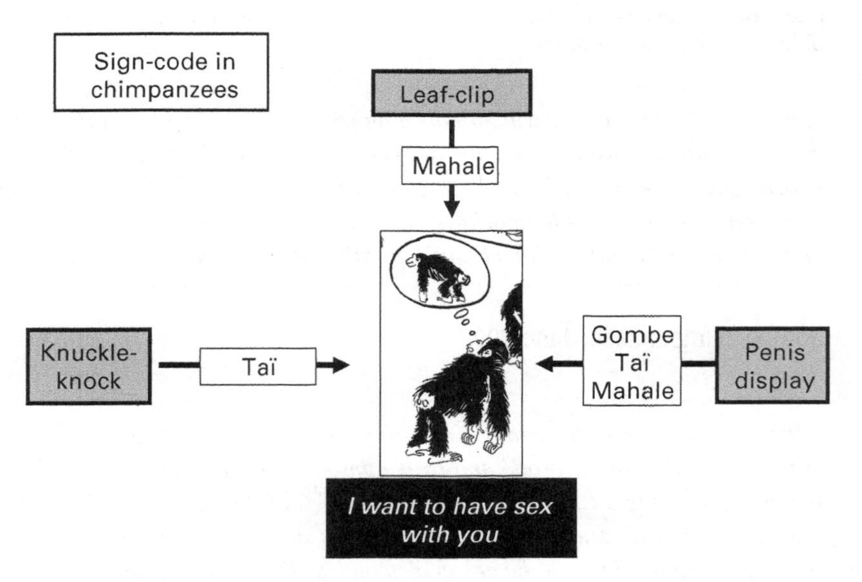

Figure 5.2 **Different populations of chimpanzees use different signs to mean "I want to have sex with you."** While dominant males in all populations simply show their interest in mating with sexually active females by showing their erect penis (shown in gray on the right side), in Mahale and Taï, low-ranking males use different group-specific signs to attract the attention of sexually active females. Low-ranking males in Gombe do not seem to use specific signs to attract female attention.

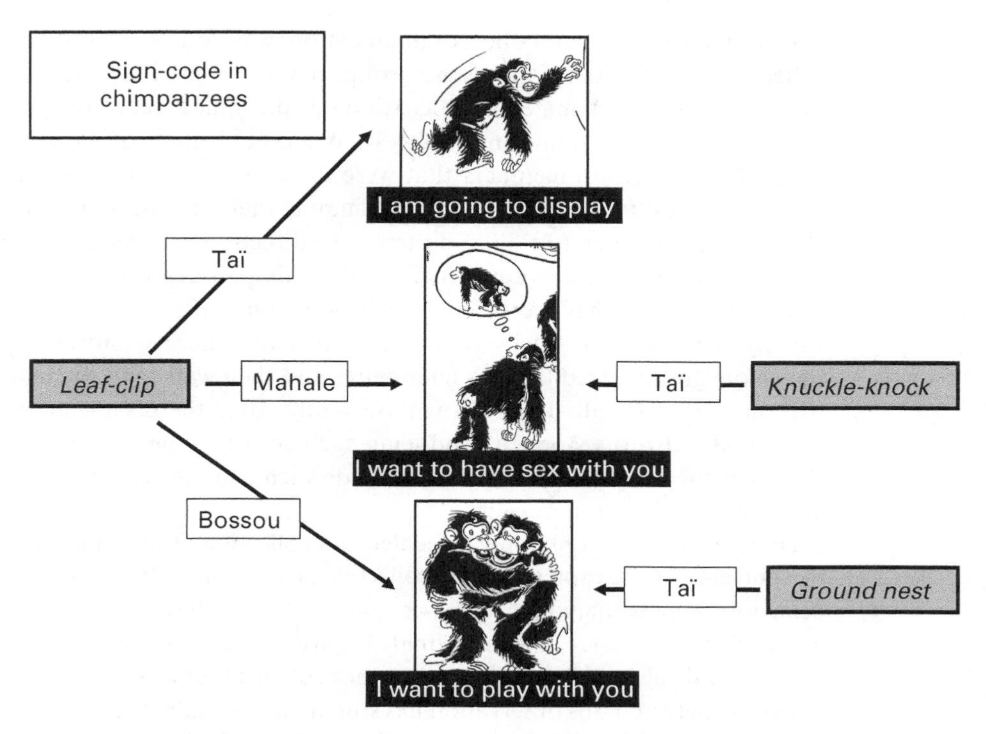

Figure 5.3 **A preliminary glossary of the sign-code used by wild chimpanzees.** Three symbolic meanings (in the center) of three behavioral elements, the signs (framed in bold) are performed by chimpanzees from three different populations (in the box on the arrows). In different chimpanzee populations, the same sign can have different meanings and the same meaning can be conveyed with different signs.

knuckles. These behaviors are symbolic signs in the sense that they stand for something that they bear no external resemblance to and this is correctly understood by most group members. For Mahale chimpanzees, the sound of ripping leaves means something that is understood by all group members, while the Taï chimpanzees all understand the same meaning from the sounds produced by knuckle-knocking. In a way, this is very similar to human sign language, where gestures stand for meanings.

At the same time, the same sign can have different meanings in different chimpanzee populations. For example, while leaf-clip means "I want to have sex with you" to the Mahale chimpanzees, to the Bossou chimpanzees from Guinea, it means "I want to play with you," and to the Taï chimpanzees, it means "I am going to display." Thus, different chimpanzee groups attach different meanings to leaf-clipping and use this behavior systematically in group-specific contexts (Figure 5.3). Further, to express "I want to play with you," the chimpanzees in Taï South group first construct a rough ground nest until the play partner joins in. The striking feature of this leaf-clip sign-code is that this behavioral pattern has no direct morphological or visual connection to its understood social meaning. Thus, the meaning of this sign is based on an *arbitrary social convention* that is shared by all group members.

Another case of symbolic communication was found in Taï chimpanzees. When Brutus was the boss of the group, it was 80 chimpanzees strong and 9 of these were adult males. He occasionally drummed very conspicuously on the buttressed trees found in Taï forest. When he drummed twice on the same tree, all of the group members that were dispersed throughout the forest were observed to rest for 60 minutes before resuming their activities. In contrast, when he drummed twice on two different trees, the direction between them represented the new movement direction, which all the chimpanzees within auditory range followed, even if they were silent for a long time after the direction change. Finally, if he drummed twice on the same tree and once on another tree, he and the whole group rested first for 60 minutes and then walked in the new direction without any new calls to confirm this.[2] Sadly, after the disappearance of four adult males, Brutus stopped producing such conspicuous drumming. It is possible that the presence of fewer males made such complex communication about group movements unnecessary.

Human culture is commonly presented as a collection of ideas and values, which, in a nutshell, can be expressed as a "collective cast of mind." In other words, cultural behaviors possess shared meanings or values within each social group, and it is this aspect that determines how individuals behave. In many instances, this "collective cast of mind" has no direct relevance to the ecological conditions prevailing in each human society and this observation has stimulated the claim that culture in humans can free the individual from its environment. Intriguingly, we seem to observe the emergence of a similar process in wild chimpanzees. Sure, there are presently only a very limited number of behavioral elements that could be considered as symbolic in chimpanzees and this remains tremendously different from what we see in humans, but in the course of this book, we will learn to be cautious about strong claims. Are symbols in chimpanzees really so rare or is this simply all new to us and we are slowly beginning to establish ideas and insights into this topic – including the fact that chimpanzees could well have symbols? Systematic studies are surely needed before we can answer this question.

Arbitrary social conventions in chimpanzees

The first step towards symbolism is the development of arbitrary social conventions that persist over time within different social groups and that may have meaning we have not yet uncovered. In chimpanzees, a number of behavioral patterns have been observed that clearly seem to be social conventions that have been adopted by many individuals within a social group and that may have persisted for generations. Here, I would like to provide some intriguing examples of them in the context of how chimpanzees handle skin parasites. Chimpanzees in general love to handle ticks and lice, move them between their teeth and lips, and look at them repeatedly as they hold them out on their protruding lower lips.

[2] Boesch (1991b, 1996).

Mahale chimpanzees, Mahale National Park, Tanzania, September 1999

After resting for a while, Alofu grooms Fanana, who is sitting with his back towards the younger male. After a while, Alofu finds a louse egg on Fanana's back, pulls it off with his teeth, takes one leaf from a nearby sapling and places the tiny egg on the leaf. As I near to get a better view, I see Alofu fold the leaf lengthwise in two, thereby covering the egg with the leaf blade. He then uses the nail of his thumb to cut a half circle in the leaf and, lifting the flap to expose the louse again, he puts it back into his mouth and chews on it (see Figure 5.4). Replacing the louse on the unfolded leaf with his lips, he then refolds it to cover the louse again and then cuts the upper part a second time. After cutting a second half circle in the leaf, he exposes the louse, puts it in his mouth again and chews on it. Fanana, who had been closely watching Alofu's actions, sits down and shows his back to Alofu again, who grooms him for another minute before laying on his back to go to sleep, shortly before Fanana does the same.

Taï chimpanzees, November 1991

Some females, including Salomé, join Brutus and Falstaff who immediately start to groom this popular female. The grooming rapidly develops into a big chain, with

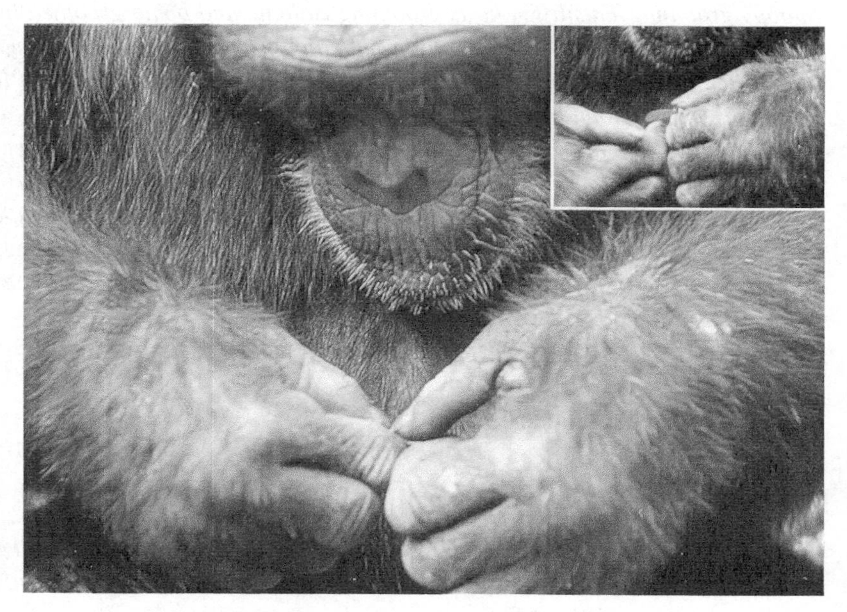

Figure 5.4 **Leaf-fold with ectoparasite in Mahale chimpanzees**. After placing a louse egg on a leaf, which he holds with his two index fingers (insert top right), and folding the leaf over the louse, Bonobo, an adult male, is shown here cutting the part of the leaf covering the louse with the fingernail of his left thumb. Once he has cut a half moon, he will lift it and recover the egg with his lips.

Figure 5.5 **Macho using his forefinger to hit a tick he has placed on his forearm**. An adult female who was attracted by the manipulation of the tick observes his actions very closely for the length of time he hits and manipulates the tick. While all Taï chimpanzees use this technique, it is very different than the one used by the Mahale and Gombe chimpanzees.

Snoopy grooming Falstaff, who is grooming Salomé and being groomed by her, who Brutus also grooms while two females groom him. The positions change regularly and it is difficult to track precisely who is grooming whom. Suddenly, Falstaff lip-smacks louder, as he has found a tick in Salomé's hair. After removing it, he turns aside, places the insect on his right forearm and starts hitting it with the forefinger of his left hand, while both Salomé and Brutus intently watch what he is doing from 10 cm away. Falstaff repeatedly picks up the tick with his lips and alternately bites and looks at it, but then replaces it on his forearm and continues to hit at it with his forefinger. I count 240 hits in all. Once the tick is nicely squashed, Falstaff swallows it and the grooming session resumes.

Gombe chimpanzees, June 1992

After dipping for driver ants, Prof is joined by Frodo, who has finished eating some of the juicy Parinari fruits that have fallen to the ground. The two rest and groom one another, searching for dirt, ticks, and lice in each other's hair, preferentially focusing on places that are difficult for an individual to access directly – the armpit, the back, the face. Suddenly, Prof turns around and tears leaves off a small sapling, places the leaves carefully in a pile in his hands, and with his lips places a louse he found on Frodo on top of the pile. Using both thumbs, he tries to squash it against the carpet of leaves, all the while making lip-smacking noises. Frodo, who stands next to him, stares intently at Prof's actions as Prof takes the louse between his lips and tries to press some juice out of it. He then replaces it on the leaves and tries to crush it again with his thumbnails. He repeats the procedure 4 times, all the while with Frodo staring at his actions from just 10 cm away. Then letting the leaves fall to the ground, both resume their grooming.

A person's first experience with chimpanzees is extremely rich and memorable and it acts as the reference point for all future experiences. In that sense, my research at Gombe and Mahale were key to my realization that population differences are a reality and not a bias of our personal experiences. Luckily, I arrived at Gombe with my Taï chimpanzee glasses firmly on. That is to say, I arrived there with a detailed knowledge of Taï chimpanzee culture and therefore knew some of the things I should look for before immersing myself in a new culture. In this sense, my two 3-month-long stays in Gombe in 1990 and 1992 were some of the most enriching experiences I have ever had with chimpanzee culture.

When I arrived at Gombe, I knew that when Taï chimpanzees use their lips to pick up and place something repeatedly on a surface during a grooming session, it means they have found and are handling an ectoparasite. As we have seen, Taï chimpanzees place them repeatedly on their forearm then hit them before putting them in their mouths and chewing on them. While leaf-grooming had been described in Gombe chimpanzees before this time, its purpose remained a mystery, as observers who had carefully watched this behavior saw neither the chimpanzees placing anything on the leaves with their lips nor any parasites on the leaves.[3] Consequently, when I was first in Gombe and I saw Fanny leaf-grooming during a grooming session with her uncle Freud and placing something she had been chewing on four leaves with her lips, I immediately suspected that it was a parasite. In a way, it was so strikingly similar to what I had seen so many times in the Taï chimpanzees but at the same time, thanks to my biased Taï glasses, it was so different! Carefully watching two more instances of leaf-grooming that same day by Frodo and Freud confirmed to me that it was lice from their grooming partner's hair that they were placing on leaves. So it appears that the Gombe chimpanzees added a new element to this cultural trait just before I arrived in 1990, some 30 years after Jane Goodall started studying this group. This could be a very nice example of "cumulative cultural evolution," by which individuals belonging to the same social group add elements to a cultural trait over time and contribute to its sophistication.

The same thing happened during my first days with the Mahale chimpanzees in 1999. They had also been observed to leaf-groom and its function also mystified observers, who assumed it to be the same as in the Gombe chimpanzees. I arrived in

[3] When I asked Jane Goodall and Richard Wrangham, who had been carefully looking at leaf-grooming since long before I visited Gombe, they confirmed to me that they had not seen the chimpanzees placing anything on the leaves and, therefore, that they considered the behavior to be aimless and a mystery. The Tanzanian observers shared exactly the same view; however, they agreed with me in 1992 that the chimpanzees were placing insects on the leaves.

Mahale, now with both my Taï *and* my Gombe glasses firmly on, and, after seeing Alofu and a consort folding a leaf that they had placed something on, I immediately knew that this behavior was strikingly different from what the Gombe chimpanzees were doing.[4] When I discussed this with Toshisada Nishida that evening, he agreed that this was something different and told me that he had seen it done this way for the 30 years he had studied the Mahale chimpanzees. Having no personal experience with the Gombe chimpanzees, he did not realize that the Mahale method was different.[5] To me, the leaf-fold is very intriguing. If it is seen as a way to smash lice and ticks, it is certainly not the most efficient method; it is very cumbersome and how precisely they cut the leaves with their fingernails requires very fine motor control (see Figure 5.4).[6] Later, researchers in Mahale confirmed this leaf-folding behavior and function. Furthermore, the chimpanzees in Bossou have recently been described to place objects in the palm of their hand and poke, push, and drag the item with their index fingers.[7] This suggests another variation in chimpanzees on the theme of handling ectoparasites.

Why do all individuals in a group conform to such arbitrary conventions? Repeatedly hitting a tick that a chimpanzee has placed on its forearm and cutting a leaf that has been folded over do not look like very efficient techniques for smashing lice or ticks. There seems to be no way to determine why one way should be preferred over the other and we are left puzzled about why chimpanzees from one group copy one another so precisely. In addition, the very detailed and fine motor control required for the task is striking – I tried to cut the upper part of a folded leaf with my thumbnail and failed miserably. When Alofu was leaf-folding, he followed a very precise order of actions – find a louse, put it in his mouth, turn to cut a leaf, place the louse on one side of the flat part of the leaf, fold the leaf so as to completely cover the louse, use one fingernail to cut the leaf above the now covered louse, lift the 5 mm-wide circle that he has just cut, and remove the louse with his lips. It seems highly unlikely that such a sequence would be learned in such an exact way without precisely copying what you have seen others do. Imitation of the precise actions of a group member seems to be unavoidable in this case. But why would one want to stick to such a complex sequence of actions for something as simple as eating a louse? To me, conformity (in the form of belonging to a social group,

[4] During my 2-month stay, I saw leaf-fold performed 116 times and by as many as 28 individuals.

[5] A similar situation occurred when a researcher from Gombe came to visit Mahale and saw the Mahale chimpanzees scratch the back of a partner before grooming him there. This behavior, which is common in Mahale chimpanzees, was never seen in Gombe chimpanzees and pointed to a cultural difference between these two populations (Nishida pers. comm., Whiten *et al.* 1999). Without a Gombe observer visiting Mahale, this "social scratch" difference would have remained unnoticed. A systematic study of the leaf-folding behavior was undertaken more recently that confirmed my observation and found that the chimpanzees indeed place lice on the leaves (Zamma 2002).

[6] Some skeptics could argue that those differences are too finely graded to be reliably documented. However, following my stays in Mahale and Gombe (in 1990–92 and 1999), I published detailed accounts of these differences in several reports (Boesch 1995, 1996, 2003, Boesch and Boesch-Achermann 2000), and other researchers have since described these differences independently for Mahale and Gombe (Nishida 1987, Zamma 2002, Goodall 1968, 1986, Whiten *et al.* 1999).

[7] Nakamura and Nishida (2006, p. 38).

Table 5.1 Symbolic cultural traits and social conventions observed in eight chimpanzee communities. Preliminary list found when comparing a sample of eight different chimpanzee communities ordered from West to East Africa. ("+", behavior is present; "–", behavior has not been observed; "blank", no data found about this behavior)

	Chimpanzee communities							
Pattern	Bossou	Taï N	Taï S	Gombe	Mahale	Kanyawara	Ngogo	Budongo
Knuckle-knock	–	C	–	–		–	–	–
Branch-shake	C	C	C	C	C	C	C	C
Leaf clip	P/F	D	D	–	C	+	C/P	+
Ground nest		+	C					+
Index-hit		+	+	–	–			
Index to palm	+	–	–	–			–	–
Leaf-groom	–	–	–	+	–		+	+
Leaf-fold	–	–	–	–	+		–	–
Leaf insect inspect		–	–	–				+
Ground groom		–	–	–	+		–	–
Hand-clasp groom	–	+[1]	+[1]	–	+		+	+[1]
Water splash		–	–				+	+
Shrub bend	+	+	+	+	+		–	
Throw splash	–	–	–	+	+		–	
Social scratch	–	–	+[1]	–	+		+	–
Leaf strip	–	–	–	+	–		+	–
Leaf-pile pull	–	–	–		+			
Branch slap	+	+	+	–	–	–		+
Mutual genital touch	+	–	–		–			
Stem pull-through	+	–	–	–	+	+		–
Shrub bend	–	–	–	–	+	–		+
Rain dance	–	+	+	+	+	+		+
Groin tuck		–	–			+	+	+

For knuckle-knock and ground-nest context: C = courtship. For leaf-clip context: P = play initiation, F = sexual frustration, D = display. For groom hand-clasp and social scratch: 1 = seen only when one specific individual involved.

References: **Bossou**: Nakamura and Nishida (2006). **Taï North and South**: Boesch (1991b), Boesch and Boesch-Achermann (2000). **Gombe**: Goodall (1986). **Mahale**: Nakamura *et al.* (2000), Zamma (2002), Nakamura and Nishida (2006). **Ngogo**; Whiten *et al.* (1999), Langergraber (pers. com). **Budongo**: Assersohn *et al.* (2004), Zuberbuehler, Wittig and Crockford (pers. com).

identifying with that social group, and therefore "doing as they do") seems the only natural explanation as to why we observe such different ways of handling ticks and lice in three different groups and why all group members precisely follow their own group's arbitrary method.

Do we know everything there is to know about symbolism in wild chimpanzees? That I was able to find so many subtle cultural differences after only two months in Mahale and six months in Gombe strongly suggests that many more might exist and that with new and careful systematic observations, many more will be found. In another case, one researcher who had spent a great deal of time observing the Mahale chimpanzees went to observe the Bossou chimpanzees and listed five subtle differences during a two-month stay, which illustrates the potential of a comparative approach.[8] Sadly, only a handful of different chimpanzee populations have been studied until now and a systematic comparative study is still lacking. The paucity of data on symbolic culture leads me to question how well we understand them. Like deciphering an unknown language that you have had no previous experience with, we surely must miss many signs the chimpanzees use in front of our eyes because we do not understand them. Often before a hunt, Taï males scratch their bellies conspicuously while standing upright and looking at the others in their group. What are they communicating? All chimpanzee observers have been struck by how easily the chimpanzees can meet others when they are totally silent. Were such meetings planned ahead and, if so, how? I use these examples to show that there is room for more of such arbitrary communicative signs that we might have missed.

Table 5.1 presents a preliminary list of cultural traits found after comparing eight chimpanzee populations. Some of these traits could be related to specific meanings, as seen in Figure 5.3, while for some others, no specific meaning or function has yet been proposed. This illustrates how quickly such a list could increase if more cross-site observations were done.

Arbitrary social conventions have been suggested in two other primate species. First, long-term observations in a group of capuchin monkeys (*Cebus capucinus*) revealed the appearance and transmission of five social conventions, such as hand-sniffing, sucking of others' body parts, "finger-in-mouth," and hair-pulling games.[8] They were observed to be performed by about 22–45% of the group members. One of the most enigmatic traditions in monkeys is the stone-handling done by Japanese macaques; while chewing provisioned wheat and corn, these monkeys manipulate stones in various ways, including picking them up, scattering them, and sometimes rubbing, carrying, clacking, or cuddling with them. This solitary behavior that happens after feeding is typically used for playing or grooming when other food types have been consumed.

Innovation and transmission of inventions

How do such differences originate within groups of chimpanzees? Fossey, a young adolescent female, gave us a clue. When she migrated into the Taï group, she immediately followed the males for protection from the resident females, who can be aggressive to female newcomers. Brutus, Falstaff, and the other males welcomed new females and were perfect guides for them to discover this new part of the

[8] For the capuchins, see Perry and Manson (2003); for the macaques, see Huffman (1996).

forest. We never learned how far she originally came from and this is still one of the big mysteries in chimpanzees, as we have not been able to track the dispersal movements of the females between chimpanzee communities. However, Fossey was quickly integrated into her new group and became used to the presence of the weird human bipeds that were always following them and it was with her that I observed an unusual type of tool use. Chimpanzees in the Taï forest love to eat a diverse diet, not only with many types of fruits and leaves but also with many types of insects, mushrooms, and meat. They regularly inspect *Cephalotermes* termite mounds to see if the mushrooms that grow inside have emerged (this only happens once a year, when the fungi sprout in lovely white hats) and if they see any signs, they scratch the surface to remove the surface layer and then dig into the thick mushroom mass to eat bits and pieces. These mushrooms are very hard and I had to use my Swiss Army knife to cut off a piece to taste it. After she had found such a mushroom one day, I saw Fossey cut a twig from a sapling, remove the leaves attached to it, and shorten the end with her teeth. She came to the mound and used the stick instead of her nail to scratch the mushroom and remove little bits to eat. That was the first and only time I saw this innovation in this group.

Innovation in a group of chimpanzees is fairly common; for example, I saw 20 new behavioral patterns emerge in the North group during a two-year period,[9] but others picking up these innovations is rare. A similar pattern has been seen in Mahale, with 32 new behavioral patterns appearing in one community but only 11 of them spreading to other group members.[10] Fossey was not alone in demonstrating new ways to solve ecological challenges and her invention, which seemed like a very good one to me, was never picked up by the others. In another example that happened back in December 1990, Kendo, the highest-ranking male in the community at the time, turned on his side while he was napping on the ground and started to leaf-clip without sitting up. When he finished, he went back to sleep. It was the first time in the 10 years I had been following them that I saw a chimpanzee leaf-clip in this context. In the following days, I noticed that Kendo continued to leaf-clip while resting, and, remarkably, some of the other group members began to imitate him. Within a few weeks, most of the males and some of the females were leaf-clipping in this new context and, thus, a new fad was born! I could not identify any incentive to leaf-clipping in this context, but it is worth noting that the possible inventor occupied a much more prestigious position in the group than little Fossey did and this might have played an important role as to why he was imitated and Fossey was not.[11] This observation suggests that the identity of the inventor might sometimes be more important than the usefulness of the invention.

[9] Boesch (1995); for innovation, see also Biro *et al.* (2003).

[10] Nishida *et al.* (2009) observed these 32 innovations over a period of 27 years. Only new behaviors occurring after 20 years of observation qualified as innovation in this study.

[11] Prestige has been proposed to be important in explain the acquisition of a cultural trait in both humans and animals (Durham 1991, Boesch and Tomasello 1998). Recent experiments with a captive group of chimpanzees and free-living humans have confirmed that, with all else being equal, the model with more prestige will be more frequently copied (Horner *et al.* 2010, Harris and Corriveau 2011).

In chimpanzees, successful innovations could spread to other social groups by migrant females, who would become the carriers of culture. In all studied chimpanzee groups, females transfer between groups once they reach sexual maturity but before they begin reproducing, when they are around 10–12 years old. At that age, they have already acquired all the behavioral patterns typical of their natal group and when they transfer, they could potentially spread cultural traits between groups. Male chimpanzees, on the other hand, spend their entire lives in the group they were born into and therefore cannot fulfill this function.[12]

When Diva arrived in the Taï South group in February 2007, she was faced with the sometimes aggressive resident females and spent most of her time with the males. However, she integrated quite rapidly into her new group and was seen more and more frequently with the other females. She remained low-ranking for a very long time, however, and when she occasionally performed an unusual behavior, she was not copied.[13] For example, during the first nut-cracking season after her arrival, she used mostly wooden clubs to crack the nuts in trees, while South group members almost exclusively use small stones to do so, even though stones are much rarer than wooden clubs. The resident individuals did not appear to show any curiosity or interest in her behavior. On the contrary, a year later, Diva was seen to select stone hammers to crack nuts in trees in exactly the same proportion as the resident individuals of that group. So, in that case, it was the immigrant female who adopted the prevalent behavior of her new group and not the other way around.

However, stone selection is not the only cultural difference between neighboring communities of Taï chimpanzees. I detected many differences (see Figure 5.6) between the three communities we follow; for example, North group chimpanzees put their whole arms into termite nests to extract many handfuls of grubs, while South group chimpanzees put only their forearms into the nests and extract one handful at most. Similarly, only the South group chimpanzees eat the piths of the herbal plants that are found in swamps throughout Taï forest, and only the North group eats the abundant *Thoracotermes* termites. In addition, the South group eats the *Strychnos* fruits only when it is fresh while the North group eats them only when they are rotten. More differences like these are now surfacing with the habituation of the East group, which seems to also have its own subculture within Taï.[14] For

[12] In chimpanzees, males remain in the groups they were born into for their entire lives. The very rare cases of males transferring between groups were only observed in special cases where a mother with a dependent son transfers away from her new group as it begins to disintegrate. This happens, for example, when groups that have fewer than two adult males are too weak to resist larger neighbor groups. In other species, including most monkey species, it is the males, not the females, who transfer between groups, and so males would be the potential carriers of cultural traits. The majority of human groups are like chimpanzees and the other Great Apes in this respect.

[13] Lydia Luncz, a researcher at Taï who spent a lot of time in the forest following new immigrant females, documented the differences in hammer choice between neighboring groups and how the new females adapted to the prevalent group preference (Luncz *et al.*, 2012). I documented the first differences between neighboring groups in Taï chimpanzees in 2003 (Boesch, 2003).

[14] We habituated four neighboring communities of chimpanzees in the course of the Taï Chimpanzee Project (Boesch 2009). We started with the North group and, as an attempt to understand more about intergroup interactions as well as dispersal of individuals, we habituated their neighbors to the south.

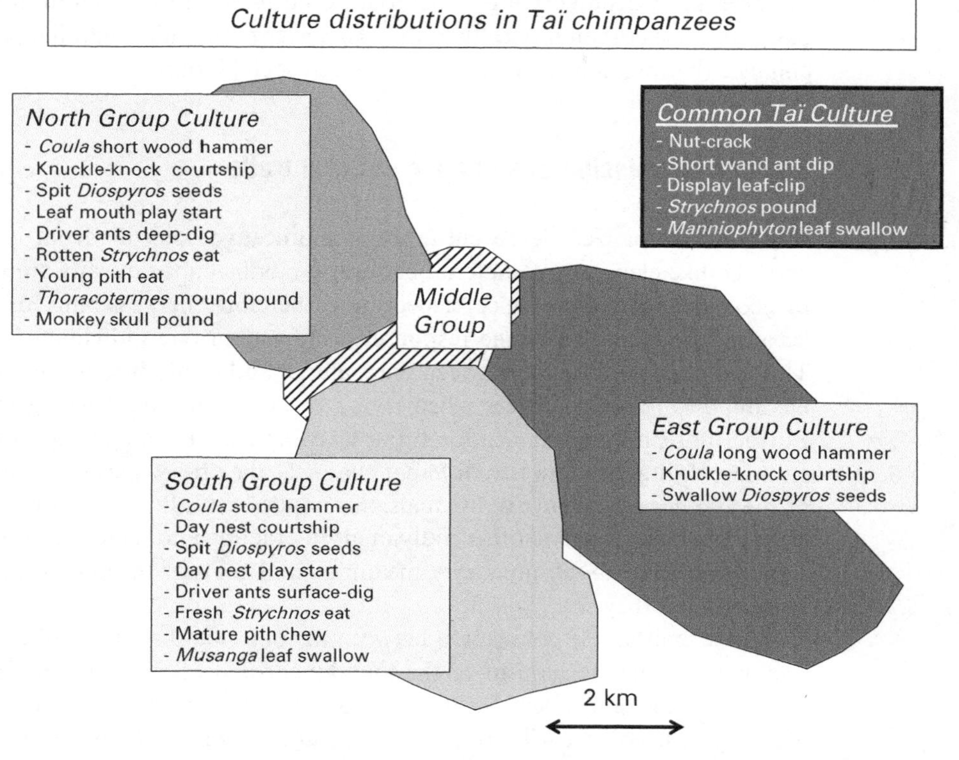

Figure 5.6 **Cultural differences between three neighboring communities in the Taï forest** (expanded from Boesch 2003). Some of the common Taï chimpanzee cultural traits shared between all three communities are listed. The territories of the communities are represented with different gray patterns and the list of specific cultural traits that distinguish them from one another is listed. The middle group remained exceptionally small during the entire observation period and we have only limited observations on it.

example, they use wooden clubs as hammers, like in the North group, but they select much longer wooden clubs than the North group and they knuckle-knock like the North group but unlike the South group.

Cultural differences persist between neighboring communities in Taï, despite the regular transfer of females between them. For large communities with many adult males, which are attractive to young females, we observe an immigration of a new young female about once or twice per year. From the example of Diva, we can assume that even if they showed different behaviors to those of their new community, it is unlikely that it will lead to cultural changes. However, if a high-ranking female transfers between groups, as we have seen for communities that are in

However, it turned out that we had actually habituated two communities there, the Middle group and the South group. With a continuing decline in the number of chimpanzees in the North group, we then started to habituate the East group, which neighbors the South group. All groups are neighbors and have regular intergroup encounters (see Boesch 2009).

the process of disintegrating,[15] then one could imagine that a cultural trait could pass from one community to the other. However rarely this might happen, it could enhance the spread of cultural traits between social groups.

Social transmission of symbolic cultural traits

How do chimpanzees learn cultural communicative behavior from group members? Unlike elements of material culture, symbolic cultural traits are not limited by ecological or technological aspects and therefore the possibility of individual learning producing the same response in more than one individual is less likely. This is a major difference between the material and symbolic culture and one that has important consequences when trying to determine the learning mechanism. Furthermore, for communication traits, learning the meaning of the cultural element must be practiced on two different channels, the one of the emitter and the one of the audience. For naïve individuals, this implies a triadic dimension as they must learn from observing two other individuals interacting and therefore need to understand something about intentions, meanings, and goals of the interactions to make sense of what they see.

For the emitter, the connection between the object and sound could be based on a simple learning mechanism, as the sound made by leaf-clipping, for example, can only be produced by a leaf being ripped and no other object. Thus, trial and error with any other object will not produce the correct sound. However, this does not explain why all males use exactly the same sign to mean the same thing. In addition, we have not seen young males make mistakes as would be expected if individual learning was taking place. The only sign young Taï males will use to attract the attention of estrus females is the knuckle-knock, and they continue to do so even in the rare cases when the females ignore their signs.

For the audience, the auditory or visual perception of the behavior must be related to a particular shared meaning for the communicative function to be effective. For example, when you hear the sound associated with knuckle-knocking by Taï chimpanzees, it means a male wants to mate with a sexually active female. Those who hear the sound (the audience) have to infer from the context that the male knuckle-knocker (the emitter) is expecting a specific behavioral response from the female. Interestingly, this behavior is mainly done by young adolescent or adult males who want to remain inconspicuous when they invite females to mate, in order to avoid larger dominant males from possibly interrupting their mating. In chimpanzees, males start mating with females when they are as young as 2 years old and everyone is extremely tolerant with them to the point that old females will get up while resting to present to such males. However, infant and juvenile males have

[15] We saw this once with a prime female migrating with her infant son from the Middle to the South group as the Middle group had only one adult male left (Gomes, Deschner and Luncz, pers. obs.). A very similar process has been observed in Mahale chimpanzees (Nishida *et al.* 1985).

never been seen to knuckle-knock as they are not really competing with the big males. However, females as young as 4 years of age, such as Simone, the daughter of Salomé, already understand the meaning of the knuckle-knock, although it has never been directed at them. I saw her and her friends regularly rush to a knuckle-knocking male and present to him, while he had clearly intended to attract someone else. The reaction of the young infant females shows that this is learned in ways other than through social punishment of a male whenever females did not react appropriately. Similarly, observations have been made of Mahale chimpanzees using the sound of leaf-clipping to locate out-of-sight estrus females that had been courted by other males or by rival males and to interfere with their mating.[16] Thus, learning the communicative meaning of signs is also done by group members not directly involved in the interactions, and, therefore, this requires some understanding of the motivation of others.

Symbolic culture and social conventions in other animal species

Song birds have long been proposed to possess population-specific songs that males acquire as they mature from listening to neighboring males sing. In this way, song cultures become clearly established so that the exact region of the males can be determined from the specifics of the song. If the learning period is limited to an early maturation time in some bird species, which suggests a genetically determined learning period, in other species, learning is much more flexible and extends throughout life. Vocal copying for songs that have not been shown to possess any special meanings or specific communication values would seem to have relatively less relevance in terms of our discussion of symbolic culture.

Dolphins and whales including killer whales, have also been proposed to possess cultural vocalizations in the sense that they possess population-specific clicks or songs that, in some cases, seem to be learned socially. For example, the songs of humpback whales (*Megaptera novaeangliae*) show interesting aspects of culture for a very large mammal.[17] All males in a population share the same song at a given time, but the song changes gradually throughout the singing season. In one population of humpback whales found off the coast of Bermuda, one-third of their song changes on average each year, which is too rapid a rate than could be plausibly explained by genes. However, off the east coast of Australia, the entire song changed within two years to match that found off the west coast, apparently triggered by the movement of a few individuals from the western to the eastern side. Here again, the symbolic or communicative values of the songs need to be clarified.

Orangutans have also been shown to possess behavioral elements that have a symbolic cultural dimension.[18] Of the 19 likely cultural variants that have been

[16] Matsumoto-Oda and Tomonaga (2005).
[17] See Payne and Payne (1985) and Noad *et al.* (2000).
[18] See Van Schaik *et al.* (2003).

identified in a comparison of 6 different populations of orangutans in Kalimatan and Sumatra, 3 potential communicative sounds – 2 kiss-squeaks and a raspberry sound – have been proposed to be cultural. Suaq Balimbing orangutans emit the raspberry vocalization during the final phase of nest building, while the Lower Kinabatangan orangutans make this sound just before they start to build nests. The communicative meaning of this sound is not yet totally clear. Nevertheless, this shows a potential similarity of the cultural ability between chimpanzees and orangutans. In bonobos, cultural traits have been proposed recently, but since very few populations of this species have been studied, the evidence is sparse.[19] Intriguingly, 10 behavioral elements that have been proposed to be cultural traits in chimpanzees have also been observed in bonobos. This includes leaf-clip, which, similar to the Bossou chimpanzees, the Lomako bonobos seem to perform with the mouth in the context of soliciting play, and the mature females seem to perform with the hands to attract males, which is the opposite of what has been seen in Mahale chimpanzees. More data would be needed in both cases, but overall, here again we see a pattern that could be similar to that seen in chimpanzees.

The observations on bonobos illustrate the main limitation we face when attempting to answer questions about culture in other animal species – namely, the paucity of data from more than one population in any species. In cases where such studies exist, they rarely achieve the depth and duration equivalent to the few long-term studies on wild chimpanzees and this makes it difficult to answer questions about meaning and communicative value. However, we should remember that absence of evidence is not evidence of absence. On the other hand, it could be argued that symbolic culture will remain rare as, by definition, it requires some symbolic abilities and these have been demonstrated in only a few species, such as chimpanzees, orangutans, gorillas, and a few others.

Contribution to the culture debate

The hardcore experimental psychologist, although intrigued by the sign-codes described here, will undoubtedly remind me that human language is much more than a small collection of symbolic signs, but is instead a very complex symbolic communication system with unlimited possibilities thanks to its grammar and semantics that allows humans to communicate about the past and the future and share ideas and beliefs. He will then specify that human language is systematically taught by parents and teachers and it is this teaching that allows us to acquire such complex languages so rapidly. Finally, the hardcore psychologist will reiterate that, through a process called "ontogenetic ritualization," it is known that individuals can develop signs that look as if they were symbolic but that, in reality, were progressively modified into their final form as interactions developed so as to eventually make the gestures appear in the final shape we see.

[19] See Hohmann and Fruth (2003).

No one is arguing against the unique complexity and power of human language, but what I find so fascinating is that an animal species without speech spontaneously develops such a group-specific symbolic communication system. Our observations of Simone and other youngsters in the Taï forest show that, contrary to the ontogenetic ritualization argument, such signs can be fully acquired through social interactions without any practice or prolonged development. Furthermore, because ritualization develops within specific dyadic interactions, this process is unlikely to lead to group-specific traits used by all group members.

Animals should be expected to use the most economic or least demanding means of reaching their goals. Too often, psychologists assume that once you have acquired a cognitive skill, it should be used systematically to learn all behavioral patterns an individual possesses in its repertoire. Following this, some psychologists tested for the presence of imitation by using such diverse and unnatural tasks as sand-throwing, raking out-of-reach food, opening boxes, tongue protruding, picking mirrors, and extracting food from opaque or transparent boxes. Unsurprisingly, the results were very mixed and confusing. Recently, an experiment has shown that captive chimpanzees will imitate only if they have no clue as to how to solve the problems from their own perceptions, while otherwise they will use their perception to guide their attempts to reach the goal that they have seen others achieve.[20] This stresses how chimpanzees combine individual learning with social aspects as a function of the tasks presented. I have argued above that imitation will be more systematically used for symbolic communicative traits, like leaf-clip or knuckle-knock, than for material cultural behavior patterns.

Symbolic culture in chimpanzees possesses one of the key attributes of culture in humans, namely a total arbitrary nature disconnected from ecological constraints. The meaning of the sign-codes are based purely on "social shared conventions" that are understood by all group members. This could be viewed as the first step towards the development of beliefs, myths, arts, and discourses. I could provocatively suggest that as long as we remain incapable of communicating with chimpanzees at their level, it seems premature to say that they have no such abilities.

If material culture opens the door to new dimensions of the environment in which the individuals live, symbolic culture opens the door to a new dimension in the social world. In humans, especially through language, this has permitted totally new forms of communication, which have allowed individuals to share information about past and future times, as well as about internal states. Without language, communicating about these dimensions would be much more difficult

[20] The experimenter showed the chimpanzees a box with food inside that could be opened in two different ways, only one of which was demonstrated to the subject. This box was presented either in an opaque or in a transparent box. In the first case, chimpanzees faithfully copied the method used by the chimpanzee demonstrator, while in the second case they were mainly influenced by the tool being used and the goal to reach (Horner and Whiten 2005). What this experiment shows is that chimpanzees imitated when they could not obtain all the information they needed through observation alone. Since chimpanzees do not imitate blindly in all cases, experiments that failed to show this ability might also be considered as experiments where the experimenters failed to understand the requirements needed to trigger imitation in chimpanzees rather than prove an inability in the chimpanzees.

and limited. One direct consequence is that we would not expect the effect of symbolic culture to be as important to survival and reproduction in non-speaking species than material culture is to them. Nevertheless, in chimpanzees, we see many domains of social life where symbolic elements have become important and permitted the expression of cultural communication forms that are based on socially shared meanings. If it could be argued that such shared meanings comprise only a tiny fraction of what is seen in humans, we must, in contrast, acknowledge that decoding such meanings in a non-speaking species has proven to be very challenging for human observers, who are used to relying heavily on human-specific symbolic communication channels.

The learning of symbolic cultural elements requires a more precise social understanding of others so that the "how," the "when," and the "with whom" will be coordinated in an efficient way and the responses of the audience can be correctly anticipated. Cultural sign-codes in wild chimpanzees could be acquired only with a certain level of understanding of others and of their intentions and expectations – including not only estrus females but also competitive males. Such elaborate understanding of others – also called mind-reading or theory of mind by psychologists – could not be shown in experiments with captive chimpanzees and were therefore proposed to be a uniquely human ability.[21] A more realistic proposition would be that the acquisition of symbolic communication signs would bring a selective advantage to the individuals who perform them only in a rich social environment, and therefore, we should expect the acquisition of symbolic communicative abilities in the presence of an adequate social learning environment. The social conditions faced by most captive chimpanzees can be so deprived that we should not be surprised that such cognitive abilities do not emerge. In support of this idea, captive chimpanzees who have had much richer social lives and, for instance, have been members of an artificial sign-language program for many years, acquired better mind-reading abilities than less socially enriched captive chimpanzees.[22]

The introduction of the distinction between material and symbolic culture helps to clarify when different types of social learning mechanisms will be expected and that the assumption used for many psychological experimental studies – that once a cognitive ability exists, it will be used in all cases – is overly simplistic and does not take real-world requirements into account.

[21] For many years, such broad claims based solely on captive experimental studies have been accepted without much discussion (Galef 1992, Hauser 2000, 2006, Povinelli 2003, Penn and Povinelli 2007, Penn *et al.* 2008, Tomasello 1990, 1999, Tomasello *et al.* 1993, 2005), even though the social context of such experiments was totally artificial. Furthermore, in almost all cases, captive chimpanzees were actually tested for their understanding of the human mind, which we should expect to be quite different than their understanding of the chimpanzee mind (Boesch 1993b, 2007, de Waal 1996, 2001, 2009, Whiten 1989). Finally, when some experiments were done that looked at chimpanzees' understanding of other chimpanzees, they were suddenly much better than when tested with humans and many psychologists have changed their minds about what chimpanzees understand about others (Hare *et al.* 2000, 2001).

[22] We will come back to this issue later in the book but for now, see Matsuzawa (2006) and Tomasello and Call (2004) for details on performance improvements in subjects living in socially enriched captive conditions.

In the next chapter, we will concentrate on the learning of culture by youngsters. For the three domains of culture we have just reviewed, youngsters will need different skills to acquire them and the social environment will play different roles to make such acquisitions possible or more efficient.

Synopsis

- Symbolic culture in the communication domain is observed in many chimpanzee populations.
- Arbitrary social conventions have also been documented in many instances in different chimpanzees populations and other animal species.
- Symbolic culture is the cultural domain that can be totally independent from ecological influences and is purely based on the properties of the social group.
- Symbolic communication based on group-specific shared social conventions form a collective cast of mind that allows the maintenance of such symbolic culture in chimpanzees.
- The absence of systematic studies on this aspect of culture in chimpanzees and other primates makes it likely that we are still underestimating it.
- The wide-ranging symbolic culture domain in humans should not distract from the fact that some basic aspects are shared with our closest living relatives.

6 Learning culture: from pupils to teachers

Two types of education need to be distinguished, as their products are very different: the education of the individual that is common to humans and the animals, and the education of the species that belongs only to humans[1]

George de Buffon (*Histoire Naturelle des Quadrupèdes*, 1762)

Taï National Park, Côte d'Ivoire, 12 February 1986

Salomé is cracking Coula nuts with a hard wooden hammer and, as usual, 5-year-old Sartre is sitting in front of her and eagerly begging for a share. Her hesitation to give him kernels indicates to me Salome's increasing reluctance to share. Instead of starting to pound nuts himself, to my surprise, Sartre collects a load of nine nuts that he brings back and adds to Salomé's heap next to the anvil, thereby taking over the task from his mother so that she can crack more nuts without having to do all the collecting. Then after his mother refuses to share the first nut, Sartre immediately cleans the anvil of all shell remains from the previous nut, as she regularly does, and places a fresh nut for her to pound. Salomé waits for her son to finish these preparations before pounding this nut. Once open, Sartre begs for it and this time she lets him take about a third of the kernel. Salomé now concentrates only on pounding the nuts and Sartre goes on regularly collecting them for her. From this point on, nut-cracking has acquired a cooperative dimension between the two: nuts that are collected by Sartre and cracked by Salomé are partly shared with Sartre, while nuts collected and placed on the anvil by Sartre are invariably shared with him.

During the next *Coula* season a year later, Salomé remained very generous at first, sharing parts of her nuts each time Sartre begged for them. However, in January 1987, she again started limiting her sharing and he immediately reacted in the same way he had the year before by collecting and placing the nuts for his mother to pound. While this continued to work quite well for Sartre, Salomé also started leaving behind intact nuts and her hammer on the anvil while she left to collect more nuts. Sartre, who was obviously eager to eat more nuts, immediately began using

[1] « il y a *deux éducations* qui me paraissent devoir être distinguées, parce que leurs produits sont fort différents : l'éducation de l'individus qui sont commune à l'homme et les animaux, et l'éducation de l'espèce qui n'appartiens qu'à l'homme. »

Figure 6.1 **Mothers share nuts for many years with their infants**. Perla hands the nuts she just cracked to her son, Papot (top left and right). Xérès allows her daughter to take a piece of the nuts from her mouth (bottom left).

his mother's tool to crack the nuts she had left on the anvil – a rare opportunity to practice with a good hammer! Salomé and the other chimpanzees would normally never leave a hammer behind, as they are rare and there is the risk that it would be stolen by another chimpanzee. After she has collected more nuts, Salomé would come back and resume cracking nuts while Sartre waited nearby for her to share with him. Finally, two months later, she started nut-cracking under another *Coula* tree with whatever hammer she could find. Once Sartre finished eating the nuts Salomé had left for him to crack, he collected some more and continued to crack them with the efficient tool his mother had left behind. For the last two months of the season, Sartre continued collecting and placing nuts for his mother and she also stimulated his nut-cracking by providing him with nuts and her good hammers.

In order to understand why nut-cracking seemed so challenging during the many years I observed the Taï chimpanzees cracking *Coula* nuts, I have repeatedly tried to crack nuts myself using their technique, not only because the nuts are very tasty but more importantly to clarify why some hammers are worse than others.[2] Why did youngsters not succeed with one hammer when they were already successful

[2] It is important to point out here that I only tried to crack nuts when I was sure no chimpanzees could see me because I did not want to be perceived as a feeding competitor.

with another one? What I learned after practicing for over 12 years is that, despite making progress, I would never be able to compete with good nut-crackers like Héra or Salomé, who ate more than 2 nuts per minute. They could keep such a rhythm because they were able to perfectly control the force they needed to exert in order to break open the shell and leave the kernel intact. I tried hard to exert just the right amount of force but I was not successful and my kernels were usually smashed into tiny pieces. Moreover, using natural branches as hammers requires two additional skills. First, some branches are simply too soft and make very bad hammers so one must learn to select a hard wood. It is funny watching youngsters during this learning phase as they might try obviously unsuitable wood types or branches that appear to be suitable but are really too soft and break after only a few hits. In this respect, being able to use mum's hammer is a big help. In addition, branches are never perfectly straight; they are mostly curved or S-shaped and often split. Thus, besides selecting an optimal tool, it is also important to have a good understanding of how essential it is to hit the nut with the flat surface of the hammer and here again, youngsters make many funny mistakes, such as hitting the nut with their hand or completely missing the nut while pounding. Within one or two seasons, however, they come to master selecting an optimal tool and a suitable anvil.

Such complex interactions that occur between mothers and infants around learning to nut-crack are frequently observed, especially for 4- to 6-year-old individuals that are highly motivated to learn this skill. Still though, as a young baby Sartre already intently watched his mother nut-cracking and he eagerly gave food grunts and ate the ones she shared with him. He was too young to nut-crack by himself but his motivation to do so was encouraged by his mother's generosity. At the same time, I sometimes saw young mangabey monkeys watching the chimpanzees cracking nuts and waiting until they left so that they could eat the leftovers. For mangabeys in the Taï forest, nuts are a desired but hard-to-access food source. They seem unable to copy the chimpanzees' technique so they can only use their teeth to open nuts. Equally striking, although *Coula* nuts are very abundant in the forests of Loango National Park in Gabon, the nuts there are not part of the chimpanzees' material culture and they simply do not eat them. Baby chimpanzees watch their mothers' actions intently but they will not see the *Coula* nuts as food in Loango. Thus, thank to their mothers' activities, infants will be immersed in their culture at a very early age.

Human children also learn their culture from their parents and other group members and this is not done at school in most human societies but rather in the family setting when faced with the relevant contexts. A typical example of such cultural learning is provided by a Mexican Mazahua two-year-old helping to dig a corn field with her mother, observing and eagerly carrying out a part of the task that aids her mother's efforts.

The girl starts to remove soil from the bunches of grass her mother has just removed, making the exact same movement she has observed her mother making. Her mother stops digging and watches her. When the mother wants to continue, the girl tells her to stop as she wants to pick up a bunch of grass and remove the soil from it just where the mother wants to start digging. The mother lets her do this and waits … [Soon] the girl has taken over a task from the mother so that the mother can continue digging without having to remove the grass. They work together for about five minutes … When the girl sees her mother removing the soil from the grass (while the girl was chatting), the girl protests and demands her task back.[3]

As in the example with Sartre, learning and teaching in humans mix together in complex ways based on the motivation of the naïve to learn by observation. In humans, this is typical for subsistence economies where learning takes place in family settings and the forms of learning often adjust to the changes in the eco-cultural environment.

In Tahiti, a man described his early learning: "I can make [an earth oven] and cook in an earth oven because I saw my 'grandfather' make them every day. I would sit next to him, and I understood how to do it … I just looked at the way things were done, and when I got to be thirteen years old, I understood how to do the work … He taught me a little bit, but most of the time, it was by my own eyes that I would see how to do the work."[4]

Such forms of learning and teaching are indeed very different from our modern school system, but were representative of human teaching from very early on. Schools are associated with a more independent, individualistic mode of apprenticeship. The famous anthropologist Margaret Mead (1970) commented on the "striking" difference between the kinds of teaching and learning that occur in traditional societies in which children desire to know ("steal" knowledge), as compared to modern schooling societies in which adults wish to teach ("proselytize"). Mead considered it a "sorry" addition that "the emphasis has shifted from learning to teaching, from the doing to the one who causes it to be done, from spontaneity to coercion, [and] from freedom to power".[5]

Cultural acquisition in the natural world

All cultural learning takes place within the specific learning environment that prevails in that culture. The Taï culture distinguishes itself from the Loango culture by seeing *Coula* nuts as food. However, to open the nuts, a hammer is needed and has to be found. For Salomé to be able to see a "good hammer" in a dead *Coula* tree branch lying on the ground requires a certain degree of imagination, by which the reality of the direct perception of that branch is mentally "transformed" into a hammer. Thus, when looking for a good hammer, Salomé does not merely see branches lying on the ground, as would, for example, a mangabey monkey in Taï forest or a chimpanzee in Gombe or Loango. Through her mental transformations, she is able to see potential hammers and independent of their real shape, she can "see" the shape they would have once they were modified to be hammers or select out of all

[3] de Haan (1999, pp. 77–78) quoted in Rogoff *et al.* (2003, p. 190).

[4] Levy (1973, p. 451) quoted in Paradise and Rogoff (2009, p. 109).

[5] Margaret Mead (1970, p. 12) quoted in Paradise and Rogoff (2009, p. 112).

branches the one with the shape most similar to the technical requirements.[6] This *subjective reality*, which is at the base of flexible tool use, is in many respects independent of *objective reality*, the reality directly perceived, but is instead the result of the individual "transforming nature to conform to his/her subjective wants and needs." It is necessary for Salomé to predict which of the many branches in her surroundings will make a good hammer. When selecting a potential tool, Salomé and the other chimpanzees always briefly scrutinize their immediate surroundings and then typically go straight to one particular branch or stick out of the many that are available and, if necessary, quickly modify it into a tool. When you first see an adult chimpanzee make a tool, the most impressive aspect is this natural swift ease with which they make all the necessary transformations to the raw material. You cannot avoid thinking that they know perfectly well what they need to modify before they begin and this is part of what seems to be going on in the selection process before they act. In support of this view, Taï chimpanzees' tool-making is characterized by the fact that most modifications made to the raw material to produce a tool are systematically done before use.[7]

The selection of a leaf that will be used to sign "I want to have sex with you" will require more imagination, as the subjective reality of the leaf is more distanced from the objective perceived reality than is the selection of a hammer; while mentally transforming a perceived branch into a hammer remains, so to speak, in the physical mode – for example, imagining the flat shape needed to hit a nut – mentally transforming a perceived leaf into a communicative sound involves different modes – for example, relating the type and the size of the leaf with the stripping sound it can produce in front of an estrus female only when high-ranking males are close by. Therefore, the subjective reality for symbolic culture traits requires more challenging mental transformations linking different perception modes to master all at the same time. This may be part of the reason why symbolic culture is more rarely observed.

It is easy to see that the more complex the tool use is, the more elaborate such mental transformations will need to be. In other words, the more anticipatory subjective steps are available to the individual, the more complex transformations can be projected mentally, which opens the door to more complex tool use. One of these would be sequential tool sets, by which the subjective reality produced by the individual "sees" through many steps before imagining how the goal can be reached. The tool sets used by Loango chimpanzees in Gabon are a good example of such a subjective reality, but we have plenty of examples in humans as well; for example, planning to build a car requires an individual to project a vast number of future,

[6] Here, I am pleased to thank my father, the psychologist Ernest Boesch, for pointing out to me this important aspect of tool use, especially the mental transformation that this implies (see as well Boesch, E. 1991, 2003).

[7] Boesch and Boesch (1990) showed that over 90% of the modifications seen during tool use are done before use over many years of observations. Tool-making in Taï chimpanzees is more frequent for soft material like small sticks, which were modified in 91% of the observations, while wood hammers were made in 6.5% and stone hammers in only 1.6% of the observations (Boesch and Boesch 1990).

Figure 6.2 **What is happening around me?** Even before they can walk independently, young chimpanzees inspect their surroundings. From the safe haven of his mother Héra's belly, Homère, who is only a few months old here, looks at what is happening around him and his attention is constantly drawn to what his mother is doing. Through her actions in the environment, Homère will learn a lot and perceptually order the objects in the world around him so that he will be able to "see" a wooden branch as a hammer.

transformative steps. Thus, with tool sets we observe an anticipatory process in which symbolic culture directly profits from material cultural complexities, which permit a higher degree of independence from the physical constraints in a given environment.

For youngsters to acquire symbolic and material culture, we would expect that some learning of the subjective reality of their surroundings would be necessary, and that, before starting to practice any of these cultural traits, the stimulation of the mental transformations of objective perception will be required. However, this requires the appropriate social learning environment. For a city boy like me, a liana is simply a plant that grows thanks to the support of trees. I was surprised to learn that this is not how all people see lianas. For many people living in the African forest regions from Côte d'Ivoire to Gabon, a liana is also a water reserve so that when you are thirsty, you just have to look for a *Manniophyton* liana, which, in a 1-m section, contains 1 liter of clean water. How would I have ever seen this reality of a vine as a water container if I had not travelled to Africa? For young men in some parts of the Southern Pentecost Islands, a liana is an elastic rope that slows you down when you jump from high up in a tree during the traditional Naghal ritual in Vanuatu. So depending on where you live, humans see a liana as either just

Figure 6.3 **Young chimpanzees build up their subjective reality** as they watch their mother interact with her world. Chimpanzees' curiosity is legendary and each youngster experiences what their mother experiences when she is interacting with social partners (top left), eating fresh *Strychnos* fruit in Gombe (top right), extracting wood-boring bees from a nest built inside a dead piece of wood (bottom left) or cracking nuts with heavy stones (bottom right).

a plant that supports itself on other plants, or a reserve of water, or a life-saving elastic rope.[8] This subjective reality in humans is a direct function of the cultural environment we have experienced.

Sartre, like all youngsters in the Taï forest, "sees" many different things – branches, hammers, toys, weapons – in the branches lying on the Taï forest floor. This is the result of being constantly exposed to the different ways these objects are used by older individuals in his social group (see Figure 6.3). As a consequence, when chimpanzees are old enough, they will start to manipulate stones according to the subjective reality they have acquired from their passive but intense observations. And these actions will in turn lead to their perception of some stones as good hammers and other stones as bad hammers. On the other hand, Gombe infants will only see branches as toys, as, to my knowledge, adult chimpanzees there have never been seen to use them as hammers. Some captive infants may have never

[8] Many studies have shown that the way humans categorize objects, animals, and plants in their environment is a direct function of the way these things are used in their culture and that, very often, how westerners look at our surroundings is very different than many other societies (Medin and Atran 2004, Atran *et al.* 2005, Henrich *et al.* 2010a).

even seen branches. All chimpanzee infants "see" leaves as food and sponges, but, in addition, Taï infants "see" them as a display sign, while Gombe infants "see" them as a material to groom parasites and Mahale infants "see" them as a material to fold parasites in and a signal to use when asking for sex. By growing up in such different socio-ecological environments, infants' perceptions of these objects will unconsciously be shaped and they will perform different activities with them and so discover different properties about them.

A similar process has been suggested for orangutans, in which members of the Suaq Balimbing population in Sumatra have been seen to use tools to extract insects from tree holes with different frequencies. Interestingly, in this rather solitary species, the females from the north are far less gregarious than the central females and the former use tools much less often. Researchers there argue that young northern females who see this tool use performed by experienced foragers much less often might not "see" when a tree hole is good for tool use and, therefore, use tools less often.[9] In classic psychology, this process, called "prior knowledge," guides our perception of the world and is heavily based on past experience and previous actions that we have performed in our surroundings.[10]

A recent experiment illustrated the influence of subjective reality on the way individuals solve a task in a novel situation.[11] When encountering a log with honey inside in a human-made hole, the Kanyawara chimpanzees that occasionally use sticks in their daily life made a stick to access the honey, while the Budongo chimpanzees that have never been seen to use sticks but regularly use leaves did not use any stick at all, but when using a tool used leaves to access the honey. So prior experience directly influences the subjective reality of the chimpanzees and the solutions reflect their local culture.

Recent work on the topic in child development has shown that a very similar process is at work in humans. If children are known to be ubiquitous imitators, the question is how to decide which actions to imitate. So it has been shown that children are much more likely to learn from trustworthy informants – with conflicting information, they endorse the actions of adult individuals who have proven reliable in the past – and tend to favor members of the same culture – they select adults who abide by the norms of their own group.[12] This provides a rich view of social learning whereby the pupil will evaluate the task to learn and the different models available and through a selection process based on prior knowledge will decide "what", "how," and "who" to copy. This process suggests that less optimal solutions proposed by ingroup members might be favored over outgroup models and this

[9] see van Schaik *et al.* (2003).

[10] In his book, *Making up the mind*, Chris Frith (2007) explains at length how important the prior knowledge we gain of our world is for the development of the brain and our cognitive abilities, and how we are constantly using our past experience of the world to develop our knowledge and predictions about objects and group members.

[11] See Gruber *et al.* (2009).

[12] Harris and Corriveau (2011), Corriveau and Harris (2009), Buchsbaum *et al.* (2011).

could therefore explain the persistence of culture despite group member migrations between different groups possessing different cultural traits.

The learning and transmission of symbolic cultural traits will be quite different than that of material cultural traits, which are at least partly constrained by technical requirements. Consequently, if a material cultural trait like nut-cracking can be learned with a mixture of individual practice guided by some social element, the learning of a symbolic trait will require the use of a much more faithful copying mechanism. Copying the effect on objects, like a nutshell breaking when it is pounded by hard hammers, and using the same objects, like stone or wooden hammers, will be important guides for the faithful reproduction of material skills with specific technical constraints. However, this would not be enough to produce an "I want to have sex with you" leaf-clip, for the goal is not directly perceivable, or a leaf-fold, where the manipulation of the leaf are very complex and detailed. A precise imitation of the movements that make up the leaf-fold would be mandatory, especially as the goal – eating the louse or tick – could be attained with more direct methods than the Mahale leaf-fold and the Gombe leaf-groom.

Learning symbolic communicative cultural elements will add further complexity to the learning by naïve individuals. To acquire an "I want to have sex with you" leaf-clip, a purely imitative process is required in which the movements with the leaf must be precisely copied as a leaf can be ripped in many different ways. Furthermore, the "context of production" of the behavior is a shared group-specific character essential for communication to function. Thus, symbolic communicative sign learning requires the naïve individual to learn not only "how" to perform it, but also "when" and "with whom." The learning of "when" and "with whom" will be particularly strongly facilitated by a better understanding of others and their intentions. Insofar as the knuckle-knock in Taï or the leaf-clip in Mahale has no direct connection to the message it conveys, the communicative value of these signs can only function if a certain understanding of the intention of the emitter is understood by the potential audience.

Mothers as the main culture vehicle

Like all youngsters in Taï, Sartre enthusiastically begged for nuts from his mother. Nevertheless, as seen above, Sartre would regularly try to crack the nuts his mother would leave behind while she was collecting more nuts under the *Coula* trees. Little Sartre would then try his best until his mother would come back with new nuts to crack. Salomé did not leave nuts behind for Sartre to crack when he was 2 or 3 years old, but regularly did so when he was 5 years old. Thus, it looks like she was reacting to the ability level that her son had reached. Moreover, every now and then she would even arrange the nut-cracking anvil by leaving the hammer balanced on the root and placing an intact nut in the roundish trace of wear on the anvil before she went to gather more nuts. This seemed to work well because in the majority of cases, Sartre would take the hammer and hit the nut with it, and at 5 years of age, he was strong enough to successfully break them. In this way, Salomé *stimulated* her son to learn about what makes a good hammer and how to use it.

A year later, Salomé, possibly adapting to Sartre's increased strength and improved nut-cracking, stopped making such arrangements for him but regularly left him her own hammer, thereby *facilitating* his nut-cracking. At 6 years of age, Sartre had become quite successful with his mother's hammers and we saw him crack nuts by himself more frequently, although he limited himself to doing so only with good hammers, mostly those of his mother. The cost for Salomé was that she had to search for a new hammer each time and as there was not always another good tool nearby, she sometimes had to hit a nut more times for it to open than she would have using the good hammer she had left behind for Sartre. But all in all, Sartre's tool use improved and she did not need to share as many of her own nuts with him as she was before this period.

Salomé is the rule and not the exception, as all chimpanzee mothers teach their infants in these ways.[13] In fact, mothers stimulating nut-cracking in their offspring peaks when infants are between 3 and 4 years of age, as mothers are seen to stimulate them like Salomé did with Sartre in over 80% of the observed cases while they were collecting nuts, or once every 10 minutes of nut-cracking (Figure 6.4). This active support by mothers occurs in addition to the nut-sharing she does with her offspring from the first year of life until the youngsters become expert nut-crackers themselves.

On the other hand, when the youngster is regularly trying to pound nuts by himself later on, maternal *facilitation* peaks between the age of 5 and 8 years and this involves a mother providing her offspring with whole nuts or good hammers an average of 6 times per 10 minutes of nut-cracking. Furthermore, Taï mothers seem to have a precise notion of the level of skill that their youngsters have reached so that they intervene and assist them with those skills that are appropriate for their skill level. For *Coula* nuts, for example, they emphasize the correct placement order of hammer-on-nut-on-anvil to 4 year olds, but continue to facilitate learning how to precisely pound with good hammers to 6–8 year olds (Figure 6.4).

A year later, Salomé systematically shared her hammers with Sartre, who was now 7 years old, so that he could crack the much harder *Panda* nuts. Cracking these nuts requires very heavy stone hammers, which are rare in the Taï forest, and I saw the two of them sitting and facing each other at the same anvil, with Sartre waiting with a bunch of nuts that he had collected; when Salomé was finished cracking her nut, he would use her hammer to crack his nuts on the same anvil

[13] When Sartre was 5 years old, Salomé stimulated her son 42 times by leaving nuts behind for him during 8 hours of observations and facilitated his nut-cracking by providing him with 5 nuts and 11 hammers. When he was 6, she stimulated him 30 times but facilitated his nut-cracking by providing him with 33 nuts and 41 hammers during the same amount of observation time. At the same time, another female, Ella, stimulated her 5-year-old son 7 times and facilitated his nut-cracking by providing him with 144 nuts and 19 hammers during 5 hours of nut-cracking. Similarly, Aurore facilitated her 5-year-old daughter with 57 nuts and 6 hammers in 2 hours of nut-cracking (Boesch 1991a).

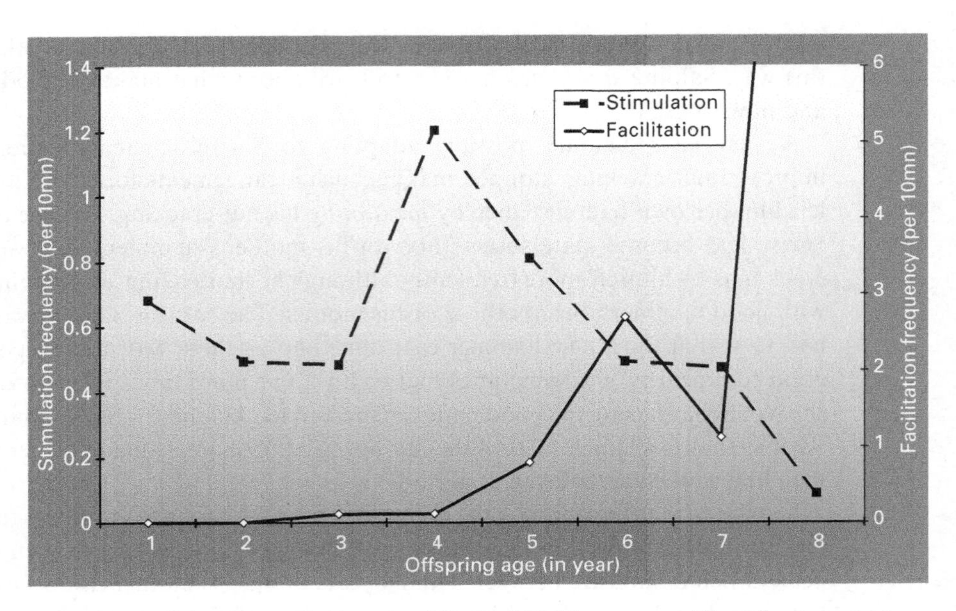

Figure 6.4 **Teaching interactions for *Coula* nut-cracking by Taï chimpanzees.** All mothers regularly intervene in the nut-cracking of their offspring and try to stimulate their attempts and facilitate their nut-cracking. They do this for many years after weaning, which occurs when infants are 5 years of age. Stimulations are mainly done for 4 to 5 year olds, while facilitations are concentrated on juveniles between 6 and 8 years of age (for 8 year olds, facilitations occur an average of 14 times per 10 minutes).

Figure 6.5 **Salomé and Sartre share the same hammer,** a 12 kg stone, to crack *Panda* nuts. Salomé gave birth to a new baby, Simone, when Sartre was 7 years old. By this time, he had become a good nut-cracker himself, although he still preferred to use his mother's hammer. As useful stone hammers are rarely found in the forest, they are a limited but necessary resource for nut-cracking. In this picture, Salomé has just opened and is eating her nut as Sartre is using her hammer to open one of his own *Panda* nuts.

(see Figure 6.5). Salomé, who was holding her newborn infant, Simone, would wait for Sartre to successfully crack his nut before she would crack her next nut. They passed her hammer back and forth to crack nuts like this for long periods of time.

Box 6.1: What is teaching?

George Buffon suggested 350 years ago that the social dimension of education belongs only to humans. This idea has remained prevalent in large circles of human sciences, such as in pedagogy, psychology, and some subfields of anthropology. Except for some agreement that teaching involves an expert directly guiding the learning of a naïve individual, there are many divergent opinions about what teaching is. The simple and abundantly cited three criteria definition of teaching, which was proposed by Caro and Hauser (1992), requires the following:

- the behavior only occurs in the presence of the pupil,
- the behavior is costly and does not provide any immediate benefit to the teacher,
- the behavior facilitates acquisition of knowledge or skill learning in the pupil.

The last criterion of this definition seems questionable as much of human teaching would not satisfy it, as seen in the examples of stone knapping in India and Indonesia (see below) and in the failure rates in western schools.

Part of the success of this definition was that it excluded any direct reference to humans and therefore allowed for a comparative approach. Despite this, however, reviews of evidence for teaching in animals have used different behavioral milestones as positive evidence (Caro and Hauser 1992, Hoppitt *et al.* 2008, Thornton and Raihani 2008). Sometimes they are quite relaxed in accepting some synchronous activities to show teaching (examples include ants walking in tandem, bees waggle dancing together, or attention-getting behaviors, such as making calls when food is brought). I will be using a stricter definition based on the consensus that teaching should include an experienced individual modifying his or her actions with the specific aim of encouraging learning in a pupil.

In a sense, a definition of teaching should concentrate on the information that is made available to the pupil. Hence, developmental psychologists suggest that scaffolding, defined as modifying the environment to support individual learning (Wood *et al.* 1976), could easily increase learning in a pupil. For some, teaching implies a "type of social learning that transmits generalizable (semantic) knowledge from the teacher to the pupil through communication." Thus, providing information about the location of a specific food source, which might not generalizable, would not be considered to be teaching (Csibra 2007,

Premack 2007). Without any data to prove this, this seems a very subjective criterion.

Following their observations of the Bossou chimpanzees, Tetsuro Matsuzawa and his team have suggested a special type of scaffolding which involves "education by master-apprenticeship" (Matsuzawa *et al.* 2001). This is characterized by observational learning, which is in turn facilitated by the long-term mother–infant bond, an absence of active teaching by the mother, a strong intrinsic motivation by the infant to produce a copy of the mother's behavior, and a high tolerance of the mother towards the infant during bouts of observation. For Frans de Waal, social learning is guided by social relationships and extrinsic rewards are not essential; he calls this process "bonding- and identification-based observational learning" (BIOL) (de Waal 2001).

To anthropologists and cultural psychologists, teaching in humans as it is done in schools is a very peculiar way of transmitting information to children and one that is very different from the forms of natural pedagogy that is observed in most human societies without formal schooling (Rogoff 1990, 2003, Greenfield 2009, Keller *et al.* 2005). If some have opposed active to passive teaching, for many both are important components of teaching. This has led some to propose different developmental models in humans, such as the "eco-cultural model" (Berry *et al.* 2002), which sees the child's behavioral development and the acquisition of culture as resulting from the interaction between human biological potentialities and environmental conditions (e.g. Berry *et al.* 2002, Greenfield *et al.* 2003b). Alternatively, the "sociohistorical approach" emphasizes processes of social construction seen as a set of situation-specific activities, including, in particular, participation in cultural activities or practices, the use of cultural artifacts, including tools, and the historical dimension of these processes (e.g. Rogoff 1990, Greenfield *et al.* 2003b). Such approaches view teaching or pedagogy, therefore, in a very open way and tend to include many different processes as long as they contribute to improvements in learning, such as scaffolding, observations, stimulations, facilitations, molding of the actions by the teacher, correction of errors, and demonstration of specific techniques.

For discussions about teaching in animals see: Boesch 1991a, 1992, Caro and Hauser 1992, Tomasello *et al.* 1993, de Waal 2001, Thornton and McAuliffe 2006, Franks and Richardson 2006, Radford and Ridley 2006, Csibra 2007, Premack 2007, Hoppitt *et al.* 2008, Thornton and Raihani 2008, de Waal and Bonnie 2009, Csibra and Gergely 2009, 2011.

For discussions of teaching in different human societies see: Hewlett and Cavalli-Sforza 1986, Rogoff 1990, Berry *et al.* 2002, Rogoff *et al.* 2003, Greenfield *et al.* 2003b, Keller *et al.* 2005, Paradise and Rogoff 2009, Hewlett *et al.* 2011.

Figure 6.6 **Héra sharing nuts with Eros.** As Héra gave birth to a new son when Eros was just 5 years old, she tried to reduce the number of nuts she shared with Eros much earlier than Salomé had with Sartre so that he would gain more experience at cracking nuts on his own. Nevertheless, when Eros continued to pout over not receiving nuts from his mother, she would either let him have a piece of the nut she was eating (top picture), or move her wooden hammer out of the way so that he could take the entire nut that she had just opened, even though he was still eating one he had previously obtained from her (bottom picture).

In the context of tool use, each mother develops a set of solutions to ensure that her infants will quickly and efficiently learn the necessary skills.[14] When Sartre was 5, Salomé had time to teach him to nut-crack because she did not yet have a younger infant. In contrast, Héra gave birth to a new infant, Homère, when her older son Eros just turned 5. Therefore, Héra had some important reasons to make sure Eros quickly learned the technique before Homère arrived. Thus, Héra began reducing the amount of nuts she shared with Eros when he was 4, which was two years earlier than Salomé. Eros noticed this and, like Sartre, began to directly help his mother. Unlike Sartre, though, Eros first simply cleaned the nutshells from the anvil after Héra had opened a nut and placed a new nut on the anvil so that Héra essentially just had to crack the nuts. At first this was quite efficient as Héra shared more of those nuts with Eros, but this lasted only a few weeks, so Eros began to also collect fresh nuts for his mother to crack. Thus, for several weeks, we observed

[14] For those readers who would like to know more about the abundance of interactions between mothers and infants in the context of nut-cracking, see Chapter 9 in our book, *The chimpanzees of the Taï Forest* (Boesch and Boesch-Achermann 2000).

a division of labor between the two – Eros would collect and place the nuts and clean the anvil, while Héra simply had to crack the nuts. This system worked so well that during the next nut season, after the birth of Homère, Héra still shared some *Coula* nuts with Eros, even though by that point he was already a very proficient nut-cracker himself (see Figure 6.6).

Teaching styles differ across chimpanzee and human cultures

As shown above, Salomé, Héra, and others regularly assist in the learning processes of their youngest offspring so that they remain motivated to strive to crack nuts. They do this *by sharing* an average of 25% of the nuts they open with them for over 7 years, *by stimulating* the nut-cracking attempts of their 4–5 year olds more than 4 times per hour, and *by facilitating* the nut-cracking of their 6–8 year olds more than 8 times per hour (Figure 6.4). As youngsters are keen to eat what their mothers are eating, they sometimes want more than their mothers are willing to give them. However, conflicts over the amount of nuts shared happened in only 25% of cases where infants had begged for *Coula* nuts and the mothers refused to share in less than 10% of those cases.[15] This frequent and important maternal assistance seems to result from the fact that the nuts have become so important in the diet of the Taï chimpanzees that mothers try to ensure that as much as possible of the technique is properly acquired before a younger sibling is born and will need their full attention.[16] In their own way, both Salomé and Héra showed how they could help their sons to improve rapidly.

Gombe chimpanzees must also learn to use tools to fish for termites and learning this technique has been shown to take a few years. Gombe mothers are very tolerant towards the natural interest their offspring take in their activity and they let them watch at very close proximity, even if they are sometimes disturbed by their presence. However, in strong contrast to Taï mothers, Gombe mothers *never* share their tools with their infants, and they *never* actively aid access to the termites they have caught. As one observer writes, "In no instance did I see a mother actively facilitate her offspring's learning in any way. Mothers were oriented completely to the task of termite fishing and rarely even made eye contact with offspring. A mother never offered a termite to her offspring, never handed her offspring a tool and never

[15] Nut-sharing and potential differences between maternal willingness and infant request are detailed in Boesch and Boesch-Achermann (2000, pp. 205–221). In addition to the fact that conflicts about sharing are rare and mostly won by the youngsters, most nut-sharing is actively favored by the mother as she allows access to the nuts or turns herself towards her infant, and in 10% of the cases, she actively holds the entire opened nut out for her infant. Until the age of 3, all of the *Coula* nuts that infants eat are shared with them by their mothers, and by the age of 5, when they are normally weaned, half of all *Coula* nuts that they eat still come from their mothers (p. 205). This is even more extreme for *Panda* nuts; until the age of 6, they only eat *Panda* nuts that have been shared with them by the mother (p. 206).

[16] I am not suggesting that mothers know when they are going to have a new infant, but in evolutionary terms, mothers who make sure that their offspring can crack nuts efficiently before the arrival of a new sibling will be more successful and have more surviving offspring than those who have juveniles that have not acquired enough of the technique to acquire enough nuts by themselves.

Figure 6.7 **Xérès sharing *Coula* nuts with her daughter, Xindra.** Xindra is seen begging with her outstretched left arm for the nut that her mother is cracking (top left). Once opened, Xérès facilitates Xindra's access to the cracked nut by twisting the hammer away and letting her take the whole nut (top right). Xérès immediately takes a new nut from the small reserve she has on the ground near her hammer and places it on the anvil (bottom left). After opening it, both Xérès and Xindra eat nuts side by side (bottom right). Mother–infant pairs can repeat such sequences for over 3 hours per day during the 3–4-month *Coula* season.

molded the offspring's behavior while fishing. Chimpanzee offspring were simply allowed to observe and occasionally 'steal' a tool or a termite, sometimes resulting in a negative reaction from the mothers."[17] This could not be more different than our observations of Salomé, Héra, and the other Taï mothers. What could account for such dramatically different maternal teaching styles?

Westerners often consider the active teaching that goes on in our modern schools to be the only way for humans to teach others. I have only seen two instances of active teaching by Taï chimpanzees: once Salomé corrected an error in the placement

[17] Detailed studies done with the Gombe chimpanzees about the acquisition of termite fishing have been published (Lonsdorf *et al.* 2004, Lonsdorf 2005, 2006). This quote comes from Lonsdorf (2006, p. 44). As these studies were done well after the published observations of the Taï mothers sharing nuts and teaching youngsters, they specifically controlled for the presence of such interactions and they explicitly discussed such aspects in their reports. Therefore, we must accept that the absence of the maternal intervention in Gombe chimpanzees is real and reflects an essential difference between the two populations.

of a nut by Sartre and she used his nut to demonstrate how to open it to him, and another time, another female, Ricci, demonstrated to her daughter the correct way of holding a very complex wooden hammer.[18] As a result, some have denied that any such teaching exists in chimpanzees and if you compare this to the modern western school system, I can understand why. However, considering how teaching happens in some human societies should not lead people to simply brush aside all the other important and complex interventions of chimpanzee mothers in the learning of nut-cracking (see Figure 6.4). It is true that in modern western societies, a more active teaching form dominates in schools and in family circles, where the parents and family members have themselves gone through the school system; such a form includes demonstrating things to kids and correcting their errors. However, even in the western world, most skills are learned outside of schools, where we either try things by ourselves or improve them as we realize what does and what does not work, or we learn by watching what others are doing. If you now try to remove the ethnocentric glasses we all wear, you will realize that the situation is radically different in many human societies, including in South America, Africa, the Middle East and Asia, and imposing things on others is often considered to be wrong or inappropriate. Margaret Mead characterized many small-scale cultures as "learning cultures" because children in these cultures acquired culture easily and quickly without teaching. Social anthropologists used the term "osmosis" to describe how easily children in small-scale cultures acquire a wide variety of knowledge and skills without active teaching; it was automatic, without effort, and nobody failed.

Hadzaland, Tanzania, July 2011[19]

Nine-year-old Abiba is searching for underground tubers with her small digging stick in the thorny bushes that are so typical of this region. Her mother Madi is in front of her and uses the blunt end of a stick to hit the ground; from the sound that emanates, she judges if it is worth digging in the area. Abiba uses the same technique 1 m away from Madi. Abiba starts to dig but stops immediately when she notices that Madi has moved to another bush and starts to dig there. Abiba starts to dig closer to her mother. She quickly encounters a large stone and struggles to get it out. After Madi removes her first tuber, she watches Abiba struggling with the stone. Without saying a word,

[18] See Boesch (1991a). Some authors have totally ignored that these two instances of active demonstration were embedded in a whole suite of teaching activities provided by the mothers (e.g. Tomasello and Call 1997, Csibra 2007, Call and Tomasello 2008, Hoppitt *et al.* 2008, Csibra and Gergely 2011). In addition, they ignore the fact that active teaching in our societies is associated with a great deal of much more passive learning.

[19] Here, I would like to thank warmly Frank Marlowe and Colette Berbesque for introducing me to the Hadza people, to their land and traditions in the summer of 2011. I would also like to thank the Hadza people themselves for having been so welcoming to us and letting us experience their material and social culture directly. Discovering how efficiently their simple material culture has allowed humans to survive during an extreme drought in one of the driest habitats in Africa was extremely impressive and confirmed to me the richness and resourcefulness of human cultures.

Figure 6.8 **Learning to dig for tubers in the Hadza women**. Variable interactions prevail between mother–daughter pair when digging: Abiba and Madi are digging for tubers side by side in the same hole (upper left), mother Madi looks at how Abida is extracting a stone (upper right), daughter Abida looks at how Madi is extracting a tuber from the hole (below left), and Madi indicates with her left hand where Abiba should continue to dig (below right).

Madi sits next to her and places a wedge stone under her digging stick, which she uses to lever the stone out of its place. Thanks to her mother's intervention, Abiba can continue digging without impairment and after removing some more stones, she extracts two small tubers from the hole she has dug. When she makes it clear that she wants to rest, Madi tells her something and Abiba continues digging. After working together for 30 minutes, Madi indicates a spot in Abiba's hole with her forefinger and Abiba immediately begins digging there and finds more tubers. During two and a half hours of work, Abiba extracts about 2 kg of tubers. She always digs close to her mother, who watches her, helps her to remove stones three times, encourages her verbally twice, and directs her once (see Figure 6.8).

Teaching styles differ throughout human societies. In some human societies, active shaping and guiding will dominate only when the costs of making errors during the learning process are high; whenever costs of making errors are low or do not matter, trial-and-error learning prevails. Depending on the task one is trying to acquire, learning-by-doing followed by facilitation and stimulation from an expert, as seen in the Hadza example, supplemented with pure observations is seen in many societies.[20] Contrary to the habits of the western world, adults in many societies only rarely speak to their infants, who must passively listen and watch in order to understand and make sense of what the adults around them are doing. A key human activity for many millennia has been the production of stone tool artifacts, which represents the bulk of our knowledge about the technology produced by our ancestors. Observations of modern humans in India and Indonesia that produce such stone artifacts have revealed how the learning of these skills emerged. Stone knapping is a social activity with experts and novices working side by side and, in some cases, numerous instructions and discussions occurring. Interestingly, despite what looks like a great deal of teaching and guidance by the experts to the novices, the products made by these adult apprentices remain of a low quality for years, with shorter and thicker tools produced by the novices than by the experts. Furthermore, the novices' use of technical strategies is more varied and some of the experts' techniques are strikingly never used by novices. Typically, learning to stone knap happens very slowly and requires an average of 10 years of practice for a novice to become expert. Thus, for stone knapping, imitation is limited to only some aspects and a lot of trial-and-error learning goes on, with teaching or guidance by others having a very limited effect for years.

To learn to weave blankets and cloth, the Dioula men in Côte d'Ivoire first practice with the aid of some scaffolding – such as when they set up the warp and produce their first large cloths – but they then learn through observation and trial and error. The son does a bit of weaving when his father is out, and when the father returns home, he corrects his son's errors without saying anything about them, and so on. In Japan, the traditional way of learning to make sushi consists of the apprentice watching the sushi master at work for hours, and it is only after watching him for many months that he will be allowed to practice, but when he does, he is expected to become proficient very rapidly without any monitoring.

Moreover, teaching styles can vary over time. Zinacantecan girls in Mexico learn to weave magnificently colorful and elaborate carpets. During recent times when innovation is valued, young girls learn the technique predominantly by trial and error; however, during previous periods when the maintenance of traditional ways was more important, learning by observation, shaping, and especially scaffolding prevailed. Thus, we see that in human societies, teaching styles include many different forms of intervention by experts and the most common one across all of them is tolerance of the apprentice observing the expert's technique, while the use of more active forms of intervention depends on the difficulty of the technique.

Active teaching is often useless with very active apprentices and relatively easy techniques. At schools, pupils sit and passively listen to what they are being told. This form of teaching cannot be more different than learning to nut-crack in

[20] For general reviews of teaching in humans, see Rogoff (1990), Rogoff *et al.* (2003); for the Dioula blankets weaving, see Tanon (1994); for the Japanese sushi mastery, see Morris and Peng (1994); for the Indonesian stone knapping, see Brill and Roux (2003), Stout (2003); for the Mexican carpet weaving, see Greenfield *et al.* (2003b).

chimpanzees, learning cornfield preparation in Mexico, or learning to make earth ovens in Tahiti (see above). In particular, youngsters learning to nut-crack are like sponges, watching their mother's actions at a very close range, constantly trying to obtain nut pieces, manipulating hammers before they can even lift them, and trying out the technique very early on. Being very flexible, they quickly detect difficulties and immediately try to solve them so that when they start attempting to nut-crack even before they have enough strength, they will already make all kinds of corrections to improve their performance, such as changing their sitting position, the angle they are hitting the nut, the position of the nut, the way they hold the hammer, or the hammers they are using, and so on. Thus, youngsters will have experienced all kinds of possible adaptations and corrections within the nut-cracking context, often before their first serious attempt to nut-crack. One gets the feeling that whenever they can make progress, they will. Under this form of *active participation* by the apprentices, active teaching would be useful only for tasks in which the skills will be understood by the apprentice only later as he matures, while the forms of teaching involving stimulations and facilitations will immediately be useful to active apprentices. However, in *passive participation* learning forums such as schools, active teaching is possibly the best way to improve the learning process.

In chimpanzees, the socio-ecological challenges encountered during the learning process could affect the level of maternal intervention; termite fishing is progressively acquired during a two-year period between two and a half and four and a half years of age, when neither strength nor availability of tools are limiting factors. Nut-cracking, on the other hand, requires the understanding that nuts, a hammer, and an anvil must be brought together in a certain order, this is first reliably understood by three-year-olds; enough strength must be developed – the strength required to crack *Coula* nuts is only reliably reached in 4-year-olds; and a good hammer should be used, which is an important limiting factor in the forest. Thus, the skills to crack *Coula* nuts are acquired during the four-year period between two and six years of age. This process is further delayed by two years for the extremely hard *Panda* nuts due to the additional strength required to open them (in fact, they are the hardest nuts found on the African continent). If we remember that interbirth intervals in Gombe and Taï chimpanzees are about five years, we understand that there is no pressure on Gombe mothers to accelerate the learning process of their youngsters, which will naturally have acquired these techniques by the time their younger siblings arrive, while Taï mothers have much to gain by actively stimulating the learning processes of their young for the more demanding nut-pounding.

Interestingly, not all nut-cracking chimpanzee mothers have been seen to teach their youngsters this skill. Bossou chimpanzees in Guinea are famous for cracking oil palm nuts with small stone hammers.[21] However, while learning to nut-crack by young Bossou chimpanzees is similar to what we have described for the Taï chimpanzees in that mothers allow their offspring to take many of the nuts they open, similar to the Gombe mothers, the Bossou mothers have *never* been seen to actively

[21] Matsuzawa (2007), Inoue-Nakamura and Matsuzawa (1997), Ueno and Matsuzawa (2004).

provide good hammers or good nuts to their infants or to place nuts on an anvil to make nut-cracking easier. This is strikingly different than Taï chimpanzee mothers. Why do we see such a difference? The explanation seems similar to the one we saw earlier for termite fishing in the sense that the oil palm nuts are much softer than any of the nut species cracked in the Taï forest and young chimpanzees are able to crack them at an earlier age (by the age of four, most have mastered the technique in Bossou). In this respect, it does not conflict with the reproductive schedule of the mother. Thus, we again see that "teaching style" is influenced within each chimpanzee population by the local technological challenges of the task and the reproductive imperative of the mother.

So we see a convergence in chimpanzees and humans, with both altering their teaching styles depending on the type of task to be learned, as well as some more general cultural attitudes towards apprentices within each population. Complex tasks, such as carpet weaving, stone knapping, and sushi making, can be learned mainly through observation without much expert intervention in some cultures, while they would require formal teaching in western societies. In chimpanzees, teaching styles differ dramatically between populations. For example, nut-cracking is taught differently in Taï compared to Bossou, while learning to termite fish in Gombe does not differ from learning to nut-crack in Bossou, but is very different from learning to nut-crack in Taï. These differences seem to be explained by how difficult the task that is being learned is and the length of the learning period. In other words, more active teaching styles will be more likely to occur either for tasks that require long apprenticeships and for which mistakes are costly, or in more passive participation learning forms.

Finally, teaching has been proposed in many different animal species. First, cheetah mothers have been shown to bring live prey back for their young to kill, which is, in a way, similar to the stimulations seen during nut-cracking in Taï chimpanzees. Callitrichid monkeys, domestic cats, and meerkats have been proposed to use similar stimulation techniques for teaching their young to find food and prey. Meerkat mothers have been observed to present progressively less disabled scorpions to their young according to their skill level so that more-skilled young would be able to practice more of the killing skills with less-disabled scorpions. However, in this species the decision to modify the prey before presenting it to the pupil was not dependent on the mother's capacity to evaluate the level of the pupil, but was a response to changes in the pups begging calls with age. Nevertheless, as pups very rarely find mobile prey themselves, providing partly disabled prey to them decisively facilitates skill acquisition.[22] These observations of specific behaviors directed towards apprentices in primates and felids nicely show that, first, teaching is not exclusive to humans and many forms are found in different animal species. Second, they also reveal that mothers have the ability to consider the different skill levels of their young and try to react specifically to them. Third, the tasks for which

[22] See Thornton and Raihani (2008) and Hoppitt *et al.* (2008) for a discussion of teaching in some animal species.

such maternal teaching was seen were generally much simpler to learn than those seen in chimpanzees. Whether teaching occurred was more affected by how essential it was for the young to feed independently before the next generation of youngsters arrived. Thus, an interaction between how time-consuming it is to learn a task and the reproductive biology of the species seems to provide a general explanation for the use of teaching intervention in many animal species.

Learning one's own social culture

Ali, the young male who was adopted by Brutus when he was 5 years old, spent his juvenile years following this very special foster mother and was exposed at a much earlier age than usual to the "male model." Young Ali would even follow Brutus when he went on patrols and attacked the males of the neighboring groups. This seemed very dangerous to me as Ali could not run as fast as the adult males and if they were defeated, this might have been risky. Brutus was a very generous adoptive father; he shared the nuts he spent hours cracking with Ali and gave him large pieces of meat, much to the frustration of the many females who had been patiently begging from him. In this unique social learning environment, Ali developed into a premature and gifted hunter. Brutus was the male who made most of the complex anticipations during hunts and by the time Ali was 14 years old, 6% of his hunting movements were full anticipations, which is about 10 times more than those seen in his age mates. I remember a hunt where Ali, who was 11 years old at the time, fully and correctly anticipated the effect that Brutus' actions were about to have on the red colobus and he ended up capturing a large adult male. During this time, a hunter who was 12 years older than Ali was waiting on the ground to be sure the colobus would enter the correct tree. In general, Ali's learning environment allowed him to learn hunting techniques six years earlier than the other males.

The social learning environment is essential for the development of socially normal behavioral traits. Not only was Ali immersed in such an environment from birth onwards, but all youngsters are exposed to the social patterns that are characteristic of their group. It is through such social partners that youngsters come into contact with important external objects like sticks, hammers, nuts, leopards, bees, monkeys, and so on, and over many years, they have the opportunity to see how others behave with them. So during the hunting seasons, for example, a Taï infant will "see" a red colobus monkey as meat and not simply as a monkey in the forest, contrary to a forest antelope, which will be "seen" as an animal that Taï chimpanzees do not eat. In addition, Taï infants' experiences teach them that monkeys are hunted by groups in which each individual performs a different role and that the meat is shared extensively with others (see Figure 6.9). Later, the Taï male will naturally seek others when hunting and this special characteristic of hunting will be quite different from what a Gombe or a Mahale infant experiences. Such complex examples of the physical world being "transformed" by the social culture to which

Figure 6.9 **A young chimpanzee learning about meat-sharing**. By repeatedly seeing that the meat of red colobus monkeys is shared between the big males and some of the females, young Taï chimpanzees (bottom left) will "see" that meat-sharing is the norm. Brutus (top left) and Macho (top right) were both active during the hunt and are now seen sitting and eating the prey, while other non-hunter males (bottom right) are displaying around them but have only limited access to the meat. Such social influences will predispose the infant to abide by the meat-sharing rules at a later age.

the infant belongs is what is so characteristic of the social learning environment of young chimpanzees in the wild.

The social learning environment is also very important in humans and has long-lasting, complex effects on children. A previously mentioned study compared the development of Greek, Costa Rican, and Nso (Cameroonian) infants.[23] The upbringing styles of these groups systematically vary such that the amount of bodily contact between infant and mother increases from Greece to Costa Rica to Cameroon. In contrast, the frequency of stimulation with objects by the mothers decreases in the same order. Since mutual gaze between mother and infant increases with less body contact, it was shown that Greek infants developed the ability to recognize themselves in the mirror much more rapidly than the Nso infants. Nso infants, on the other hand, much more readily complied with social obligations than the Greek infants. This example illustrates what are now recognized as the two main learning environments in humans: the cultures that are focused on "interdependence"

[23] For more on this study of human children, see Keller *et al.* (2004b, 2005). A more recent study making similar comparisons with seven other human societies found exactly the same kind of influences of parenting style on the cognitive development of infants (Broesch *et al.* 2011).

emphasize social integration, obligation, and responsibilities and are seen in most traditional cultures throughout Africa, Asia, and South and Central America, and the cultures that are focused on "independence" emphasize the individual and technical intelligence and are typical of the western world.[24] Children growing up in such different cultures develop differently in a variety of ways. For example, children develop different perceptions of space: in interdependent cultures, the notion of space is more absolute and based on geometric coordinates and is therefore easier to share with others, while those from independent cultures acquire a more egocentric notion of space, one that is based on proximity to oneself (left, right, or above, for example). So, in humans, the importance of the cultural learning environment has been amply documented.

Chimpanzee social learning environments will be critical at some important developmental phases of an individual's life. The first is during early infancy, when infants or children are totally dependent upon their mothers and their absence has long-lasting negative consequences. Another period that is especially crucial for the development of cooperative ability is during a juvenile's socialization; this is the period after 5 years of age when a youngster, with his or her mother's support, moves more independently from her but is still constantly associated with her. However, as we have already seen, the development that takes place during the juvenile period is not independent from the development that took place during the infancy period, as the notion of self that is so important for the understanding of others is decisively affected during the first period. The general point is that development is an inclusive process and cannot be considered as a cupboard with different, independent drawers for learning periods or different abilities.

In a cooperative group, youngsters are influenced by the social solidarity that prevails between all group members so that individuals will often behave in terms of "group solutions" rather than "individual solutions." The same kind of subjective reality we mentioned about "branches being seen as hammers" applies here, where a male competitor is "seen" as an element of the team. Learning to "see" oneself and others as "elements of a team" is the result of growing up immersed in a cooperative environment. Eight-year-olds Ali and Sartre had seen Brutus and Falstaff work together for years when hunting for monkeys and when they themselves attempted to hunt, they were able to put the older individuals' actions into perspective and "see" how to maneuver the monkeys in the trees. But learning to become social is not only limited to a cooperative style, it also involves a sharing and tool-use style. The "cooperative hunting-meat sharing–tool use learning environment" in Taï can be expected to mold infants in a different way than the "large group hunting-meat sharing–no tool learning environment" that prevails in Ngogo. This in turn is different than the "no group activity–no tool–human presence learning environment" of most captive chimpanzees.

How can young chimpanzees in captivity learn about social skills? This brings us back to the "culture outside of culture" paradox. Captive infant chimpanzees

[24] Greenfield *et al.* (2003b), Rogoff (2003), Keller *et al.* (2004a).

Figure 6.10 **The natural social learning environment for chimpanzees.** The mother provides a safe haven from which the infant observes group social life. Seen here, a grooming interaction involving the mother (top left), a fight in the distance between adult males (top right), a reassurance gesture given by the mother to a dominant female (bottom left), and finally, the infant himself trying to interact with social group members from the safety of his mother's lap (bottom right). Such experiences will form the basis on which the youngster will develop his social knowledge and interactions.

rarely experience any complex group activity.[25] Should we expect them to understand such an activity without any experience and training, or should we expect it to be a purely genetic ability? The Cartesian approach which has inspired many generalizations from captive experiments does not fit well with the observations of important population differences in wild chimpanzees. The care invested in experiments by psychologists is remarkable and a number of the results obtained certainly point to some intriguing limitations in the social cognition of captive chimpanzees. However, some of these results mirror what has been found in orphaned human children who have been brought up in socially deprived orphanages in Romania.

[25] Captive chimpanzee groups can be relatively large and in a few cases even contain more than one male, but to me, just because a group is large does not mean it is complex. In contrast, wild chimpanzees groups are especially complex because of their fission–fusion nature. In fission–fusion grouping systems, group composition is continually changing, which leads to constantly changing dominance relationships, continually new and different social challenges, and constant external challenges (for the cognitive challenges of fission–fusion systems, see Aureli *et al.* 2008).

Contribution to the culture debate

Our hardcore experimental psychologist, although very interested in the detailed descriptions of the interactions between mothers and infants while learning to nut-crack, will reiterate that in whatever human society you look at, verbal exchanges and verbal teaching occur and this is what makes learning in humans so special. Furthermore, he will argue that genetics plays a more important role in chimpanzees and other animals compared to humans and, therefore, we cannot directly compare the effect of the learning environment in the two species. The hardcore psychologist will further argue that teaching is totally different in the human school system or when learning to do a task as complex as restoring a car, so that the maternal interventions seen in chimpanzees and other animal species that involve much simpler techniques cannot be considered to be of the same quality.

Personally, I totally agree that speech has opened up a world of possibilities in humans that did not exist before. I therefore suspect that part of what we see between mother and infant chimpanzees could become more complex in speaking individuals. I also acknowledge the fact that verbal exchange will happen even in very young human children, like I mentioned in the example of the Mexican girl learning to plant corn.[26] However, as seen in the stone-knapping case, verbal exchanges may not always lead to better performance, but may have more importance in terms of motivation and concentration.

The importance of the learning environment and prior knowledge are certainly underestimated by contemporary experimental psychologists. Comparative psychologists have traditionally been interested in the behavioral and cognitive effects of deprivation and many studies have demonstrated the strong negative effects this has on chimpanzees as well as on other primate species.[27] Recently, psychologists have simply ignored these results and tended to assume that deprivation in captivity is currently less important, as most captive animals are now kept in social groups. But we need to be careful here. As some psychologists[28] have recently said, "it is a disservice to psychology and to our understanding of the minds of chimpanzees (as well as other species) to hold any one individual or small group of individuals responsible as representatives for the entire species. This is especially non-productive in the case of reports of negative performance in various tests of cognitive skills." Another recently said, "Sound comparative psychology and modern evolutionary and developmental biology emphasize powerful effects of developmental conditions on the expression of genetic endowment … Most modern psychologists would expect caged human children to lose rather than develop cognitive ability. Indeed, the longer chimpanzees live in cages, the lower they score on cognitive tasks."[29]

[26] Paradise and Rogoff's (2009) paper was specifically intended to correct some cultural psychologists' impression that verbal exchange was absent and they argue that it is present in many traditional societies and used to support motivation and interest in youngsters, to occasionally emphasize aspects of what was being shown and practiced, and to reward attempts by youngsters.

[27] Harlow and Harlow (1962), Davenport *et al.* (1973), Maki *et al.* (1993), Gardner and Gardner (1989).

[28] Rumbaugh *et al.* (2008, p. 101).

[29] Gardner (2008, p. 135).

The fascinating aspect of teaching in humans is that it is so varied and presents so many different strategies, not only depending on the society but also depending on the tasks. Intriguingly, we see a very similar situation in chimpanzees. The link that seems to emerge from all this comparative work between the importance and frequency of teaching, in the form of direct demonstration and imposition of forms, and the difficulty of the tasks being learned might also explain the higher propensity of modern humans in western societies to rely on such forms of teaching. The higher reliance on complex and opaque techniques makes imposition and demonstrations more and more necessary.

Learning symbolic cultural traits will possibly involve less teaching than material cultural traits, as they are technically and physically less demanding. As a consequence, individual practice, so prevalent in material culture, will be less important in symbolic culture, and therefore there will be less scope for teaching interventions. How can you facilitate an "I want to have sex with you" leaf-clip? It will be the social partners to whom such a communicative sign is directed who will, through their reactions, either positively or negatively reinforce the predictions that a youngster could have made about their meaning. Thus, we should expect more imitative learning and less teaching for symbolic cultural traits than for material cultural traits.

In the next chapter, we will expand on one specific social domain to see if chimpanzees have a notion of death. We have seen that psychologists tend to argue that, contrary to humans, chimpanzees have no ability to perceive and reason about the unseen. However, over the years, we have made some very intriguing observations about the behavior of chimpanzees when confronted with death. This will allow us to address some questions about the chimpanzee mind and its similarity to our own.

Synopsis

- The learning of material culture relies on much individual practice and copying of other social group members.
- Subjective reality in youngsters permits them to structure their environment and guide their learning of the culture they share with others.
- As in symbolic culture, the learning of social cultural traits is very dependent upon imitation and the socially important influence of group members.
- The maternal role is much more limited in social culture than in material culture, as social traits are often learned at a later age than material traits.
- Teaching of youngsters by mothers extends over many years in chimpanzees and is more frequent for cultural tasks requiring extended learning periods that collude with the reproductive agenda of the mother.
- Teaching styles in chimpanzees vary extensively according to how challenging a task is, like in humans, and how it conflicts with the maternal need to invest in a new offspring.

7 Dead or alive? Towards a notion of death and empathy

The human species is the only one that knows that it must die and it knows it only by experience.[1]

Voltaire, *Traité sur l'homme* (1774)

So death, the most terrifying of ills, is nothing to us, since so long as we exist, death is not with us; but when death comes, then we do not exist. It does not concern either the living or the dead, since for the former it is not, and the latter are no more.

Epicurus, letter to Menoeceus

Taï Forest, 8 March 1989

Tina, a 10-year-old juvenile female, lost her mother just a few months ago. Tarzan, her 5-year-old brother, actively sought to be and was adopted by Brutus, the dominant male. He is very caring and even regularly shares meat with Tarzan. Tina is more cautious with Brutus and always follows a certain distance behind him. Thus, Brutus is regularly seen being followed in a row by Tarzan, then by Ali, who he adopted some five years earlier, and finally by Tina. At 7:45 this morning, after hearing unusually loud calls nearby, we rush to find Brutus near the motionless body of Tina, whose intestines are visible through a long cut on her belly. We later concluded that a leopard had caused this injury and she apparently died as a result of it biting through her second cervical vertebra. Brutus is quickly joined by several other individuals of both sexes. At first, the males behave aggressively and display nearby and drag Tina's body over short distances. Ulysse carries Tina's body 2 m but Brutus moves it back to where it had been, just about 5 m away from where the attack had occurred. Ondine and two other high-ranking females sniff Tina's wounds and some nearby leaves on the ground. Ulysse inspects and holds one of Tina's hands. Four more females arrive and very carefully approach the body, which is now guarded by the males and Ondine, the alpha female. One of the females smells the wounds on the body but does not lick them, while her nearby infant is chased away. After an hour, one of the males lies down beside Tina and begins to groom her and Brutus does the same on the other side. A low-ranking female approaches and sniffs the body but Ondine and Brutus chase her away. Over the next hour and a half, Brutus and two adult males groom Tina's body for a total of

[1] *L'espèce humaine est la seule qui sache qu'elle doit mourir et elle le sait que par experience.*

55 minutes. This is a remarkable sight, as none of these males were ever seen grooming her while she was alive. Salomé, the beta female, smells Tina's wounds and genitals. Nearby, subadult and low-ranking females carefully inspect the place of the attack and the blood stains on the ground. In contrast to when Ella was wounded by a leopard, no one licks a single drop of blood on Tina's body. Héra and another female approach and Ondine and Brutus allow them to smell the wounds while their two infants are chased away. Tarzan, Tina's younger brother, is the only infant allowed to come near Tina and he gently smells different parts of her body and inspects her genitals. Tarzan then very briefly grooms her and pulls gently on her hand many times while watching her. During this time, Brutus chases away a female with her juvenile daughter for a second time and she screams and flees. At 9:07 a.m., Brutus gently taps Tina's chin while looking at her face. Ulysse later shakes one of her hands and legs softly while also watching her face. It looks as if they are testing for some sort of a reaction. Also for the first time, Brutus briefly plays with Ulysse's hands before both start to groom Tina again. The dominant individuals play like this several more times, always very close to Tina and generally grooming her afterwards. Two and a half hours after the attack, the chimpanzees begin to be noticeably bothered by the ever-increasing swarm of flies on the dead body and after guarding the body for over 6 hours, the chimpanzees silently leave Tina's body.

Figure 7.1 **Three adult males guarding Tina's dead body**. Brutus, Macho, and Ulysse (from left to right) are guarding Tina's dead body after she was fatally attacked by a leopard (her perforated belly with intestines protruding is visible here). Macho is gently shaking Tina's right leg, possibly to test for a reaction, while Ulysse, who had been grooming Tina's body, is watching him. Along with many other members of the group, these three males will guard Tina's body for over 6 hours following the leopard attack.

Did Brutus and his group realize that Tina was dead? Did he have an understanding of death? Was he simply reacting to the fact that she was no longer moving, or did he associate not moving with being dead? Several aspects struck me as very special when I observed the events around the time of Tina's death. First, the dominant individuals seemed to impose a system of respect on the others by chasing the infants, juveniles, and low-ranking individuals away each time they tried to approach within 3 m of Tina's body. Second, it was unusual how the adult males simultaneously cared for Tina's body and played with one another around her body. Third, no one licked the blood from her wounds or the ground, which is what normally happens when chimpanzees are injured by leopards. I was puzzled that no one tended to Tina's fresh and bleeding wounds immediately after she had died. Did they know that doing so would be useless in her case? Why did they lick the blood from the ground only when individuals were wounded and not when individuals were killed?

Before addressing these and other questions, let me first give you a second example of the Taï chimpanzees' reactions to death.

Taï Forest, 23–24 March 1991

We came across several individuals mid-morning, including Bijou, a young female who we had heard giving loud alarm calls for 10 minutes. She is holding her 2-year-old son Bambou against her chest and when we arrive, she climbs into a tree to avoid the arrival of the displaying males, who climb up the tree behind her and smell Bambou's motionless body. Bambou has no visible wounds and since he had been in perfect shape yesterday, I assume he must have died after falling from a high tree. When the situation calms down after 30 minutes, Bijou returns to the ground and follows the group. Kendo, the new alpha male, guards her for close to 3 hours. An hour after Bambou's death, Bijou starts laying him on the ground for short periods of time, but continues to carry him along with her throughout the rest of the day.

The next morning, we find Bijou still carrying the swollen and smelly body against her chest. At around 8:00 a.m., Bijou places Bambou in a nearby branch while she eats in a tree when Kendo starts warming up to display against her. Brutus rushes to take Bambou's body in his arms. While Bijou tries to avoid Kendo's display, Brutus brings the body to the ground, where he rests with it near two males who both smell the infant's body. By 9:20 a.m., Bambou's body is so swollen that the skin tears in several places and flies starts to swarm. After a long rest later in the day, the group starts to move off and Bijou looks hesitantly between the group and Bambou's body. Seeing this, four females with their infants, one of whom was Bambou's closest playmate, return to Bambou's body and some climb a small tree above the body and look down at it. Three females and Bambou's playmate each make a few soft "hou" calls and then they all silently leave. At 2:56 p.m., Bijou leaves Bambou's body to catch up with the group. Brutus and four males are sitting and waiting silently for her 80 m

Figure 7.2 **Castor carrying her fatally ill infant, Congo**. Castor's 14-month-old infant died from an unknown disease a few hours after this picture was taken. Before he died, Congo was too weak to hold onto his mother and she had to constantly carry him and brush flies and dirt from his body.

away. Shortly after meeting these males, Bijou returns to Bambou's body on her own and carries him 20 m. She continues to hesitantly move him in this way for another 80 minutes and then she leaves him behind for good.

Mothers clearly experience the death of their infants as a loss and they hesitate to leave them behind. As seen in Figures 7.2 and 7.3, Castor also faced such a dramatic experience when her infant died from an unknown disease. However, they quickly stop behaving towards them as they would if they were alive. Gombe chimpanzee mothers have long been described to behave like Bijou and carry the bodies of their dead infants around for some time[2] and there have been three cases of Bossou chimpanzee mothers carrying their dead infants for even longer than has been observed in Gombe and Taï.[3] However, group members in Gombe have been

[2] Similar behavior by mothers whose infants had died was mentioned by Jane Goodall as long ago as 1968. At that time, there was a polio outbreak at Gombe that killed many individuals, including some dependent infants.

[3] The mothers of both infants (1.2-year-old Jimato and 2.6-year-old Veve) carried their bodies for 68 and 19 days after death, respectively (Biro *et al.* 2010). In general, all members of the community demonstrated high tolerance levels towards the corpses. That these bodies were carried for such excessively long periods is especially surprising because the corpses seemed to be totally mummified and therefore looked very different than when alive. The authors suggest that the mothers did not realize what was

Figure 7.3 **Castor cracks nuts with her dead son on the ground**. Castor placed Congo's body next to her on the ground (just behind the hammer near Castor's right foot) while she was cracking nuts and she picked him up again before moving on. Congo has been dead for one day and Castor stops carrying him later this same day.

observed to react violently against the dead infants, even sometimes cannibalizing the dead bodies after snatching them away from their mothers shortly after death.

More puzzling in the case of the dead Bambou was how hesitant the other females were when they started to move on and leave him behind. Why did these females come back? Why did they make these soft calls before finally abandoning him? Were they aware they would never see him again?

However, it is not only mothers that seem to react in this way with dead infants. On 4 November 1987, two months after Falstaff had been badly injured during a leopard attack that left him with a perforated lung (see Chapter 4), Snoopy, Salomé's eldest son and Falstaff's hunting partner, was alone with him and both groomed each another intently. They stayed together for 2 hours while we heard the rest of the males of the community moving towards the north of the territory. Snoopy, who clearly wanted Falstaff to follow him, moved a few meters then waited for him, but Falstaff seemed too weak and did not move. Snoopy came back to groom Falstaff, but Falstaff still did not move. Snoopy then moved about 200 m north and drummed repeatedly on a large tree buttress. Falstaff did not answer or follow.

happening but "because lactation ceased once the infants had died, the mothers' reproductive cycle returned; such hormonal changes, which prepare the mother for the arrival of a new infant (normally around weaning), may have contributed to a gradual 'letting go' of the dead infants' remains." For accounts of how the Mahale chimpanzees react towards dead infants, see Hosaka *et al.* (2000) and Kooriyama (2009).

Snoopy then screamed loudly for a long time as he ran towards the north. This was all a very unusual behavior for an adult male chimpanzee and it was even more unusual for a highly self-confident individual like Snoopy. Falstaff was never seen again and most likely died from the injuries he sustained during the leopard attack. Snoopy's reaction as he left Falstaff for the last time is similar to the females' reactions when they returned to Bambou for the last time. Do chimpanzees realize that a dead individual will never "come back" and that badly injured ones will die and then never be seen again?

Understanding of death in humans and other animals

The understanding of death has an important place in human cultures. In all human societies, death is accompanied by elaborate cultural rituals and prohibitions. Such rituals have a long history in our past and archaeologists have found traces of them as far back as the emergence of modern humans, *Homo sapiens sapiens*, 100 000 years ago.[4] All human cultures today have methods for treating and disposing of dead bodies, which all have the effect of isolating living individuals from contact with decomposing corpses. Another essential human reaction to death involves various types of formal and informal mourning. Again, while these differ between cultures, it seems that there is a universal need for some sort of mourning behavior in response to the death of close friends and acquaintances. Our understanding of death allows us to recognize something as "dead" because we have an internal representation of death that is distinct from the direct perceptual information that we receive from an inanimate body. It is this separate representation that explains our ability to reason about death rather than merely respond to death in the environment.[5]

What is meant by "an understanding of death"? This is not an easy task to explain because, in addition to the many different descriptions of rituals related to the death of a group member in different human societies, it is difficult to know exactly how individuals spontaneously react upon first encountering a dead person and what they understand of this phenomenon in different societies. Developmental studies intended to uncover the concept of death in children have distinguished different components, but the majority recognize five major aspects of understanding death. These five stages are mastered by children in a relatively fixed sequential order between 5 and 10 years of age. The first is "irreversibility," which refers to the recognition that once a living thing dies, it cannot be brought back to life (e.g. a dead body will not move later on), and is acquired by 5 years of age. Next is "inevitability," which refers to the acknowledgement that living things eventually die, followed by "universality," which refers to the understanding that all living things die, followed by "cessation," which refers to the understanding that all life-

[4] See Mithen (1996), Baron-Cohen (1999).
[5] See Allen and Hauser (1991).

defining functions cease at death (e.g. a dead body cannot be thirsty). These three are acquired between 6 and 7 years of age. Finally, "causation," which refers to the understanding that death is ultimately caused by a breakdown in bodily function, is mastered between 7 and 10 years of age.[6] How much children fear death seems to be related to how complete their understanding of death is, especially in relation to the cessation and universality components. Thus, independent of age, the better a child's understanding of death is, the less fearful they are of death.

Furthermore, some have suggested that a notion of death requires an ability to recognize the continuity of self and others through time. Self-recognition, possible only when the difference between "self" and "other" is developing, would be a requirement for understanding the irreversibility of death. Researchers have studied the development of self-recognition by recording how children react to their reflections in a mirror. These studies have found that it is only after having experience with mirrors that children learn to recognize their reflection. Furthermore, this seems to correspond to the time it takes these infants to develop an increase in empathic reactions towards others. Thus, one proposed scenario suggests that the development of self-recognition in human children would open the way to a certain understanding of others that allows the development of compassion and empathy. This, in turn, would permit an understanding of death. Following such a developmental scenario, some psychologists have proposed that animals could not possibly understand death because they are not self-aware and do not have the capacity to attribute mental states, both abilities that are necessary for such an understanding. Furthermore, they have proposed that animals lack moral emotions and senses and therefore have no capacity for empathy, sympathy, shame, guilt, and loyalty.[7] In contrast, Frans de Waal rejects such claims and has spearheaded an inclusive approach that incorporates observations and anecdotes from animals; from this, he claims that empathy has a long evolutionary history and can be seen in birds, dogs, cats, elephants, dolphins, and many primates.

Is it true that animals have no understanding of death? Field studies with different monkey species have suggested that mothers carry their dead infants with them but react towards them as if they were alive but passive, thereby showing no recognition of death.[8] Descriptions of yellow baboon mothers in Kenya provide a good illustration of this general phenomenon.

> Mothers persist in the apparently automatic embracing of their infants even after the infant's death. They continue to carry the decomposing and increasingly dehydrated corpse, despite the fact that this usually means that they walk three-legged, setting the corpse down whenever they stop to feed and then retrieving it again. … After about three days … mothers leave the corpse on the ground for gradually increasing periods of time while they forage at greater

[6] For examples, see Speece and Brent (1984), Slaughter and Lyons (2003), and Barret and Behne (2005). Slaughter and Griffiths (2007), in particular, look at the relationship between fear and an understanding of death in children.

[7] See Hauser (2000, 2006). For an alternative, more bottom-up approach, see de Waal (2001), de Waal and Ferrari (2010).

[8] Zuckermann (1981), Cheney and Seyfarth (1990), Fashing *et al.* (2011), Anderson (2011).

distances away, eventually either lose it or leave it, looking back at the corpse with repeated signs of conflict and ambivalence and sometimes giving alarm barks.[9]

Mothers in other primate species have also been observed to carry their dead offspring around for days without showing any specific reaction to the fact that they are dead rather than just temporarily inanimate.

Elephants have long been known for their very intriguing behavior around dead individuals and it has even been suggested that family members are able to recognize the dead bodies or even the bones of deceased group members.[10] And this interest in elephant carcasses is not due to a generally high level of interest in dead bodies because elephants do not show any interest in the dead bodies of other species. Elephant researcher Cynthia Moss described how she once discovered the fresh carcass of a dead female elephant in a clearing just when the family of the dead individual was arriving. The elephants became silent and approached, smelled, and touched the dead body. Then, scratching the ground with their feet, they threw earth on the body. Some also gathered branches and palm leaves and placed them on the dead body. Other similar descriptions of elephants covering dead bodies exist. In addition, Cynthia Moss thinks elephants can identify dead individuals. As evidence of this, she notes that 3 weeks after the natural death of one of the old matriarchs, she collected the lower jaw and brought it back to her camp and placed it with the other jaws she had collected from dead elephants. Three days later, the family of the deceased elephant crossed the camp clearing. The elephants made a detour to inspect the jaws of their deceased family member and the 7-year-old son of the dead matriarch remained behind for a long time, gently touching the jaw and turning it over with his trunk and feet. Cynthia Moss was convinced that he recognized his mother's jaw. She also mentions Agatha, the daughter of Annabella, who for years regularly returned to the place where her mother's skull was and carefully touched and manipulated it. In a similar case, another observer witnessed an elephant family coming across the bones of a female family member, Jezebel. The family, which included Jolene, Jezebel's daughter, silently and gently manipulated the bones for an hour.

Definitive proof of an understanding of death in non-speaking animal species will be hard to collect, but the striking reactions seen in chimpanzees and elephants towards individuals who suddenly do not move and might not be seen alive anymore should move us to entertain this possibility.

Mourning and compassion around death in chimpanzees

Following Bambou's death, Bijou gave birth to another infant, Baloo, in April 1992. Sadly, Bijou died 2 years later, possibly from a fall from a tree but without showing any external signs or injuries, and left Baloo an orphan.

[9] For the baboon descriptions, see Altmann and Altmann (1970, pp. 129–130). Maternal transport of dead infants has also been reported in mountain gorillas (Warren and Williamson 2004), and in ring-tailed lemurs (Nakamichi *et al.* 1996).

[10] Douglas-Hamilton and Douglas-Hamilton (1975), Moss (1988), Poole (1996).

Taï Forest, 31 May–1 June 1994[11]

While observing a group of five chimpanzees in the morning, they came across the very recently deceased body of Bijou. They immediately stood upright to look at it, gave loud alarm calls, and climbed a few meters up the surrounding trees to look intently at the body. Before he began climbing, Brutus, the oldest male of the community, took little Baloo against his belly so as not to leave him behind. They were soon joined by many others and they stayed in the trees near the body for nearly 5 hours, alarm-calling most of the time and making fear screams and aggressive pant-hoots. They spent 93% of this time in the trees looking down at the dead body, but a few eventually came down to the ground. Brutus did so twice to eat in a nearby fruiting tree, but each time he returned to sit 5 m from the body. Only five individuals touched the body and only one individual – Kendo, a male – did so aggressively. Many individuals moved around in the trees, breaking off leafy branches and letting them fall so that they cover the body. When they started to leave, Brutus repeatedly tried to take Baloo along, but Baloo resisted and Brutus eventually left him alone near his mother's dead body.

The next morning, Baloo remains alone with his dead mother until 2:43 p.m., when Brutus, on his own, appears and immediately embraces and grooms Baloo. Finally, an hour later, Baloo allows Brutus to carry him on his back away from his dead mother. Brutus carries him for only a short time but the two remain together for the rest of the day. That night, Brutus and Baloo share a night nest. Sadly, Baloo disappears only 5 days after his mother died. Despite Brutus' care, he was simply too young to survive without milk.

Brutus' reaction towards Baloo, like his care for Bambou's body just before Kendo's display, is moving to us because it seems to reveal a genuine concern for those who are helpless. It is difficult to avoid anthropomorphizing Brutus' intentions and motivations in this situation.[12] Over many years, Brutus was observed to adopt two very young orphan males, to rescue alarm-calling individuals, to tend to the injuries of Ella and many other individuals who had been attacked by leopards, to ensure that young group members respected the bodies of dead group members, and to protect Bambou's body during social squabbles. So we cannot simply dismiss his reaction as merely an intriguing anecdote. Like all chimpanzee hunters, he was regularly exposed to death. Was Brutus aware that Bijou lying motionlessness on the ground meant that she would stay motionless "forever," that she was "gone" and it therefore made no sense for Baloo to stay near his dead mother's body? Did Brutus realize that Baloo would be safer with him rather than alone

[11] The observations of Baloo and Brutus around the time of Bijou's death were made by Annemarie Fränkl, Honora Kpazahi, Grégoire Nohon, and Claudia Steiner.

[12] As Frans de Waal (2009) might say, this situation so appeals to our natural empathy that it is very hard not to immediately interpret this observation in terms that attribute complex mental reasoning. However, conversely, just because this observation is so appealing does not mean that Brutus did not mentally reason when he decided to carry Bambou's corpse away.

Figure 7.4 **Brutus, the oldest male of the Taï north community**. Brutus, who was born around 1951 and died on 1 March 1997, was the alpha male of the North Group for over 10 years. During this time, there were nine adult males in this community and he was a very ardent fighter against neighbors. At the same time, he was the best hunter and the main meat-provider to the females of the community. Long after losing his alpha position, he continued to be the primary travel direction leader and showed much empathy towards weaker group members. He adopted two male orphans, Ali and Tarzan, and cared for them for many years.

near his mother's dead body, which a leopard might come to scavenge on? Daily observations tell us that Brutus would normally be unconcerned about leaving a baby sleeping alone on the ground with only their mother. What is impressive is that Brutus, even after spending a night far away from both Baloo and his dead mother's body, bothered to return and take Baloo away with him. What was so special in his mind about this case that made him intervene?

Covering the body is another impressive and puzzling behavior associated with death in Taï chimpanzees. In all five cases where chimpanzees died without external signs to indicate a possible cause, we saw group members cut leafy branches of different sizes and, without eating anything from them, let them fall on the dead body. In the case of Ondine, the highest-ranking female of the group, her dead body was discovered rather late in the afternoon and the chimpanzees slept in the trees above the corpse. When the observers arrived the next morning, the chimpanzees had already left, but Ondine's body was found almost completely covered under two large leafy branches. The branches placed on Ondine's body were exceptionally large, unlike in most other cases, where the leafy branches that were used were rather small and thus covered only a small part of the body. Why would chimpanzees want to cover the body of a dead group member? Is this in any way an attempt to bury the body?

Here, however, we need to be cautious before drawing any conclusions. A few months before this episode, I was with the chimpanzees when they came across a freshly dead bushpig that a leopard was feeding on. The chimpanzees made many alarm barks and chased the leopard away. During the 2 hours that they stayed with the bushpig's body, mainly above it in the trees, they also let small leafy branches fall on it. Thus, to me, letting small branches fall on a dead body seems to be an attempt to elicit a reaction from it to determine if it is still alive rather than an attempt to bury it. And in this case, they only dropped small objects and nothing close to the size of the branches that we found on Ondine's body. Taï chimpanzees do not prey on adult bushpigs either and none of the chimpanzees fed on it.

Guarding the dead body of group members for hours, immediately helping orphans who remain close to their dead mothers, testing for a reaction by shaking or covering the body with leaves, not licking blood or wounds as is usually done with injured individuals, showing signs of "sorrow" when leaving the dead, showing signs of "respect" by keeping youngsters at bay, and carrying the corpse to safety – these unusual behaviors are only seen in association with death! If chimpanzees had an understanding of death, these behaviors would make perfect sense to us. If not, they make you wonder, to say the least.

About empathy and compassion

While driving in the Côte d'Ivoire many years ago, my wife and I prepared for the worst as we approached a police roadblock. Earlier that day, I had badly injured one of my fingers by slamming it in the car door. The policeman approached our car with his gun and the expression on his face told me he meant trouble. Seeing the bandage on my finger, he asked me what had happened. When I told him about my accident, he replied, "Ouiii ououaaa, that must have hurt. Pardon, pardon!" and he shook his hand as if he himself felt my pain and let us drive on without any further delay. Compassion for others comes spontaneously to us and we immediately identify with the pain others feel (in other words, we attribute a mental state to the other). And we can feel this compassion even if the injured person does not make any sound to indicate he is in pain or if he or she is a total stranger to us, like in my experience with the policeman. Compassion requires not only a sense of self that ties into imagining what it would be like if I would have hit my thumb, but it also represents an emotional fusion of self and other – because the other hit his thumb, he must feel the same as I would if I had hit mine – and this is independent of the fact that, right now, I have not done so. Therefore, this is only possible if one adopts the other's perspective.[13] Because he was empathetic, the policeman not only realized that I was in pain from hurting my finger, but also that I could not perform a simple action like extracting my driver's license from my pocket. Hence, under such a definition, compassion requires the attribution of feelings to others, while empathy

[13] In Frans de Waal's book, *The age of empathy* (2009), he uses marvelous examples to describe how widespread such compassion is in the animal kingdom and how this ability develops.

Figure 7.5 **Fitz licks a wound on Darwin's right foot**. Such tending to injuries is systematically seen in Taï chimpanzees and is not limited to close relatives, as has been seen in the Gombe chimpanzees. Saliva has a strong antiseptic function and its regular application when the wound is fresh helps speed up the healing process and might explain why wounds often heal so rapidly in Taï chimpanzees.

requires the attribution of competence to others.[14] Since, as we saw, some psychologists have proclaimed that animals do not have self-awareness or a capacity to attribute mental states, then compassion and empathy should not be possible to them.[15]

In the 3 hours and 15 minutes following the leopard attack on Falstaff, 9 different individuals groomed him without interruption and licked his wounds for almost 3 hours. While he was resting the next day, other group members tended to him for almost the entire day. Similarly, when a leopard wounded Ella (see Figure 4.4), she was constantly surrounded for the rest of the day by other chimpanzees, who licked her wounds for over 6 hours. I saw similar care provided to all of the numerous individuals we saw injured by leopards. Such treatment was provided as long as necessary, with some wounds being tended to for longer periods than others. For example, Gitane had many cuts on her back that she was unable to reach herself and other group members licked her wounds for as long as one month after the attack, by which time the wounds were almost completely healed. Similarly, group

[14] See Cheney and Seyfarth (1990, 2007) and Hart and Karmel (1996).

[15] Just to remind the readers, such claims have been made solely on the basis of experimental results with captive individuals that have never experienced any of the challenges that their wild counterparts face on a daily basis. For compassion and empathy, it would be necessary to regularly experience events such as encounters with leopards, attacks from violent neighbors, falls from high trees, and other events that could trigger such feelings. Surely I am not suggesting that we should regularly confront those chimpanzees with leopards, but we simply need to acknowledge that captive chimpanzees are not like wild chimpanzees. Interestingly, because captive chimpanzees without prior experience do not show empathy, it is safe to assume that this ability is not purely genetically based in this species.

Figure 7.6 **An adult female cleaning a wound near her partner's eye**. Tending to others' injuries is a social activity that attracts the attention of social partners and is performed in front of youngsters who are learning how to behave in the group. In this photo, the infant of the injured female watches several females take turns tending to the cut over his mother's eye.

members tended to Falstaff's hardly visible wound for two months after the attack that probably perforated the lung under his right arm. Rousseau's left testis was injured by a leopard's claws (see Figure 4.5), and other individuals still occasionally licked his ruptured testis as much as two years later. Group members tended to wounds on easily accessible body parts, like the front of arms and legs, for a much shorter period of time, as the injured individuals were able to attend to such injuries themselves. This suggests that the tending given to wounded individuals is proportional to the severity and the location of the wounds (i.e. the needs of the individual), and not just to the external aspects of the wounds.

Wounds remain painful for long periods of time and other chimpanzees seem to be aware of this. Gérald, Ella's son, and Goyave, Gitane's daughter, were still dependent upon their mothers for transport when they were injured, but the two youngsters adapted their positions when riding on their backs so that they would not touch the wounds. Freshly attacked and injured individuals are normally in a state of shock. Dominant group members waited close to Falstaff for more than 3 hours until he was able to move. Similarly, group members waited for Ella at the site of the leopard attack for more than 2 hours, with Kendo, her eldest son, remaining close to her the entire time. Why do these chimpanzees behave in such a way when they are with freshly injured individuals? Is it a reaction to the fact that the wounded individual does not move or are group members aware that they are in pain and handicapped, or both? Waiting like this at a given location in the forest is unusual, and in Ella's case, many of the individuals left to feed on nearby trees before returning to her. Kendo chased away all infants that came near Ella during the first few hours after the attack and he was regularly supported in this by other males. However, they tolerated most of the adult group members who approached carefully and kept at a certain distance from Ella. After being attacked by a leopard, Fossey, a young immigrant adolescent female, had many cuts on her head and shoulders and she was quite afraid of the resident females, who are known at times to be quite aggressive towards immigrants. Two males took turns protecting her when adult females approached, which permitted Fossey to rest quietly during the hours following the attack. Protecting injured individuals by keeping certain adult group members at a distance and chasing away infants is unusual in chimpanzees and tailored to outstanding situations.

Sadly, a new context where empathy has become very important is when an individual is caught in a snare. Human populations have been steadily increasing in many isolated regions of Africa and this has resulted in an increase in human encroachment into chimpanzee habitats. Farmers everywhere protect their crops by placing snares around their fields to capture invading animals, mainly forest antelopes and monkeys. However, snares represent a fatal danger to chimpanzees as they are made out of very strong metal cables (bike brake cables are sold in almost every small village throughout the continent) that cut deep into the flesh of a trapped hand or foot. If the cable is not removed quickly, this can lead to gangrene and necrosis of the body part that has been deprived of blood circulation to the point that the limb dies and falls off. Many chimpanzee populations live in areas where people set out snares to capture animals for food and this is happening more and more in protected areas as the fauna is vanishing outside such areas, and quite frequently chimpanzees become trapped.

Taï Forest, 20 June 1985

At 3:00 p.m., I am following a large group of chimpanzees in the western part of their territory. Suddenly, right in front of me, Véra, an adolescent female, catches her left wrist in a snare trap. Screaming, she spins around over and over again until the cable finally breaks from the tension. She then sits and looks quietly at the four fingers of her left hand, the cable tightened so strongly around them that it completely disappears into the flesh. It must be pure agony. I see no blood, which probably means that the cable has not cut the skin. Schubert, the beta male, approaches her slowly and Véra pant-grunts softly and stretches her injured hand out in his direction. As he is turning his back to me, I only see him bend forwards, head down at the height of her wrist. He must be placing one of his long canines underneath the cable and with a jerky movement of his head, the cable is removed. Véra utters a scream and dashes away followed by the whole group, all of them silently changing their initial direction and walking away from the trap without stopping for 90 minutes.

Figure 7.7 **Nino stepped in a snare when he was a 7-year-old juvenile** and three of the fingers on his left hand were trapped. As you can see in this picture, a piece of the cable remained deeply embedded in the flesh of his fingers and prevented him from using his hand until the cable rotted away and fell off 7 months later. He licked it regularly, which prevented an infection, but he was cut so badly that the flesh from his second and third finger fused and remained this way for his entire life. Luckily, though, Nino recuperated very well from this event and developed into a strong and impressive male.

Although she must have been in excruciating pain, Véra did not move at all while Schubert was trying to help her. In addition, Schubert seemed to have understood immediately that Véra needed help and he provided it without hesitation. Not only did it seem that Schubert felt empathy for her and reacted appropriately and immediately, but Véra understood this and stayed still as long as he was trying to free her despite the pain she must have felt.[16] Despite the fact that it was most likely the first time that Véra was trapped in a snare, she seemed to immediately understand that she would profit from Schubert's help and she let him provide it. Spinning on the ground like Véra did tends to cause cables to be break loose from the tree they are fixed to, but once an individual has freed themselves, the loop of the cable is usually deeply and tightly embedded in the flesh and touching it must be extremely painful. This is an important point because, while Taï mothers remove snares from their dependent infants within minutes and adults remove their own rather quickly as

[16] I can hear the hardcore psychologist call out, "Anthropomorphism, anthropomorphism!" The point is that I can only evaluate this observation as a human and not as, say, a penguin. I could suggest that Véra did not move because she was scared about this highly dominant male. This would actually be equally anthropomorphic, as young female chimpanzees retreat in front of high-ranking males. Suggesting that Véra did not move because she just wanted to see what Schubert intended to do does not agree at all with how juveniles typically react. Finally, suggesting that Véra did not move simply because she was in too much pain does not really explain why she stretched out her hand towards Schubert.

well, the main problem is with juvenile individuals who do not yet understand how to remove snares. I have seen five juveniles with snares on their hands whimper and avoid the slow approaches of their mothers who seemingly wanted to help. These individuals possibly did not yet understand that help was right there. The sad consequence of this is that snares remain on juveniles for much longer, up to 12 months in some cases (see Figure 7.7), than they do on infants and adults, which often leads to death. For example, a 5-year-old female died from gangrene a few months after stepping on a snare and another lost her entire hand.

Are these observations of selectively caring for and relieving others' pain expressions of compassion and empathy by chimpanzees? Or do they simply associate wounds with some types of behaviors because they have seen others behaving this way? An explanation based on association would have to take into account that care was provided only for certain kinds of wounds, like those in hard to reach places or when injured individuals were too weak to care for themselves. The rarity of serious wounds (some 20 individuals in the 3 communities were seen to be wounded by leopards in 25 years) makes it very unlikely that a simple association process could elicit such a precise response by group members. In addition, in these cases, care was provided to all group members and not only to close kin, thereby excluding the possibility that they were simply helping their relatives. In this context, it is relevant to note that when chimpanzees injure individuals who they do not empathize with, they do not tend to their wounds. I observed seven cases of chimpanzees attacking and biting females from neighboring communities and in none of these cases did they lick blood from the wounds, the body, or the ground. Thus, chimpanzees do not simply lick blood because they see it dripping or because they like the taste of it. These aspects suggest that chimpanzees possess a certain form of empathy toward group members.

Are population differences observed in such domains? As seen in Chapter 4, striking differences have been observed in how frequently chimpanzees tend to wounded individuals. For example, the Gombe and Mahale chimpanzees limit their tending to close kin. A very vivid description from Gombe illustrates this well. An adult female from that group, Little Bee, was seen collecting and bringing fruit to her mother, Madame Bee, who had been so badly injured by neighboring males that she could barely move. Little Bee was the only group member who provided food to her mother, who died a few days later.[17] Obviously Little Bee strongly empathized with her mother. Does the more restricted empathetic tendency seen in Gombe towards injured individuals have something to do with the less frequent occurrence of large injuries due to the paucity of leopard attacks? This might suggest that empathy is not a rigid species-specific behavior pattern but emerges out of a specific social environment where a high level of within-group solidarity provides important survival benefits.

Snaring has sadly become an even bigger problem for other populations of chimpanzees than Taï. When I visited the Sonso community of chimpanzees in

[17] Description from Goodall (1986).

Figure 7.8 **Skull of a young chimpanzee in the Taï forest**. What goes on in the chimpanzee's mind when they see such a sight in the forest? We are able to draw some conclusions based on their behavior when they encounter dead individuals, but deciphering their actual thoughts remains speculative.

the Budongo Forest Reserve in northeastern Uganda, I was shocked by how many individuals had badly crippled hands. Some adult females have such deformities that they could barely use their hands for climbing in trees, which is a real handicap for a chimpanzee. In fact, 40% of the adult members of one of the communities have extreme handicaps from snare injuries, like missing fingers, toes, feet or hands, or suffer from badly crippled hands.[18] Chimpanzees in Kibale National Park, Uganda, also suffer from snares, although to a lesser extent. If the snares would have been removed quickly, these injuries would have recovered completely with time. However, in both Budongo and Kibale, no case of support towards trapped individuals has yet been reported. Similarly, the emblematic mountain gorillas of the Virunga Volcanoes in Rwanda also suffer injuries from snares, which have left some badly crippled and others dead. In one instance, an adult male gorilla was seen removing a snare from the wrist of an adult female, similar to what I saw Schubert do with Véra. So empathy seems flexible and is possibly a function of the socio-ecological environment.

Observations of such empathetic behavior in other wild primate species are very rare, but not because injuries have not been observed. On the contrary, injuries resulting from social fights or from predator attacks occur quite regularly. Instead, such behavior is rarely seen because there appears to be no support provided to injured individuals (see also the reaction of vervet monkeys to injuries in Chapter 4). Monkeys seem to look at disabilities as anomalies or as objects of interest rather

[18] Quiatt *et al.* (2002)

than as handicaps.[19] This absence of reaction to injuries has been suggested as resulting from an absence of compassion and empathy.

The origin of a notion of death

Death in modern western societies, with the prevalent role of medical intervention, has become a scientific fact that can be easily measured with precise apparatus and the cause of death is almost always quickly identified. However, this has certainly not always been the case in human history. For farmers from the forests of Africa who still live a rather traditional life, like the Wobé, Oubi, Kroumen, and other Bantu, a body without any external, obvious signs to indicate what the cause of death was could have died from supernatural forces and this would be a threat to the entire society. Therefore, before the body can be buried, the person responsible for the death must be found.[20] It is common in this case to ask the body to tell the family who was responsible for its death. This is done when the family members carry the body in the village during the funeral ceremony and the burial is delayed until the body has indicated his murderer. The guilty person must then be punished before order can be re-established in the community. Thus, for humans, understanding the cause of death is a challenge and often a distinction is made between a body that has visible wounds and one that does not.

Do chimpanzees differentiate between different types of death like humans do? The most striking sign of such a distinction comes from the observation in Taï chimpanzees that group members produced many more signs of fear towards dead bodies without wounds than towards dead bodies with obvious wounds.[21] In the case of Tina, who was visibly killed by a leopard, some of the adults directly touched, groomed and manipulated her body, whereas for diseased-related deaths, which result in dead bodies with no obvious signs of violence, like in the cases of Bijou and Ondine, chimpanzees barely touch them. Furthermore, chimpanzees stay on the ground near dead bodies with wounds or injured survivors, while they stay mostly in the trees and were observed to spend only 8% of the total time on the ground near individuals who died from disease. The vocal reactions to these two types of dead bodies are also strikingly different. When the Taï chimpanzees find a body that died from unknown causes, they seem to be dominated by fear and give many alarm calls; in contrast, when they find wounded dead, they give no alarm calls.

[19] Cheney and Seyfarth (1990, 2007).

[20] For the Kroumen and Oubi, see Holas (1980); for the Wobé, see Schwartz (1975); for the Bantu, see Turnbull (1961). In general, death of very young infants and very old people are accepted as natural and the villagers do not look for a murderer in the way described in the main text.

[21] In Taï forest, we observed 5 individuals who had recently been injured by leopards and where the group reactions could be recorded in detail (for a total of 233 hours of group members present), one case of Tina being killed by a leopard (for a total of 55 hours of group members present), and 10 cases of individuals killed by a disease that the group discovered when we were present (for a total of 285 hours of group members present).

Death following an illness is complicated by the fact that, depending upon the disease, it can be extremely sudden in a way that looks totally incomprehensible. Léo, the dominant male of the Middle group in the Taï forest, was a young, healthy, and impressive full-grown male. On the morning of 14 February 2002, he was observed behaving absolutely normally – eating, drumming, and displaying. At around 10:45 a.m., he started to become quieter and vomited four times. Then at 1:03 p.m., he climbed a small tree and lay on in its branches. At 1:38 p.m., he suddenly fell abruptly to the ground, where he lay with very shallow breathing sounds for 10 minutes, and then, without any other signs, he stopped breathing and was dead. It took some time for the field assistant[22] who was present to realize what had happened and he could hardly believe his eyes. Similarly, the three chimpanzees who were with Léo did not seem to realize what had happened and continued to rest around him. Within 10 minutes, flies started to swarm over the body and Bob, another male, groomed Léo but only for a few seconds and then started to chase the flies away and left rapidly.

A similar reaction to death was also observed in a small group of captive chimpanzees in a safari park in the US. After Pansy, a female in the group there, died, the three remaining group members, including the daughter of the deceased female, showed several behaviors that, according to the authors, were similar to human responses to the death of a close relative.[23] Most notably, they were seen to closely inspect the body, test for signs of life by manipulating her limbs at the moment of death, and remove straw from her body the next morning. None of them groomed her after death. At Gombe, when an adult male died after falling from a tree, the other chimpanzees present erupted in aggressive displays and alarm calling, with much mutual embracing and touching. They frequently stared at the body and some appeared to sniff it, but nobody touched it in the 4 hours before they left.

The possible difference found in the expression of compassion and empathy between chimpanzee populations suggests that the experience of living in an environment with a high level of predation forces preyed upon individuals to react. So far, the Taï chimpanzees are the only known well-studied chimpanzee population that suffers high predation pressure from leopards. Help to wounded individuals directly improves the survival of group members and, through reciprocity, will indirectly benefit the individuals providing help. Thus, compassion and empathy for wounded individuals would be selected for in an environment with a high threat of predation. Likewise, human ancestors survived in environments where they were subject to predator attacks from saber-tooth tigers and a number of other predators, and support to wounded individuals might have been very important for survival.[24] Once empathy and compassion has developed in the domain of predator defense, it will impact other social domains. We could expect that similar forces

[22] This observation was done by Louis-Bernard Bally, who has been following the Middle Group for many years. It turned out that Léo died of anthrax (Leendertz *et al.* 2004).

[23] For the captive death, see Anderson *et al.* (2010), for Gombe's death, see Teleki (1973b).

[24] See Boesch (2009) for a more extensive discussion of the effect of predation on our ancestors and its possible effect on human evolution.

had a similar impact and that compassion and empathy appeared quite early in our history.

To return to the development of a concept of death in humans and the five different components considered by psychologists (see above), the shaking of a dead individual's extremities by group members is reminiscent of the notion of irreversibility (i.e. that dead things cannot be made alive again). Did the chimpanzees test this specific aspect in Tina? More puzzling are the distinctive reactions of chimpanzees to death due either to illness or wounds, as this goes in the direction of the notion of causality of death that is acquired so late in humans (see above). Do these chimpanzees understand that death can have different origins and because of this, they express more fearful reactions towards those they do not understand (illness-induced deaths)? If this were the case, it would suggest that chimpanzees might have a rather developed notion of death.

The presence of compassion and empathy, as seen in some domains in chimpanzees, opens the way to having an understanding of death. Many observations described in this chapter raise intriguing questions about chimpanzees' understanding of death, injuries, and needs. Until now, only elephants, dolphins, gorillas, chimpanzees, and humans have been observed to care for and support wounded or badly handicapped group members. Further, only elephants, chimpanzees, and humans have been seen to treat dead individuals in a special way. More observations are certainly needed to clarify these different aspects. However, because death is luckily a rather rare event, only observers who have observed chimpanzees for several hundreds of hours will be able to detect such clear but subtle differences in their behavior when interacting with injured and dead individuals.

Contribution to the culture debate

Although impressed by the detailed observations presented here, our hardcore experimental psychologist will nevertheless warn us about giving too much weight to anecdotes. Impressive anecdotes exist for many animal species, but only a few have had conclusions drawn from the anecdotes confirmed with controlled experiments. Simply too many factors could explain such anecdotes. Furthermore, to him, understanding others in the special case of the notion of death leans too much on the anthropomorphic side and without experiments to support them, he cannot accept my interpretations. Finally, he would suggest that not all humans have the same notion of death and, therefore, trying to find a notion of death in chimpanzees similar to that of western societies might simply be beside the point.

However, most knowledge about humans rests on daily observations and much anecdotal evidence. If we were to exclude anecdotes, we would know a great deal less about humans as often the most interesting and informative behaviors are the rarest ones. No experiment has given us any direct information about how humans react to death, to life-threatening situations, and to deadly violence. Similarly, the reactions to death or major injuries in chimpanzees are too rare to be subject to experiments, which is an even more important reason to consider anecdotes on such topics very seriously. Luckily, even though they are rare, they still occur repeatedly; as a result, we could confirm that Brutus did react in an empathic way in many dramatic instances and the way he reacted corresponded to how other Taï chimpanzees have reacted in similar situations. This builds into a striking set of observations requiring an explanation.

Anecdotes were decisive in showing that elephants have a very special and intriguing relationship with elephant bones and this resulted from the dedicated observations by biologists who had spent years studying them. Similarly, striking special behaviors performed towards needy group members have been reported in whales trying to prevent others from becoming beached by pushing them away from the shore. Furthermore, dolphins have been reported to help injured individuals by supporting their bodies at the surface level, allowing the injured individual to breathe even if it is unable to swim. All of these observations in chimpanzees, gorillas, elephants, whales, and dolphins add to the picture that many animal species show behaviors compatible with a certain notion of death and a strong empathy towards individuals in need. Despite the fact that humans certainly show far more elaborate rituals around death than seen in any nonhuman animal species so far, the widespread presence of a notion of death and empathy towards others emphasizes that it must have originated quite early in the course of evolution.

Synopsis

- Death elicits reactions in Taï chimpanzees that strongly suggest that they realize dead individuals do not move or need help anymore, and that they will remain inanimate.
- Large injuries elicit tending, support, and cleaning behaviors in Taï chimpanzees and, to a lesser degree, in other populations, which suggests empathy.
- Elephants, dolphins, whales, and gorillas show some similarities to chimpanzees in their reactions to death and injuries.
- In chimpanzees, a notion of death seems to coincide with how much experience a group has with predation pressure leading to regular injuries and death.
- In humans, a notion of death takes many years to fully develop and seems to include several components, including irreversibility of death, that have now also been suggested in chimpanzees.

8 Wild culture – wild intelligence: cognition and culture

If one were to order all mankind to choose the best set of rules in the world, each group would, after due consideration, choose its own customs; each group regards its own as being the best by far.

Herodotus, *The Histories* (p. 185).

Our climb to the top has been a get-rich-quick story, and, like all nouveaux riches, we are very sensitive about our background.

Desmond Morris, *The Naked Ape* (1967).

Loango National Park, Gabon, 15 October 2009 (extract from a video sequence)

Ida, an adult female from the Rekambo community, arrives at an underground nest of Melipone sweat bees that, judging by its state, had initially been discovered by elephants and where we had also seen another chimpanzee trying in vain to get at the honey. Without any hesitation, Ida goes towards a sapling that is 5 m from the nest entrance, stands upright to seize the sapling, bends it towards her and breaks it 1 m above the ground. Then turning the branch upside down, she uses her teeth to break off a 120-cm long stick. She then goes towards the nest and inserts the thinner end of the tool into the ground. Holding the stick with her left foot and right hand while steadying herself with her left hand at the base of another sapling, she uses all her strength to insert the stick 50 cm into the soil, all while her 4-year-old son watches closely nearby. She pulls out the stick and smells the tip, then pushes it back deeper into the soil. Once the tip of the stick becomes too bent from this heavy use, Ida breaks it off with her hand. Two minutes later, she makes a second longer tool from a sapling 10 m away and uses this one to dig deeper. After working like this for more than 4 minutes, she reaches the top of the honey chamber about 70 cm under the surface. She starts licking the tip of the stick and removes the loose earth around the hole with her right hand. Alternating between removing earth with her hand and digging with her stick, Ida eventually pulls out her honey covered hand 3 minutes later. Her son immediately puts his hand into the hole as well and both feed on this delicacy for 15 minutes.

Figure 8.1 **Ida pushing the stick into the ground to access the honey chamber**. She will work for 5 minutes to reach the honey, which is in a chamber about 70 cm below the surface. Her son watches her activities very closely and is seen here bending forward to concentrate on the stick being pushed deep into the ground (picture extracted from a video sequence).

Hadzaland, Tanzania, 17 July 2011

I follow 38-year-old Moda and her 4-year-old son, Saïdi, from the Dedawako camp on a trip to find underground tubers. With them are two other mothers and their infants. When we reach a tree with one of the large vines that produces Ekwa tubers, Moda pounds the ground with the thick end of her digging stick around the vine; the sound this produces is used to detect the presence and location of the large tubers. Seemingly satisfied, she sits at this spot and starts flattening the ground by pounding it with a stone a few times. Then she begins digging a hole with the sharp end of her stick, but rapidly encounters a stone in this rocky soil. Using the stick as a lever, Moda dislodges it from the surrounding earth and brings it to the surface. She works for 5 minutes before bringing out the first tuber. By this point, the end of the digging stick is dull and Moda sharpens it with a large knife, after having quickly lit a fire with some dry grass and twigs lying nearby. She places the sharpened end as well as the tuber into the fire. She turns the stick around a few times in the fire until she considers it to be hardened enough and then rubs the tip against a rock to remove the blackened surface. She divides the grilled tuber in two, gives half to her son and both eat. She then resumes digging, removing four more stones from the hole, some of which are impressive in size.

Figure 8.2 **Moda digs for tubers with her stick while her son Saïdi watches**. She uses a strong digging stick that she has sharpened with her knife and she regularly uses fire to harden the tip before digging and removing the stones that are in the way of the tubers.

Moda and the others work for 45 minutes around this vine and she extracts over 3 kg of tubers, most of which she will carry back to camp to share with the others.

The Hadza are known to possess one of the least complex technologies of all human societies, including known present-day foragers.[1] Thus, Moda's technique to extract tubers reveals some fascinating similarities and differences in the use of sticks compared to what we see in Loango chimpanzees. In both cases, underground resources are sought after and can be quite arduous to extract, either because they are in very stone-rich soil or because they are quite deeply buried. In both cases, the main tool used is a stick that has been reduced in length and part or all of the bark has been removed to facilitate the work. Furthermore, direct visual indices – such as presence of the vine or bee nest entrance tube – are complemented with indirect cues – like the sound of pounding with the digging stick as a guide to the location of tubers or the traces of other's attempts – to locate invisible food resources.

If the digging stick wears down during use, both Moda and Ida modify it. However, while Moda uses a knife, fire, and a rock to sharpen the end, Ida only makes corrections by cutting the end away or making a new tool. Using a tool, such as a knife, a rock, or a fire, to make, modify, or improve another tool has not yet been seen in chimpanzees, but this ability has been observed in all human societies. However, chimpanzees do modify sticks directly with their teeth or hands (i.e. sharpening an end or cutting off a broken extremity). Furthermore, both Moda and Ida share the product of their efforts with their offspring, but Moda takes this

[1] Marlowe (2010, p. 71) citing Murdoch and Provost (1980).

one step further and carries some of the products she has extracted back to camp in a container to share with others. Containers, which have been seen in all human societies, are absent in chimpanzees. As chimpanzees do not have a base camp that they return to daily, there is certainly a difference in motivation that might play a role in explaining this difference. Furthermore, a digging stick was used to reach the goal in both cases, but while Moda used it to fulfill three different functions – to detect tubers by sound, dig holes, and lever out stones – Ida used the sticks only to dig into the ground.

These differences are reflected when one compares the general technological level reached by the Hadza foragers (Table 8.1) with that reached by the Taï and other chimpanzee groups (Table 8.2). Using the same classification method, the Hadza possess approximately 39 tools,[2] while the Taï chimpanzees possesses 13 and all known wild chimpanzee groups combined possess 21 tools. While these numbers are somewhat similar, some of the Hadza's tools are much more complex; for example, the arrows they use are made up of a combination of up to nine different material elements. Following this classification system, the Hadza possess 92.5 technounits in their tool repertoire, while the Taï and all chimpanzee groups together possess only 16 and 30, respectively. This produces an average of 2.4 technounits for the Hadza, with men using more complex tools – an average of 2.9 technounits – and women using generally less complex ones – an average of 1.3 technounits. Chimpanzees collectively use tools with an average of 1.42 technounits, a number that is undoubtedly increased as a result of the complex sequential tools used by Central African chimpanzees (in bold italics in Table 8.2). In a previous similar comparison, it was shown that the Tasmanian aborigines, who became extinct over 100 years ago and are famous for possessing the simplest technology of all known human foragers, possessed 13 different tools with a total of 17 technounits.[2]

The flexible use of tools seen in the example with Moda can also be seen in a higher level of multi-functionality in the tools used in human foragers, so that for the Hadza, a stone, for example, can fulfill up to eight different functions, while all chimpanzees have been thus far seen to use them for three different functions (see Table 8.3). Similarly, branches are of a more general use to human foragers than they are to chimpanzees. For example, the Hadza have been seen to have 10 different functions for them, while wild chimpanzees have been seen to use them for only 6 different functions.[3] To be fair, Bossou and Taï chimpanzees use stones for only

[2] In Oswalt's taxonomy of technology, a tool that is used to impinge on masses that are incapable of motion is called an "instrument," a tool used to kill or maim species capable of motion is called a "weapon," and a tool used to control the movement of prey or protect the tool user is called a "facility." Unmodified material is called a "naturefact," but after modification, it is called an "artifact". Finally, a "technounit" is a count of the number of different material elements that are combined to make a functioning tool, so that a hammer, which is comprised of a wooden handle and an iron head, is 2 technounits, and a wooden arrow, which is comprised of a wooden stick, feather end, ligament material, bard, and twine, is 5 technounits (Oswalt 1976, Marlowe 2010). The study comparing Tasmanian aborigines with Tanzanian chimpanzees was by McGrew (1987).

[3] Contrary to some claims (Hauser 2009, p. 194), this table shows that chimpanzees do not restrict the use of one object to only one function, even if they seem to be less flexible than humans.

Table 8.1 Material culture in Hadza foragers (adapted from Marlowe 2010). This classification follows Oswalt's taxonomy of technology (1976), which was developed to compare human forager groups and allows a qualitative categorization of the tool kit used to get food in any culture, including chimpanzees

Category	Form	Artifact/ naturefact	Use	Number of technounits
Hadza foragers				
Instrument				
Simple	Stone (hammer + anvil)	N	Pound Baobab/nuts	2
	Porcupine quill, thorn	N	Extract nut	1
	Thorn	A	Picking out nut, teeth	1
	Stick	A	Pull down bird nests	1
	Stick	A	Knock off baobab pods	1
	Stick	A	Pull down bird nests	1
	Stick	N	Dig tubers	1
	Twig	A	Extract honey	1
	Hammer of iron	A	Make arrow heads	2
	Knife metal	A	Cutting meat, bark	2
	Leather scabbard	A	Hold knife	2
	Quiver of skin	A	Carry arrows	5
	Axe	A	Chop tree for honey	2
	Stakes of wood	A	Climb tree for honey	1
	Chisel	A	Make arrow heads	1
	Stick whitted	A	Apply poison	1
	Baobab pod as cup, bowl	N/A	Carry honey, drink water	1
	Gourd with handle	A	Carry water, honey	2
	Gourd with handle and rope	A	Get water in tree	3
	Leather pouch	A	Carry objects	3
	Wood pole	A	Carry meat	1
	Basket	A	Carry berries, birds	3
	Ceramic	A	Cooking	1
	Needle	A	Sewing	1
	Skin, cloth	A	Carry, food, nursling	1
Weapon				
Simple	Iron head arrow (5 types)	A	Kill prey	7–9
	Wooden head arrow (3 types)	A	Kill prey	1–3
	Bow	A	Kill prey	5
	Stick	N	Club animal	1
	Stone	N	Throw to kill prey	1
Facility-tended				
Simple	Blind of grass, branches,	A	Conceal hunter	4
	Fire drill of wood	A	Make fire	2
	Torch of wood	A	Smoke and stun bees	2

Total: 39 tools comprising 92.5 technounits.

Table 8.2 Material culture in Taï chimpanzees with additional simple instruments as seen in Tanzanian (in italics) and Central African chimpanzees (in bold italics)

Category	Form	Artifact/ naturefact	Use	Number of technounits
Instrument				
Simple	Twig/shoot	A	Fish for ants, honey	1
	Leaf/grass	A	Investigative probe	1
	Stick	A	Dip for ants	1
	Leaves	A	Sponge for brain, water	1
	Stones (hammer + anvil)	N/A	Crack nuts	2
	Stones/twig (previous + twig)	N/A	Extract almonds	3
	Stick as collector	A	Extract honey	1
	Vine	A	*Fish for termites*	1
	Stem/stalk	A	*Investigation probe*	1
	Bark	A	*Fish for termites*	1
	Stick as pounder	A	***Break open bee nest***	1
	Stick as perforator	A	***Locate bee/ant nest***	1
	Sticks (Pounder/Perforator + enlarger + collector)	A	***Extract honey***	3–5
	Sticks (enlarger + collector)	A	***Reach and extract driver ants***	2
	Sticks (perforator + collector)	A	***Reach and extract termites***	2
Weapon				
Simple	Stone	N	Throw to drive prey	1
	Branch	N	Throw/stab predator/ prey	1
	Stick	A	Maim wood-boring bees	1
Facility-tended				
Simple	Stick	N	Lever open hive entrance	1
	Stick	A	Stir up ants	1
	Sapling	A	Elevated site for ant dipping	1

Total: 13 (*21*) tools comprising 16 (*30*) technounits.

Table 8.3 Multi-functionality of tools in Taï chimpanzees and Hadza foragers

		Chimpanzees	Hadza foragers
Stones			
	Hammer to access food	+	+
	Weapon	+	+
	Wedge	+	+
	Hammer to peel bark off sticks	–	+
	Cut roots	–	+
	Repair beehive	–	+
	Whetstone to sharpen tool	–	+
	Hammer to flatten ground	–	+
Branch			
	Digging stick	+	+
	Weapon	+	+
	Perforate for food location	+	+
	Lever	+	+
	Open bee nest	+	+
	Hammer	+	+
	Walking help	–	+
	Food detection by sound	–	+
	Stakes of wood	–	+
	Stick knock off fruits	–	+

two different functions and Taï and Loango chimpanzees each use branches for only three different functions.

These similarities and differences in material culture seen between the Hadza and chimpanzees illustrate the outcome of comparisons done between human foragers and chimpanzees in general. Parts of these differences are clearly influenced by ecology, as tubers are particularly abundant and rich in savanna regions like those where the Hadza live. Still, the ability to use tools to make and improve other types of tools, to use containers, to combine different material elements to produce tools, and to use fire are general abilities that are seen in most human societies but, as of now, have not been seen in chimpanzees. What could have led to such differences and how have cultural differences contributed to this?

My proposition is that cognition develops differently in different cultures in order to resolve the specific challenges that individuals face. The socio-ecological conditions faced during the lifetime of each individual will select for a culture-driven cognition that has been amply documented in humans living in either different cultures or under different socio-economic conditions.[4] As for humans, I argue that a

[4] Segall *et al.* (1999), Carpendale and Lewis (2004), Rogoff (2003), Greenfield (2009).

similar process is at work in animals that will lead to different cognitive abilities and performances as a function of the cultural conditions faced by the individuals.

Culture-driven cognition

As mentioned in Chapter 6, Sartre had to progress while learning to nut-crack if he wanted to catch up with Salomé's performance. Only a stone hammer will crack open the very hard *Panda* nuts, but stone hammers are rare in the Taï forest. How do chimpanzees find these hard stones? The most abundant stones in the forest, laterite blocks, are too fragile to crack *Panda* nuts but are the most frequently used stones to crack the softer *Coula* nuts. Much harder granite stones (see Figures 8.3 and 8.4) are perfect for cracking *Panda* nuts but they must weigh at least 2 kg – but preferably around 5 kg – to effectively crack these nuts, and stone hammers this size are very rare. If you tried searching for such stones at random, it might take days to find a good one. To further complicate things, *Panda* trees are also rare and quite dispersed throughout the Taï forest. Thus, in order to efficiently crack *Panda* nuts, the Taï chimpanzees need a precise mental map of where to find hammers and trees dispersed over a 10 km^2 large forested area or they might as well forsake cracking them altogether. By marking all granite stone hammers and *Panda* trees in a section of the forest, we were able to precisely follow the stone transports done by the chimpanzees and from this we reconstructed how they selected them before transport (Figure 8.3). Since visibility in this forest is, at most, 20–30 m on the ground, any transport done over longer distances require the chimpanzees to mentally plan where to transport the stone and compare distances between different stone hammers present in the region. Furthermore, our analysis showed that Salomé, Héra, and their consorts were extremely good at selecting the closest stone hammers to the *Panda* tree at which they intended to crack nuts, and they therefore must have a precise mental map that allowed them to compare the distances between a minimum of five out-of-sight stones and different *Panda* trees.[5] After 12 years of being in the forest almost every day, I had gained enough experience to know where to find the marked stones with regard to a given *Panda* tree at which it had last been used. However, I remained unable to compare the exact distances between stones and different *Panda* trees.

Orientation in flat, dense tropical rainforests is much more demanding for all animals than orientation in more open habitats, where landmarks – such as valleys, hills, cliffs, rivers, and especially large trees – can be seen over long distances. In addition, forest chimpanzees have very large territories, between 15 and 25 km^2,

[5] By precisely following all individually marked stone hammers within a 5 km^2 area with some 20 productive *Panda* trees, we could determine which criteria chimpanzees used to select their hammers (Boesch and Boesch 1984a, Boesch-Achermann and Boesch 1994). We found that the chimpanzees possessed a very precise memory in that they knew where the hammers were, and despite a distance of more than 180 m between a *Panda* tree and the stones around it, a distance that far exceeds the 30-m ground visibility in the Taï forest, they were able to select the closest stone within an optimal weight class to transport to the tree. Thus, the selection was done using some mental representations of the object properties and their localization in relation to the tree in which they wanted to crack nuts.

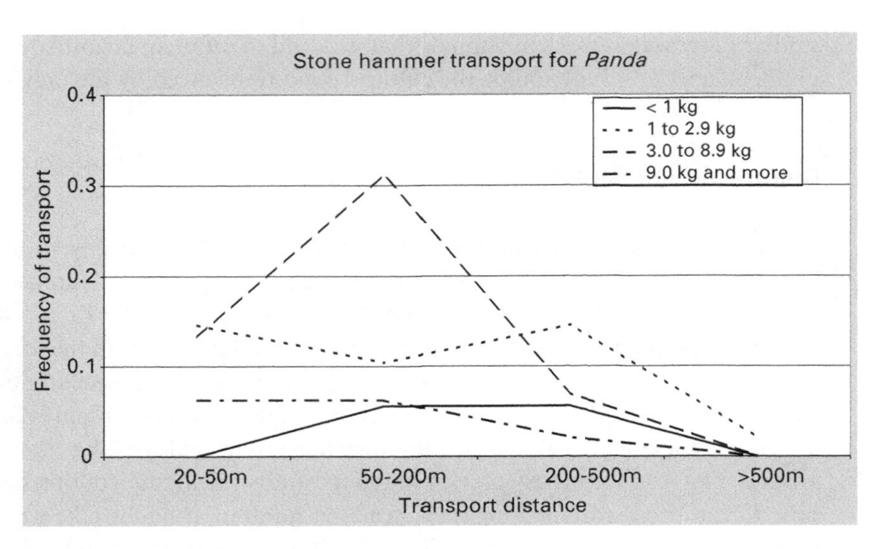

Figure 8.3 **Stone hammer transports in the Taï forest for cracking nuts.** As visibility can be as low as 20–30 m, chimpanzees need to have a precise memory of stone and tree locations to plan and execute these transports.

Figure 8.4 **Héra cracks *Panda* nuts with a heavy stone hammer.** Héra is using a rare large granite stone as a hammer. Hard, heavy stone hammers allow chimpanzees to gain 44% more energy from *Panda* nuts than if they used very hard wooden hammers (Boesch and Boesch 1983). Chimpanzees remember the locations of these stones and, in most cases, select the one nearest to the tree they want to crack nuts in. As the stones are always out of sight of the *Panda* tree, the chimpanzees show sophisticated mental mapping abilities to plan such transports.

Figure 8.5 **Loukoum and her son Lefkas eat *Coula* nuts.** Using a good hammer significantly increases nut-cracking success. Chimpanzees invest a great deal of effort in selecting good hammers, and as good hammers are rare in the Taï forest, they are often transported over long distances to the nut-cracking sites.

which include many thousands of fruit-bearing trees that produce several times a year.[6] To illustrate the challenge of orienting in the forest, if female chimpanzees did not possess a precise mental map of the trees and they walked in a straight line until they saw or smelled a fruiting tree (walking in a straight line has been shown to be the best way to randomly find trees), they would have to walk 2 km before they would come across one of the rare fruiting Parinari trees (these produce a sweet, fleshy fruit that the chimpanzees frequently eat in January). This is 10 times longer than the distance these chimpanzees actually walk when they are feeding on Parinari fruit! Furthermore, females would be able to revisit such popular trees without a mental map only every fifth day, while in actuality, they revisited them every second day on average. As we know that the fruiting season of a given tree

[6] Emmanuelle Normand mapped all the trees of the 17 species whose fruit are eaten by the chimpanzees in a 15 km^2 area to precisely follow the spatial abilities of the chimpanzees. She found 12 499 trees and for each month, fruit could be found in between 2000 and 3000 of them. Knowing that chimpanzees eat fruit from about 50 different species of large trees, this means there are about 30 000 trees from which the chimpanzees need to find the good fruiting ones (Normand and Boesch 2009, Normand *et al.* 2009). With the use of detailed mathematical simulations, she could show that the observations of the chimpanzees' daily forays in the forest coincided precisely with the use of a mental map that allowed them to find trees in straight lines and to frequently revisit especially large and rare fruiting trees. This demanding study, the first of its kind on a wild chimpanzee population, allowed us to detect some of the spatial skills used by chimpanzees on a daily basis, which would have been impossible to do in captive conditions. Together with Simone Dagui Ban, Emmanuelle succeeded in following three different females each without interruption for 28 consecutive days, and thanks to such unique observations, they were able to determine how often the females revisited the same trees to feed on them (Normand *et al.* 2009).

species lasts an average of one month, it is essential that they regularly update their mental map to the changing fruiting patterns in the forest.

Nut-cracking is a good example of a cultural behavior that selects for specific cognitive skills. *Panda* trees, like many other preferred fruit tree species in the Taï forest, are rare and finding them is a real challenge for the chimpanzees. The number of trees of different sizes and the diversity of different species of trees are such that chimpanzees more than double their efficiency at finding food with the help of a mental map and long-term memory of those trees.[7] Stone hammers are essential for optimal *Panda* cracking and are therefore regularly transported to the cracking site by different adults from the community, thereby forcing each of them to constantly update their knowledge of the locations of the stones in relation to the fruiting *Panda* trees. Thus, nut-cracking represents a special cognitive challenge requiring some sophisticated spatial and technical skills to be competently performed.

To understand how important culture is on the development of intelligence, I am first going to discuss the effect of culture on human cognition before comparing similar aspects in chimpanzees. I will then return once again to the issue of captivity that has led to much confusion and misunderstanding in the study of cognition and culture.

Cognition and culture in humans

Understanding human cognition requires cultural sensitivity. Comparing the performance of individuals from very different cultures can rapidly become meaningless as objects that make obvious sense in one culture might be totally unfamiliar in another and individuals might not know what to do with them. A cup, a glass, or a pencil might make perfect sense to westerners but are foreign to some people from Liberia or Papua New Guinea in certain contexts. To take one example, conservation of liquid is a classical Piagetian test to study the development of intelligence. Typically, to test for conservation of quantity, a certain amount of liquid is presented in a glass and then transferred in front of the child into another glass with a different shape that thereby makes the liquid look either higher or lower in the glass. The child is then asked if the quantity of liquid is the same as it was in the first glass. Schooled children in Europe and America correctly evaluate these transformations only after 8 years of age. In an attempt to compare this developmental sequence with other cultures, the same test was performed in different regions of the world and it was found that only 40% of 8-year-old unschooled Okaspmin children in Papua New Guinea gave the correct answers when tested with a western Piagetian task such as this.[8] Intrigued by this result, some cultural psychologists

[7] Normand *et al.* (in press) showed that tree size differences and tree species diversity are important factors to make long-term memory beneficial, whereas sheer number does not make such a difference. So in that case, it seems that the complexity of the rainforest represents a real challenge to fruit eaters who have a large territory.

[8] Tests for the notion of conservation of liquid were originally proposed by Piaget (1945) and then widely used within the field of psychology. The studies in Liberia and Papua New Guinea with more

invented a test that included traditional materials familiar to the people. As a result of this small change, 8-year-old unschooled Okaspmin children correctly answered 97% of questions on such a test about the conservation of quantity. This illustrates the need for experimenters to have some knowledge of their study subjects to ascertain that they properly understand the questions asked during a test.

Cultural differences in cognition have been documented in many different domains,[9] with one review concluding, "Overall, these empirical patterns suggest that we need to be less cavalier in addressing questions of human nature on the basis of data drawn from – the westerners – particularly thin, and rather unusual, slice of humanity." I will restrict myself here to domains that have a direct connection to the material, social, and symbolic cultures discussed in this book.

Notion of space and culture

Following Salomé and Sartre transporting hammers to crack nuts gave us the feeling that spatial knowledge could be directly influenced by material culture. Do we observe something similar in humans? Spatial knowledge in humans is very important for orientation between different important locations. However, spatial challenges for city-dwelling Europeans are very different than for Inuits living in the great icy North or Aka pygmies living deep in the rainforests. European children acquire a representation of space by going through three different stages. In the first stage, called the *topographic space* stage, children are only able to connect landmarks as they move in this space from one landmark to another. The second stage of space allows children to include a general idea of directions and distances to the landmarks, but they still need to use landmarks and cannot make new shortcuts between known routes. Only when the last stage is reached does a *Euclidean space* develop and at this point, children can make shortcuts as landmarks are now interconnected and new directions can be found to new places. Topographic space develops in European children at 3 years of age while Euclidean space only develops at 8–9 years of age.[10]

Cross-cultural studies of the development of spatial knowledge have been conducted in different human groups using a Piagetian approach. Generally, this has shown that environmental and cultural factors strongly affect the rate of acquisition of the different Piagetian stages, as well as the stages that are finally attained by adults. Because of this, it was proposed that the sequence of acquisition of those

appropriate traditional material were performed by Cole *et al.* (1971 in Segall *et al.* 1999) and Saxe and Moylan (1982), respectively. Many more studies have shown this problem of using the correct methodology to properly understand the level of performance in a group of subjects before drawing any conclusions about possible differences.

[9] Henrich *et al.* (2010b) recently presented an up-to-date review of the differences in cognition found between different human groups. See also Segall *et al.* (1999), Greenfield *et al.* (2003b).

[10] Piaget and Inhelder (1947) precisely studied the development of the notion of space in European children, and his first subjects were actually his own children. The cultural comparisons are detailed in Segall *et al.* (1999). For more on the Baoulé, the Inuit, and the Aborigines, see Dasen (1982) and Dasen *et al.* (1978), for the Mayan, see Brown and Levinson (2000), and for the Balinese, see Wassman and Dasen (1998).

Table 8.4 Cross-cultural comparisons of spatial and social cognition in different human societies. For the false-belief test, results are also presented from human groups living under very different economic and social conditions within western societies (the last third part of the table)

Cultural group	Performance	Reference
Spatial knowledge		
WMC[1]	Topographic + Euclidean	Piaget and Inhelder 1947
Inuit (Greenland)	Topographic + Euclidean	Dasen 1982
Aborigines (Australia)	Topographic + Euclidian	Dasen 1982
Ebrié (Côte d'Ivoire)	Topographic only	Dasen 1982
WMC	Relative before absolute	Piaget and Inhelder 1947
Vranisi (India)	Relative and absolute	Mishra *et al.* 2003
Tzelta (Mexico)	Absolute before relative	Brown and Levinson 2000
Tzotzil (Mexico)	Absolute before relative	de León 1994
Bali (Indonesia)	Absolute before relative	Wassman and Dasen 1998
Roopchandpur (India)	Absolute before relative	Mishra *et al.* 2003
Bhimeshwor (Nepal)	Absolute before relative	Mishra *et al.* 2003
Hai‖om	Absolute before relative	Haun *et al.* 2006
Guugu Yimithirr (Australia)	Absolute only	Majid *et al.* 2004
False-belief test[2]		
WMC	Present by 4 years	Wellman *et al.* 2001
Baka (Cameroon)	Present by 4 years	Avis and Harris 1991
City school (Peru)	Present by 5 years	Callaghan *et al.* 2005
City school (India)	Present by 5 years	Callaghan *et al.* 2005
City school (Thailand)	Present by 5 years	Callaghan *et al.* 2005
Tolai (Papua NG)	Present by 7 years	Vinden 1999
Yukatek (Mexico)	Present by 7 earsy	Knight *et al.* 2004
Mofu (Cameroon)	Present by 9 years	Vinden 1999
Village school (Samoa)	Absent at 5 years	Callaghan *et al.* 2005
Junin Quechua (Peru)	Absent at 5 years	Vinden 1996
Tainae (Papua NG)	Absent at 14 years	Vinden 1999
USA-Low income	37% later	Homes *et al.* 1996
London-Low income	20% later	Cutting and Dunn 1999
WMC-more sibling present	Positive effect	Perner *et al.* 1994
WMC-older siblings present	2.4 times earlier	Ruffman *et al.* 1998
WMC-affiliative maternal style	Positive effect	Pears and Moses 2003
WMC-relation quality	Positive effect	Howe *et al.* 2002

[1] WMC = White middle-class westerner sample.

[2] Present = The ability to pass the false-belief test is considered "present" when the proportion of individuals succeeding is significantly above chance level ($p < 0.05$).

levels is universal in humans, but that the timing of its acquisition was strongly culturally influenced. For example, hunter–gatherer populations that move widely in large open spaces with limited possibilities to aid in orientation, like the Inuit of Greenland and the Aborigines of Australia, possess very elaborate Euclidean mental maps using coordinates (see Table 8.4). In contrast, traditional farmers who mainly move between villages and fields, like the Baoulés of Côte d'Ivoire, have more restricted mental maps that rely mainly on topological landmarks and do not develop Euclidean maps. Thus, the socio-economic characteristics of a group's lifestyle seem to affect the development of that group's spatial abilities.

Sometimes the effect of culture on humans can be even more dramatic and lead to inversion of the acquisition of spatial abilities. Mayan people living in southern Mexico have a distinct linguistic repertoire for talking about spatial relations that is based on an "absolute" notion of space in which, for example, objects are placed "up or down hill" rather than to the left or right of the narrator. As a result, Mayan children develop an absolute notion of space before a relative one, which is the opposite of what we observe in the western world (see Table 8.4). Balinese spatial orientation is even more absolute than that observed in the Mayan in the sense that Balinese space is directed toward the highest mountain on the island, Mount Agung, which stands 1800 m high in the center of the island. However, absolute directions point towards different cardinal directions in different villages, as these are located on the seashore of this circular island and are situated around the mountain. In addition, villages created by migration keep track of the direction of the mountain from their original village. Such an extreme absolute notion of space forces men from different villages to agree about what a direction is before playing a game together based on orientation notions. Here, as in the Mayan case, it seems that the development of spatial knowledge is inverted compared to the western one.

Spatial orientation is a highly cultural ability. The ethnocentric approach, based only on western beliefs with their impression of equal achievement in all individuals, totally neglects the fact that, in reality, spatial representation in humans is experience-based and will develop as required by the cultural traditions of the specific lifestyle.

Understanding of others and culture

The ability to understand others as independent agents with their own knowledge and beliefs (i.e. theory of mind) has become a contentious topic of research because it intersects with the quest to understand differences between humans and chimpanzees. Although it was originally described in captive chimpanzees, many have proposed this ability to be a hallmark of humanity.[11] For comparison purposes,

[11] The original description of theory of mind in captive chimpanzees was from Premack and Woodruff (1978). Claims of this being a uniquely human attribute can be found in Astington and Baird (2004), Carruthers and Smith (1996), Gopnik and Meltzoff (1994), Hala and Chandler (1996), Leslie *et al.* (2004), Meltzoff (1995), Povinelli and Eddy (1996), Povinelli and Vonk (2003), Tomasello (1999), Tomasello *et al.* (1993, 2005), Wellman *et al.* (2001), Wimmer *et al.* (1988), and other publications.

I focus on only one test of theory of mind, the false-belief test, as it is by far the most frequently used. The false-belief test requires the subject to predict where a second person will think an object is hidden when the location of that object has been changed while only the subject was present. Typically, the majority of White middle-class (WMC) children pass this test by 4 years of age (Table 8.4).

Similar developmental paths have been confirmed in other, non-western human populations in which children attend city schools. In other populations, the time when children are able to successfully understand the false-belief test can be markedly delayed to the point of not being observed in older children (Table 8.4). More studies are needed to know whether those noted as "absent" would remain so after older individuals were tested.

The socio-economic conditions encountered during childhood also affect performance on the false-belief test. Studies of people with low income in the USA and London revealed that, despite an improvement in performance with age on false-belief tasks, they generally fell short of the level of performance reported from middle-class individuals; 37% passed the test by age 4 versus 61% in WMC samples, and 57% of the low-income 5 year olds were successful compared with 89% in the WMC samples (Table 8.4). To stress the importance of the social environment, family composition revealed itself to have a major influence on false-belief test performance as the number of siblings a child has positively affects the emergence of understanding of others. Furthermore, the higher the number of older siblings, the earlier an understanding of others emerges, so that 4-year-old children with two older siblings are nearly two and a half times more accurate than those with only younger or no siblings (Table 8.4). In addition, the quality of interactions within the family was also shown to accelerate the emergence of this ability, with mental stimulations by mothers and friendly relationships between family members having positive effects.

A certain understanding of the pupil by the teacher is also implied in teaching, which, as we have seen in Chapter 6, supports important elements of cultural learning. The prevalence of active teaching in our society has been attributed to schools. Only limited cross-cultural studies have been done on this important ability. As mentioned in Chapter 6, teaching in foraging societies seems to hold much less importance and most of the learning is directly done by the child and most of the adult interventions are of a scaffolding type. Two specific studies have tried to quantify that aspect. The first showed that Aka pygmy children[12] spend about 72 minutes per day observing a model, but directly imitate them for only 3 minutes per day. Furthermore, they received instruction from group members for 14 minutes per day and gave instruction for 5 minutes per day. "Systematic observational studies suggest that older children and other adults modify their behavior to help younger children learn, but that this teaching is relatively rare by comparison to observation and imitation." The second study where aspects of teaching have been

[12] Hewlett *et al.* (2011, citation from p. 1176) have studied this in the Aka pygmies of Central African Republic. For the !Kung bushmen teaching study, see Draper (1976).

documented concerns the !Kung bushmen and showed that girls and boys between 4 and 14 years of age received 1–2 adult interventions per hour in the sense of an adult "shaping a child's behavior." Such frequencies of adult interventions overlap with what we have seen between mother and infant in the context of nut-cracking in chimpanzees and suggest that, depending on the cultures under study, similar teaching interventions can be found in the two species. The forms of such interactions were developed further in Chapter 6.

Cooperation and culture

The occurrence of teamwork in many human societies has reinforced the way western scientists perceive the importance of altruism in human societies, to the point that some of them have suggested that human altruism and cooperation, with its extension to non-related individuals, is unique in the animal world.[13] Most of the evidence used to support such a proposition comes from a very limited sample of university students from western countries. Luckily, a comparison of 15 different human societies from 4 continents, including 3 hunter–gatherer groups, 6 slash-and-burn farming groups, 4 nomadic herding groups, and 2 small-scale agricultural groups revealed how strong the impact of culture is on the human tendency to share with others.[14] When given a certain amount of money with the possibility to share it with a stranger, students from different western universities systematically shared an average of 45% of it, and, when given the possibility to refuse an offer they would consider inappropriate, they rejected offers that were below 30% of the allocated amount of money.

However, the offers varied between the cultures and low offers were often accepted in some, while very generous offers were often rejected in others. The socio-economic characteristics of such cultures explain the results quite well. In the Lamalera whale hunters of Indonesia who cooperate extensively to catch large whales and divide them in a very precise way, about two-thirds of the participants offered half of the money they received. Similarly, the Ache of Paraguay generously offered almost half of the money they received in 80% of the cases and no offers were rejected. This parallels the Ache society, where meat is distributed equally among households, irrespective of who made the kill. In contrast, the Hazda of Tanzania made low offers of about 33% of the money and often rejected them. This

[13] Fehr and Fishbacher (2003), Boyd (2006), Tomasello (1999), Herrmann *et al.* (2007), Silk *et al.* (2005), Hrdy (2009).

[14] This large cooperative enterprise included economists and anthropologists in an attempt to not only follow well-established protocols that have been used with university students in the west, but also to include observers that have a real knowledge with the human societies included in the experiment and adapt the protocol to be well-received in each of those very diverse populations (Henrich *et al.* 2001, 2005, 2006, Gintis *et al.* 2003, Marlowe *et al.* 2008). It should be noted that because the money was provided by the experimenters, the study does not put the individuals in a real situation where they have to share their "own" money, and we have to assume that they all have been more generous than they would have been in a real situation. Thus, the interest of such studies is not in the level of generosity shown by the tested individuals, but mainly in the level of differences between human societies.

reflects the Hadza small-scale foraging society, in which they share meat but with a lot of conflict and attempts by hunters to avoid sharing by hiding their catch from others. Among the Tsimane of the Bolivian Amazon and the Machiguenga of the Peruvian Amazon, the offers were also very low and virtually no rejections were made, mirroring a society where little cooperation and exchange beyond the family unit is observed.

Originally, the so-called "universal tendency" of humans to share was explained by the fact that sharing was a kind of social standard and cheaters would be punished by group members. To prove this, a third partner was added to the previous experiment as a kind of referee who was free to punish whenever offers were, in his view, too low. However, here again, very large differences were found between human social groups and such third-party punishments were almost absent in some human societies, like the Hazda or the Tsimane, but quite regularly observed in large societies with more anonymous groups. A recent study compared human sharing tendencies in 16 different human populations, all including undergraduate students and, therefore, all very similar with regards to age, education, and socio-economic situation.[15] Nevertheless, they found that culture decisively influences the tendency to share in humans and even more so influences the tendency to punish others for not sharing. The main lesson from these comparisons is that humans are social animals and, therefore, we bring along our cultural norms and traditions. Thus, here again culture plays an essential role in understanding cooperation and altruism.

This review of the important cognitive variations observed in humans from different cultures illustrates one of the major challenges for understanding what humans are and about how we should perform cross-species comparisons. As two experts said concerning human knowledge about the environment, "We are all born with native minds, though some develop in a manner better attuned to their natural surroundings than others. The full expression of the folk-biology module requires environmental triggering conditions and cultural support that may be lacking for certain groups in industrialized societies, including the usual participants in most cognitive and developmental psychology experiments. From a theoretical perspective, the chief interest in studying these groups may not be to establish a baseline for generalizations about folk-biological knowledge but to explore the cognitive consequences of limited input."[16]

Cultural cognition in the laboratory versus in the real world

Before we discuss the development of intelligence in different chimpanzee cultures, we need to revisit the topic of captive studies. The reason for this is simply that almost all currently available cognitive studies are experimental studies done with captive animals. In this sense, the question I want to address in this chapter – the

[15] See Gächter *et al.* (2010).
[16] Citation from Medin and Atran (2004, p. 980).

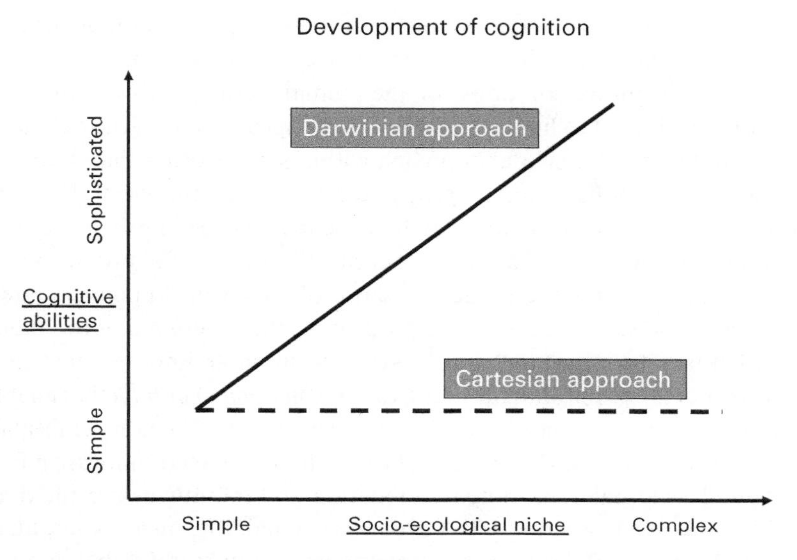

Figure 8.6 **Two different approaches to the development of intelligence** (after Segall *et al.* 1999, Boesch 2007). The Darwinian view is a biological one and considers the environmental and social conditions faced by the individual during its lifetime as decisive in influencing the development of intelligence. In particular, it predicts that the more complex the challenges of the socio-ecological niche encountered, the more sophisticated cognitive abilities will be favored. In contrast, the Cartesian view considers the effects of environmental or social conditions to minimally affect the development of cognitive abilities. Cognition is seen more as an innate ability that will simply mature with time. Therefore, to Cartesians, captive studies are applicable to the whole of the species, independent of the level of deprivation faced by the individuals.

effect of culture on cognition – has *not* yet been studied. The legacy from René Descartes' equation of animals with machines, whereby development is only marginally influenced by the environment and social conditions, remains blinding to many (Figure 8.6). This Cartesian approach essentially justifies drawing conclusions at the species level from all cognitive studies done with captive animals without considering the very special social and environmental conditions that these individuals experience. In contrast, a Darwinian approach considers that the environment as well as the social conditions an individual experiences during his lifetime will dramatically affect the abilities that he develops (Figure 8.6). As such, influences will not be uniform, some environmental aspects might have more effects on spatial abilities, for example, while social life will have more effects on aspects of social intelligence. In other words, we should not expect that the outside world's effects on the development of an individual's cognitive abilities will be systematically equal.

Opinions about the value of captive studies have diverged wildly. While experimental psychologists see them as the base of scientific progress and knowledge, comparative psychologists are more critical. Nathan Emery and Nicola Clayton, who studied the cognition of birds for years, said that comparative social cognition has failed for three reasons. First, there is too little appreciation of the ecology and ethology (which would allow us to recognize examples of natural behaviors that may require cognitive solutions) of the animal studied. Second, the comparative dimension is limited to humans or a few other species. Third, the field has adopted a very limited developmental approach within species other than humans.[17] Frans de Waal, the famous primatologist, noted that, "certain social phenomena cannot be transferred to the laboratory because it would be impossible to recreate the social and nonsocial environments that are responsible for producing and maintaining ongoing encounters, such as those observed in dominance interactions." Bekoff and colleagues stated, "one problem here is that gaining too much control over what *may* be important variables may result in an impoverished environment that does not allow the animals to use the *combination of stimuli* that enable them to make assessments of others' behavior or others' minds." The very skillful fieldworkers Dorothy Cheney and Robert Seyfarth, who conducted innovative field experiments with wild baboons, noted, "Predation and infanticide are the dark side of the Okavango Delta. Along with the open savannah, the lush islands, huge fruiting trees, and the annual floods, they set the stage for our study of baboon society and the metaphysics that underlies it."[18] These baboons have to struggle with predators such as lions, leopards, and crocodiles to survive and reproduce, and this makes them different from all baboons living in zoos. Having followed them in such a rich but dangerous world, they concluded that, "Laboratory experiments can illuminate a species' abilities only if their results can be placed within the context of an animal's natural social behavior. In the absence of such grounding, they remain difficult, if not impossible, to interpret."

Following a Darwinian approach, I suggest that the cognitive abilities of captive individuals would fall in the spectrum of cognitive abilities seen in a given species, and due to the peculiarities and deficiencies of the captive socio-ecological conditions, they would be outliers for their species in most cognitive domains and most likely at the lower end of the variation observed (Figure 8.6). This argument may partly apply to human city-dwellers as well; due to their scarce contact with nature, for example, they have been shown to possess very limited folk biology about plants and animals compared to groups with more outdoor experience.[19]

Box 8.1: What can laboratory experiments tell us about natural cognition?

In theory, experiments on social or cultural aspects are made to improve our understanding of the real world. By manipulating one key factor while controlling for the many others that may affect the specific aspect under real-life conditions, they aim to qualify the role of this key factor. Because of this theoretical advantage, some scientific disciplines, like psychology and economy, have long

[17] Emery and Clayton (2008), de Waal (2001, p. 311), Bekoff *et al.* (1994).
[18] Cheney and Seyfarth (2007, p. 49).
[19] Medin and Atran (2004).

favored the use of experiments. The central ambition of such experiments is to have broad *external validity*, so that results based on experiments can be generalized to the real world. For the purposes of this book, this issue is very thorny as no proposed cultural behaviors for chimpanzees have yet been studied in laboratory experiments.

Furthermore, experiments have struggled with the issue of *internal validity*, whereby it has proven difficult to ensure that the designed procedure effectively tests the causal chain underlining a cognitive ability, rather than just assuming it does (e.g. Guala 2005, 2012, Levitt and List 2008). And while using adult human subjects to test whether the assumptions are correct has been shown to be the best method to control for such a problem, it is surprisingly rarely used (see Table 8.5).

- Issues related to *External Validity*:

Since the whole point of making experiments is to simplify the real world so that all factors except one can be kept constant or removed, the external or ecological validity of such experiments requires extra evidence and this has proven difficult to obtain without additional experiments in the "real world" (Harrison and List 2004, Baumard 2010, Guala 2012). Furthermore, most "field experiments" are actually just artificial experiments done in the field, rather than natural field experiments that keep as much of the environment unaffected as possible by the experimenter.

In a thorough review of experimental results on reciprocity in humans, the conclusion reached was as follows: "[t]he moral to be drawn is that models and experiments can only take you so far, and the time has come for reciprocity theory to change gear and seek the test of historical and field data. This step was taken a long time ago in the investigation of related topics such as mutual insurance and collusion, and it is important to keep in mind that laboratory data – no matter how useful – cannot ultimately replace the evidence collected in the field" (Guala 2012).

- Issues related to *Internal Validity*:

Psychologists and economists have long been aware of the internal validity issue and have been using control experiments to get some measure of the bias that might affect the validity of their experiments. However, it has proven frustratingly difficult to exclude the internal validity problem, especially with non-speaking subjects, as any negative result could be a consequence of such a problem. Three major aspects could decrease the internal validity of experiments.

- *Subject understanding of the experiment*: any given subject has had a previous life, full of experiences, stimulations, and development. Such a background, which is often unknown to the experimenter, directly affects the way the subject understands and reacts to the experimental procedures.
- *Assumptions of the experimenters*: an experiment is designed by an experimenter who also has his own background and assumptions, which can be quite

different from those of the subjects, especially from those of other species, and this can lead to major misunderstandings.

- *Appropriate prior knowledge of the subjects*: there are now many studies, including with humans, that show that prior experience with a solution, an object, and a context is decisive in determining if an individual will be able to solve a test (Davenport *et al.* 1973, Frith 2007, Fragaszy *et al.* 2009). A recent study showed very convincingly that, in humans, training had a significant effect on the development of intelligence (Jaeggi *et al.* 2008).

Table 8.5 summarizes the internal and external validity of some captive experimental tests that have claimed cognitive limitations in chimpanzees and been amply cited. In all cases, serious problems with both internal and external validity have been demonstrated, and when some of these issues are resolved, the chimpanzees have performed much better on exactly the same tests.

References about the internal validity of experiments: Leavens *et al.* (2005), Frith (2007), Bulloch *et al.* (2008), Jaeggi *et al.* (2008), Levitt and List (2008), Maki *et al.* (1993), Thomas *et al.* (2008), Bania *et al.* (2009), Lyn *et al.* (2010), Guala (2012), Horner *et al.* (2011), Sternberg (2008).

Can captive experimental studies still help us to understand the effect of culture on cognition? As Box 8.1 and Table 8.5 illustrate, how generalizable captive experiments are depends a great deal upon how internally and externally valid the experiments are. An illuminating illustration comes from the different reactions of humans to the altruistic sharing experiment mentioned above. Economists, surprised to see how generous westerners are, presented the test to western students with different procedures and saw that they become much less generous if, for example, they are given a larger amount of money, given an option at the start to "not share," the identity of the owner remains anonymous, and so on.[20] Furthermore, if the recipient is not aware that a game is taking place, the owner will exit the game without giving anything in almost half of the cases. As a result of this, economists have started to question the value of experiments because the laboratory conditions systematically differ from most real-world conditions, and, therefore, the generalization of laboratory results to the real world needs to be done with utmost care.

Although they can be very damaging to the value of an experiment, problems with an experiment's internal validity can be difficult to uncover (Table 8.5). For

[20] Economists have become aware of the limits of laboratory experiments compared to the real world (e.g. Levitt and List 2007, 2008). The Ultimatum Game, where the owner of the money has to make an offer and the receiver can accept it or not, has been explored by many economists due to the discrepancies between laboratory and real-world results, and they showed that many more aspects of the experimental procedures have strong effects on how owners will share in the laboratory (see Dana *et al.* 2006, Bradsley 2008, List 2006). Some more factors that affect the amount shared by the owners to recipients are if the recipients are allowed to take rather than wait for an amount to be proposed, and those owners identified as CEOs give more than student owners (Levitt and List 2007). The list goes on, but this reminds us of how many factors in reality affect the tendency of westerners to share.

Table 8.5 Comparisons of cognitive performance in chimpanzees under different experimental paradigms where internal and external validity has been evaluated (see text for more explanations about some of the tests listed here). Note that many studies listed in the left reference list have been used to substantiate claims of "human uniqueness"

Test type	Cognitive ability tested	Results	References	Validity		Problems with validity	Results with more validity	References
				Internal	External			
Tool-making	Causality	Failed	Povinelli and O'Neill 2000; Povinelli 2003; Hanus and Call 2008	Yes	None[b]	Absence of prior experience	Success	Furlong *et al.* 2008; Bania *et al.* 2009
Trap-tube test	Causality	Failed	Limongelli *et al.* 1995; Visalberghi and Limongelli 1994; Reaux and Povinelli 2000; Martin-Ordas and Call 2009; Horner and Whiten 2007	Limited	None[a]	Many factors considered	not done	Silva *et al.* 2005; Silva and Silva 2006
Sharing test	Altruism	Failed	Silk *et al.* 2005; Jensen *et al.* 2007; Vonk *et al.* 2008	None	Limited	Complex apparatus/ Time delay	Success	Horner *et al.* 2011; Smith and Silberberg 2010
Artificial fruit test	Imitation	Failed	Whiten *et al.* 1996; Tennie *et al.* 2006	Limited	Yes	Many factors considered	not done	Horowitz 2003
Visual attention	Understanding of others	Failed	Povinelli and Eddy 1996; Povinelli *et al.* 1997; Reaux and Povinelli 2000	Yes	None[b]	Absence of prior social experience	Success	Bulloch *et al.* 2008; Thomas *et al.* 2008
Pointing test	Understanding of others	Failed	Barth *et al.* 2005; Tomasello *et al.* 1998; Call *et al.* 1998	Yes	None[b]	Absence of prior social experience	Success	Lyn *et al.* 2010
Number pointing	Working memory	Chimpanzee > Human	Inoue and Matsuzawa 2007	None	None[a]	Absence of prior experience	Chimpanzee = Human	Silberberg and Kearns 2009

Internal validity: **Yes**: No data have shown that the focus of the test was ambiguous or not natural. **Limited**: when adult human subjects considered many factors, while the experimenter designed the test considering only one factor. **None**: when test apparatus was too complex to be understood by the subject.

External validity: **None[a]**: when no equivalent of the test is found in real life, **None[b]**: when the subjects had no prior knowledge/experience due to impoverished living conditions.

example, the "trap-tube test," which was widely used for over 15 years with humans and many animal species to test for causal understanding, requires that the subjects understand which direction to push a peanut with a stick out of a transparent tube that has a trap in the middle that would prevent the subject from eating the peanut if it fell inside. Presented like this, it was fascinating to discover that captive chimpanzees and capuchin monkeys would avoid the trap, even if the trap was oriented upwards and therefore the reward could not fall into it. This suggested that the test was able to ingeniously detect limitations in the causal understanding of animal species and the test rapidly became very popular. The problems started when some variations in the way the test was presented produced incompatible results within the same species. To understand such discrepancies, psychologists Francisco and Kathleen Silva recently sought to determine if the experimenter's assumptions that the trap-tube test measured understanding of causality were correct and gave the test to adult humans. Much to their surprise, the humans, like the chimpanzees, unnecessarily avoided the trap! After questioning the subjects, it became obvious that the test was not only testing for causal reasoning, as the position of the reward in relation to the tube extremities was equally important to the subjects.[21] The main lesson taught was that "what an experimenter determines to be the causal features of the task may be very different from what the subject perceives to be the causal features of the task." This difference in perspective can easily remain unnoticed with non-speaking animal species. At a minimum, the internal validity of the experiment should be controlled beforehand with humans, but, sadly, even this is rarely the case.

Finally, experiment subjects may be cued about how to answer without the experimenters even being aware of it. This was marvelously well-illustrated by the famous Clever Hans. Clever Hans was a horse in the early 1900s that possessed the astounding ability to correctly add simple numbers by tapping the answers with his hoof. He toured Europe and entertained large audiences by correctly answering numerous mathematical questions asked by total strangers. His popularity and skill made him so famous that, in 1904, an eminent committee of behavioral scientists issued a report indicating that they were unable to detect any fraud or subtle communication between Clever Hans and his human interrogators. However, later that year, two members of that committee, Oskar Pfungst and Wilhelm Stumpf, issued a follow-up report describing their discovery that Clever Hans was indeed picking

[21] The T-shaped trap-tube test was initially developed by Visalberghi and Trinca (1989) for captive capuchin monkeys, and has since been used in numerous studies with many different species (e.g. Table 8.5, Visalberghi and Limongelli 1994, Mulcahy and Call 2006, Tebbich *et al.* 2007, Teschke and Tebbich 2011). The uncertainty about what the test truly measured allowed for many modifications and replications by different research groups to understand why chimpanzees performed so poorly. The critical test came when adult human subjects were given this task and the experimenters found that, contrary to their expectations, humans were influenced by four different aspects of the experiments and not only by the position of the trap as assumed before (Silva *et al.* 2005, Silva and Silva 2006). In other words, what psychologists had taken as proof of an absence of causal understanding in chimpanzees and monkeys now looked to be a rational consideration of more than one factor (the citation in the text comes from Silva *et al.* 2005, p. 56).

up on extremely subtle and apparently entirely unintentional cues provided by the people asking the questions and that he would fail when, for example, he could not see the person asking the question. Clever Hans was essentially a clever human eye reader and therefore stopped tapping whenever the human unconsciously signaled with his eye that he had reached the correct answer.[22] Clever Hans' tricks in solving math problems became known as the "Clever Hans effect" and psychologists started to invest more effort in trying to prevent this from happening in their experiments. Sadly, this has recently become more and more forgotten so that too many recent studies with children as subjects have allowed for huge "Clever Hans effects" (see Figure 2.7). In one video sequence of a comparison of captive chimpanzees with free-living children, the tested child faces a human experimenter who is trying, and failing, to pile books on top of each another. The father, on whose lap the child was sitting, is seen unconsciously pushing his son with his shoulder and by doing so pushes him towards the fallen book. As the son then reacts, the test was classified by the scientist as being correctly solved![23]

All this casts doubt on how valuable captive experimental studies truly are in helping us to specifically understand the cognitive ability of wild animals and even more so for our questions about the relationship between cognition and culture. Nevertheless, captive experimental studies remain of central interest in three specific domains: first, to control for possible genetic influences on behaviors, as has been brilliantly demonstrated when young captive New Caledonian crows used tools without any previous exposure to a tool-using model.[24] Often, animals have been more regularly observed in captive conditions than in the wild and, as a result, many more observations exist about them in such conditions and these can reveal a lot about the potentialities of animal species in general. Second, captive studies can help us to understand how generally some cognitive abilities are distributed among different animals. For example, tool-using skills have been tested in non-tool-using animal species – something that could not be done in the wild.[25] Third, for very well-studied species like humans, laboratory experiments can complement real-world analyses to help us understand the role different factors play in certain

[22] Besides Clever Hans, Basso, a chimpanzee in the Frankfurt Zoo, could also solve arithmetic problems and when asked "How much is 6 plus 4?" he would select a card bearing the number "10." Testing of Basso showed that he was simply very good at selecting the card that his trainer was looking at (Gardner 2005). As a result of these examples, psychologists now impose a double-blind procedure on their experiments, by which the person asking the question to the subject does not know the answer and the researcher who knows the answer preferably does not know the question. For unknown reasons, this procedure has recently been progressively abandoned, as children are again being tested in the same room or even in direct contact with their parents, allowing for Clever Hans effects (e.g. Tomasello *et al.* 2005, Warneken and Tomasello 2006, Herrmann *et al.* 2007).

[23] For the video sequence, see www.eva.mpg.de/german/filme.htm (despite such important methodological errors, this paper was published in *Science*; Warneken and Tomasello 2006). In an exchange published in the prestigious journal *Science* about the role of the Clever Hans effect in studies with children, the problem was simply totally ignored by the experimenters (comments by de Waal *et al.* 2008 followed by a response from Herrmann *et al.* 2008). The reluctance of psychologists to address this issue is unfortunate since it casts a general doubt on the value of such experiments.

[24] Kenward *et al.* (2005).

[25] Santos *et al.* (2003).

behaviors. Now, economists have expanded their knowledge of human altruism with laboratory studies where real-world aspects can more easily be manipulated.

In conclusion, to be relevant to our discussion of culture, experiments should be performed within one single culture that both the experimenters and the subjects belong to or are at least extremely familiar with. In this way, we can be sure that subjects will have the same cultural background as the experimenters. Many important studies on human cultural processes have been done in this way, with scientists testing subjects within their culture and not by extracting them from it. Our understanding of cultural processes in humans has progressed a great deal with this innovative approach. Comparative psychologists could profit from realizing the potential of such an approach.

Cultural cognition in nature

To find a good stone hammer, Salomé demonstrates remarkable mental mapping abilities. A detailed mental map proved to be very important in low-visibility habitats. Does this mean that Salomé and her consorts have more sophisticated spatial knowledge than the Gombe chimpanzees? In Gombe, for example, the open nature of the forest and the very rugged terrain with its steep ridges provides plenty of viewpoints where an individual can see out over the entire valley. Chimpanzees often stop there and may search for other chimpanzees feeding in a tree and for trees that are full of fruit. I was once with Frodo and his team when they sat for a long time on such a ridge looking intently in one direction. I searched in that direction and eventually saw a group of red colobus monkeys crossing a strip of savanna and running on the ground many hundreds of meters away from us. This is precious information for orientation. At the other extreme, captive chimpanzees have basically no such orientation issues as they receive food everyday at the same place and their cages or enclosures are so small that everything can be monitored in one glance. Thus, we should expect spatial abilities to be quite different between these three chimpanzee populations, as shown in Figure 8.7.

Figure 8.7 illustrates how some comparisons between species could easily "prove" that humans possess more sophisticated spatial skills than chimpanzees or exactly the other way around. For example, if we compare free-living humans in the forests of Ituri or in modern city schools with chimpanzees confined in captive conditions (as highlighted by arrow A in Figure 8.7), humans are likely to always fare better than chimpanzees. However, if we were to compare the spatial skills of Taï chimpanzees with imprisoned humans living in orphanages, chimpanzees will most likely show more elaborate spatial skills than humans. However, neither comparison is fair. Sadly, we have to acknowledge that the vast majority of comparisons done between human and chimpanzee cognition have followed arrow A as shown in Figure 8.7.

Studies with ravens and scrub-jays show us one solution. In an environmentally sensible approach with high external validity, the scientists first went "back to the

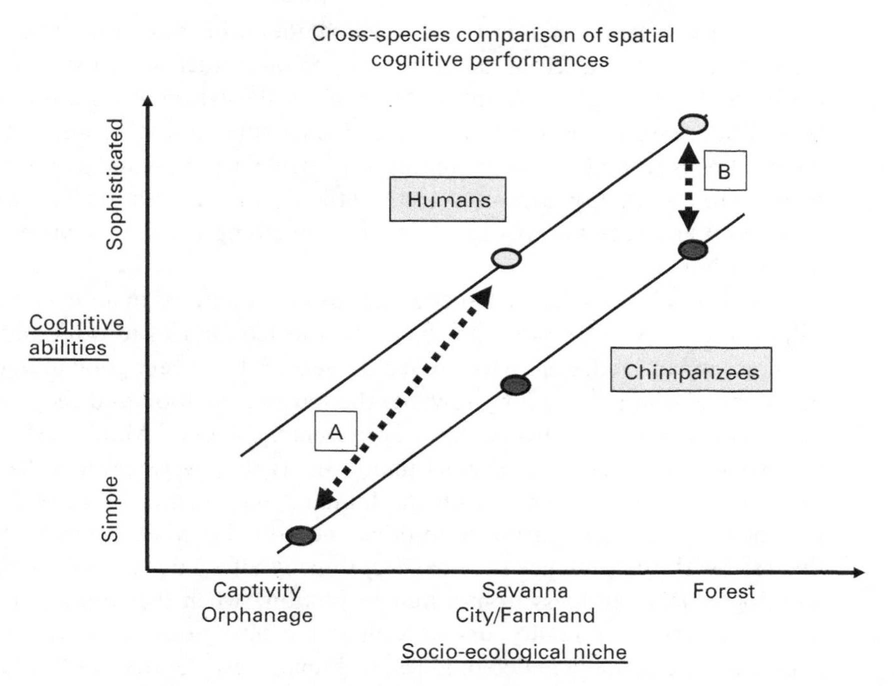

Figure 8.7 **Schematic representation of spatial cognitive skills in humans and chimpanzees.** The socio-ecological conditions prevailing in an environment are going to specifically affect the spatial orientation skills of the individuals growing up in such habitats. Cultural behavior that needs to be solved in habitats with high tree densities and low visibility, such as rainforests, are going to be more demanding than those within habitat with low tree densities, like savannas, or in environments with no spatial challenges, like captive settings and human orphanages. The human line is placed above the chimpanzee line for clarity and not because we currently have enough data to know how to place them.

wild," and there they discovered that to survive the very long and harsh winters, ravens and scrub-jays cache a large amount of food in multiple hiding places for them to eat later in the winter when food will no longer be available. How can they remember the locations of their many caches and how do they stop other birds from stealing their food? To answer such important questions, laboratory experiments were combined with observations in the wild to study the cognition of caching behavior. In addition, because the aim was to study a natural behavior, the captive conditions provided to the birds were fashioned so as to allow this behavior to express itself spontaneously. Such an approach allowed them to uncover some astounding cognitive knowledge that had been thought to be restricted to humans;[26] bird social cognition in the context of caching behavior is impressively developed in the sense that an individual will cache his food so that it will not be

[26] For the ravens, see Bugnyar and Henrich (2005, 2006), Bugnyar and Kotrschal (2002), Bugnyar *et al.* (2007), Schloegl *et al.* (2008); for the scrub-jays, see Emery and Clayton (2001, 2008, 2009), Dally *et al.* (2006), Correla *et al.* (2007), Raby *et al.* (2007).

seen by others. If a social partner is present, they will travel a long way to cache their food out of the visual sight of the potential thief, which shows that they realize that individuals learn information about the external world through sight. In addition, ravens and scrub-jays were able to differentiate between others with different backgrounds – known thieves or trustful partners – and adapt their cache strategy to the quality attributed to the other. Thus, these birds have knowledge about and attribute knowledge to others, something previously proposed to be uniquely human.

Dogs have also taught us that cognition is better studied in an environmentally valid context. "Man's best friend" has been selected for ages to work and live with humans; as expected, dogs have proved themselves to be very good at understanding humans who are pointing towards the location of food and they were much better at this task than wolves and captive chimpanzees.[27] Man's best friend was compared under better conditions than apes as they were never tested in cages but always in direct contact with the human experimenter. A careful comparison showed that both rearing conditions and how the experiments were done (in this case with the presence of a barrier) directly affected the result. Both wolves and dogs understand less from a human pointing when they are separated from him by a barrier. In reality, under comparable conditions, dogs were not better and seemed to actually understand less of human gestures than wolves. This highlights an important point I made in Figure 8.7: we need to compare only what is comparable.

Culture and cognition in chimpanzees

The best test of the role of upbringing conditions on cognition comes from studies with chimpanzees facing different captive conditions. Because humans dictate what chimpanzees are allowed to experience in captivity, researchers are free to manipulate the socio-ecological conditions experienced by chimpanzees as they feel appropriate. Whatever one may think about the legitimacy of such an approach, it represents a unique test of the impact of the upbringing conditions on the development of cognition. Before presenting the results, let me first present some of the different living conditions that the chimpanzees involved were facing. Three captive

[27] Studies on dogs have been numerous and have taught us how important a daily habit of interacting with humans is for some social cognitive abilities (Huber *et al.* 2009, Range *et al.* 2007, Hare *et al.* 2002, Topál *et al.* 2009, Kubinyi *et al.* 2007). An especially careful comparison taking into account upbringing conditions and testing procedure showed how much this affects the understanding of human pointing (Udell *et al.* 2008). Interestingly, because of their long-term interactions with humans, dogs were never kept in different rooms than the humans during experiments, unlike in all chimpanzee experiments. In addition, the fact that all of the dog subjects were pets meant that they were totally familiar with the human objects that were used during the test. Finally, it is interesting to note that many human experimenters doing dog studies are actually dog owners that have shared their lives with them for years and therefore developed a special empathy with them, which helped them to produce "dog-friendly" experiments.

chimpanzee groups were tested: the first group was at the University of New Iberia, the second was at Ohio State University, and the third was at the Wolfgang Koehler Primate Center in Leipzig Zoo.

For the first captive group, the New Iberia chimpanzees, Daniel Povinelli and his group decided to put together seven two- to three-year-old chimpanzees and kept this group in isolation for their entire lives in a small, stable, and restricted man-made environment.[28] Five of these chimpanzees were nursery-reared in a group of their peers. The other two chimpanzees were mother-reared for less than a year before joining the same social group at around two years of age. The chimpanzees had free access to five indoor and five outdoor enclosures, which contained various enrichment items, and were separated briefly for testing two times a day, five days a week. From his described methods and procedures, these chimpanzees did not have ready access to materials (e.g. sticks, foliage) or manipulanda during their early development from birth to three years. In addition, Povinelli's subjects were given only 15 minutes of exposure to a novel tool or material immediately prior to initiation of the task.

The second group, the Ohio State University Chimpanzee Center chimpanzees, included 9 chimpanzees and was comprised of 5 adults ranging from 17 to 40 years of age, 2 juveniles, and 2 infants. The animals lived in a stable mixed-age social group in an enriched environment with 24-hour access to outdoor play enclosures equipped with foraging sites and large-scale climbing structures designed to encourage species-specific behavioral activities.[29] The chimpanzees came from a variety of backgrounds, including home-reared subjects, nursery-reared subjects, and one wild-born individual. They had all spent most of their lives living under conditions of continuous human enculturation. Their daily diet included a wide variety of fresh fruits and vegetables fed throughout the day, supplemented with high-protein biscuits included in the evening meal. All subjects joined the project at a very young age (e.g. three years of age or younger). Thus, during a major portion of their physical and cognitive development, all subjects were immersed in a socially and object-rich captive setting.

Finally, the Leipzig Zoo chimpanzees live in one of the best housing facilities for captive great apes anywhere in the world. The spacious research center they live in has very nice indoor and outdoor facilities. However, all came from a medical research institution in Rijswijk, Holland, and they were so traumatized when they arrived in Leipzig that they dared not cross into the outside area provided to them at first. Rather, they all assembled on the iron mesh floor near the door, too scared to touch the grass of this luxurious free space. Some of the oldest individuals still would not walk in the free grassy outdoor area 18 months later! It is unclear what sort of a trauma they experienced before being relocated to Leipzig. Each of the 18 individuals were born in captivity and reared by humans except for the eight

[28] See Povinelli (2000) for more details and justification for this special social setting.

[29] See Bulloch *et al.* (2008), Furlong *et al.* (2008) and Bania *et al.* (2009) for more details of the living conditions.

youngest who were mother-reared.[30] The individuals in this group vary in age from infants and juveniles to adults over 30 years of age and all are subject to a very intense experimental program on a daily basis. So the Leipzig chimpanzees had arguably very rough starts in life, but since arriving in Leipzig they have enjoyed very good captive living conditions.

Tool use and culture

All known wild chimpanzee populations are more or less proficient tool users, not only systematically using tools to access food sources but using tool sets that include different types of tools of different sizes to access nuts, ants, termites, honey, grubs, larvae, and so on (see Table 3.1). Furthermore, wild chimpanzees of all populations have regularly been seen to modify raw materials *before* they use them as tools as well as during use (see Chapter 3). Due to the ubiquity of tool use in wild chimpanzees, one would expect captive chimpanzees to possess some of these abilities as well.

However, upbringing conditions seem to affect such skills negatively. First, physical properties of tools are complex and not directly perceivable. Sartre needs to consider many different functional properties of an object to ensure it is a functional tool. A wooden hammer needs to have a minimal weight, be of a certain size, and be hard enough to be suitable. In a detailed study of hammer choice in the Taï forest, we could show that chimpanzees preferentially select hard and heavy hammers, as they allow for the smallest energetic expenditure when cracking nuts. However, the farther the hammer has to be transported, the more the chimpanzees chose lighter but still hard hammers.[31] Thus, wild chimpanzees reason flexibly by taking into account hammer parameters related to the species of nut, *Coula* or *Panda*, to be pounded, including both visible ones, such as distance, and invisible ones, such as hardness and weight. Similarly, to dip for ants or honey, a stick needs to be an appropriate length and diameter. Again, wild chimpanzees select or make sticks of standardized sizes adapted to the specific purpose to be fulfilled.

Surprisingly enough, this is something that seems very hard for certain captive chimpanzees to understand. In 27 different tests, the New Iberia chimpanzees seemed unable to consider the fact that when choosing between different sticks, it made more sense to select an intact one for pulling a piece of food closer, as well as considering the fact that the stick has to be in contact with the food. On the other hand, the Leipzig chimpanzees proved to be more reliable at considering the fact that the stick should be in contact with the food, but still did not seem to consider

[30] See, for example, Tempelmann *et al.* (2011) for more details on the living conditions.

[31] Without transport, Taï chimpanzees chose the harder, heavier hammer first and the harder, lighter hammer second. If it was necessary to transport the hammer 8 or more meters to crack nuts, the harder, lighter hammer was selected first. This shows a flexible reasoning that takes into account three factors – weight, hardness, and transport distance – before selecting a hammer (Boesch and Boesch 1984a). Taï chimpanzees made 2–5 modifications on sticks before using them to produce tools for four different purposes – to extract honey, ants, nuts, and bone marrow (Boesch and Boesch 1990).

the need for the stick to be unbroken. The richer living conditions of the Ohio chimpanzees from birth on seemed to affect their performance as they could consider both of these aspects when selecting a stick.[32] They similarly properly modified tools for different purposes, which the New Iberia chimpanzees could not do as well, and they finally selected properly functional rakes to attain food, which, again, the New Iberia chimpanzees could not. The difference between the New Iberia and Ohio chimpanzees was especially illuminating as the tests were exactly the same and the evidence strongly suggests that the different upbringing conditions play a role in the development of tool-using and -making abilities.[33]

Such a deficit was also found when it came to estimating weight. To select a good hammer, Taï chimpanzees considered weight, hardness, and transport distance and combined them to select the best available stone hammer. However, the chimpanzees from the New Iberia group were unable to understand anything about the unseen properties of material, while the chimpanzees at the Leipzig Zoo were able to consider weight for choosing objects as long as it could be inferred from seeing the effect an object had on making a balance move.[34]

These recent captive studies confirmed what had already been shown for 30 years when comparing the tool-use skills of wild-caught chimpanzees and those reared under extreme deprivation and in social isolation – the latter had great difficulties in using a rake and relating it to the food. Similarly, when comparing a mother-reared chimpanzee group with a nursery-reared one, the former showed improved performance at using tools over trials, while the latter did not.[35] So all careful studies directly comparing tool-use abilities in chimpanzees with different upbringing conditions show that the "early learning environment" plays a decisive role in how chimpanzees will later perceive and think about tools (see Chapter 6 for a proposed similar effect in wild chimpanzees).

There are no equivalent formal tests for wild chimpanzees, but it seems that life would be very difficult for a wild animal if he would not be able to react appropriately

[32] This study on the Ohio University chimpanzees (Bania *et al.* 2009) presents one of the rare and thorough discussions about the key issue of early experience in chimpanzees and how this has long-lasting influences on the cognitive achievement of the individuals. They do not hesitate to explicitly discuss why the results they found disagreed with what has been found from the chimpanzees in New Iberia. The understanding of tools has been studied in the Leipzig Zoo chimpanzees (Herrmann *et al.* 2008), and they found that even limited experience with the objects increases the performance of the individuals, but they do not see the larger picture – namely, how much the chimpanzees they studied had been previously deprived and therefore how unrepresentative they are of chimpanzees as a species.

[33] A similar effect was found when a group compared Mayan children with American children on the understanding of cultural tools (Maynard and Greenfield 2003). They found that different aspects of the weaving were acquired much earlier in the Mayan versus the American children and some training did not eliminate the cultural differences even if both groups improved.

[34] The Louisiana chimpanzees have been extensively studied by Povinelli (2000, Penn and Povinelli 2007) and after having gone through hundreds of possible experiences, they have been declared unable to consider invisible physical properties of objects. In contrast, the chimpanzees from the Leipzig Zoo, who live in a multi-aged group, have been shown to be able to consider weight as long as it can be inferred from a movement due to an object's weight (a balance was seen to move more quickly if a heavier object was placed on it) (Hanus and Call 2008).

[35] Menzel *et al.* (1970), Brent *et al.* (1995).

Figure 8.8 **Taï chimpanzees gain a notion of physics about hammers very early in life**. For babies, hammers are already part of their life and they rapidly realize, as they touch them (upper left) or see their mothers using them (upper right), that each of them has a different shape, color, size, and sound as it hits the nuts. Later, youngsters will manipulate them and begin to distinguish one from the other by their weight and hardness (lower pictures). Thus, they will progressively realize how each of these aspects makes cracking nuts easier. In other words, the notion of physics about hammers develops thanks to the "nut-cracking cultural environment" they experience daily.

in unexpected dangerous situations by jumping into vines in high trees to escape attackers – and they have to be attached to the substrate – or when trying to access ripe fruits high in trees – they need to pull the branch connected to the fruit and no others. Thus, they seem to know that not only objects need to be connected to serve as support, but that they need to be steady and hard enough to hold a chimpanzee's weight. The cost of misjudging physical properties in real life is too high to allow for the type of errors that the New Iberia chimpanzees made.

Understanding of others and culture

Cultural effects can also be compared for cognitive abilities related to the understanding of others, whereby "others" are understood as independent agents with their own knowledge, intentions, and goals. One classical test has been to see whether individuals are able to understand that others see things and, through seeing, acquire knowledge. To test for this, captive chimpanzees sit facing humans, who they are separated from by a transparent barrier, and must beg or sign in order to obtain food. Sadly, the New Iberia chimpanzees, with their relatively impoverished social living conditions, performed in such a way that they seemed unable to consider the experimenter's eyes but instead only considered if the face was visible.[36] As a consequence, they would beg from a human with a bucket on his head or with a blindfold as long as he was facing them. Such a shallow understanding of others would be quite inefficient when encountering a leopard or when attacked by other chimpanzees in the wild! A faithful replication of the same experiment with the Ohio chimpanzees showed that they perfectly understood, from the first trial on, that "eyes see" and, therefore, they would only sign to humans whose eyes were open and unobstructed. Thus, as for tools, a deprived early social life without partners of different ages and natural objects to manipulate has been shown to lead to limited prior experience and prevent the development of a notion of others in the New Iberia chimpanzee group.

More focused studies that have looked at the early development of captive chimpanzees from a variety of social situations have revealed how differences like those we saw in the New Iberia–Ohio chimpanzees can emerge (see Table 8.6): the richer the social experience of newborn chimpanzees (and this was already visible with 4 hours of interactions with humans per day), the stronger an infant's ability to regulate their emotional stress when faced with a disturbance. At a later stage, early differences in maternal style, with regard to physical contact and physical stimulations, influences when a notion of self emerges in chimpanzees, as tested through mirror recognition.

The presence of different levels of coordinated group hunting (Chapter 4), of symbolic communication (Chapter 5), of different types of teaching interactions seen between mother and infants (Chapter 6), and of some of the intriguing indications of empathy (Chapter 7) all point to chimpanzees having some understanding of others. These observations suggest that once a formal test is devised to address this in the wild, we should find some evidence for this ability.[37]

Social learning in captive chimpanzees has been very rarely studied as psychologists have concentrated their efforts on the question of knowing if "captive chimpanzees can copy humans," which they then compare with the abilities of "humans to copy humans."[38] Putting aside the fact that this question has no relevance to a chimpanzee's social and cultural skills, it has turned out to incorporate a critical

[36] For the New Iberia chimpanzee results, see Povinelli (2000), for the Ohio chimpanzee results, see Bulloch *et al.* (2008).

[37] Roman Wittig, Catherine Crockford, and Klaus Zuberbuehler used playback experiments to test for the understanding of knowledge in others and they seem to have found the first evidence for this in the wild (in press).

[38] A great deal of effort, money, and energy has been invested in the marginally interesting question of whether "captive chimpanzees imitate humans" (Tomasello *et al.* 1987, 1993, Nagell *et al.* 1993, Whiten *et al.* 1996). For example, chimpanzees and children showed some propensity to imitate the way a human demonstrator opened a box, but chimpanzees did so to a lesser extent. Indifferent to the real question tested, some used this to suggest that chimpanzees were unable to imitate as they could not understand the intention of the human demonstrator (Tomasello 1999, Call *et al.* 2005, Tomasello *et al.* 2005, Herrmann *et al.* 2007). Interestingly, no comparative study has been done on the question of "humans' ability to imitate chimpanzees" to put the previous ones into perspective.

Table 8.6 Effects of different socio-cultural upbringing experiences on cognitive development in chimpanzees (only studies directly testing such experience are included)

Cognitive abilities	Upbringing conditions	Effect	Causal factor	Reference
Mutual gaze	Parenting styles	Increase	Cradling time Exercising time	Bard *et al.* 2005
Bazelton test	Vary interactive human contact	Large	Cradling time Movements	Bard *et al.* 2011
Declarative communication	Captive ↔ human-trained	Positive	Socio-linguistic prior experience	Lyn *et al.* 2010
Mirror recognition	Different mother contact	Earlier emergence	Mutual gaze Frontal contact	van Ijzendoorn *et al.* 2009
Reaching around barrier	Wild-born ↔ hand-reared	Better	More ecological prior experience	Vlamings *et al.* 2010
Tool use	New Iberia ↔ Ohio	Improved	More material prior experience	Furlong *et al.* 2008
Tool-making	New Iberia ↔ Ohio	Improved	More material prior experience	Bania *et al.* 2009
Gaze-following	New Iberia ↔ Ohio	Improved	More social prior experience	Bulloch *et al.* 2008

problem with internal validity (see Table 8.5). In a marvelous series of experiments, psychologist Alexandra Horowitz studied the ability of adult humans with at least one year of higher education to imitate human demonstrators in exactly the same test that was used for captive chimpanzees and young children.[39] In this precise comparative test, adult humans clearly performed worse than two- to four-year-old children and about as well as chimpanzees. How did these results come about? Post-experiment interviews revealed that the adult subjects did not realize they had to imitate the details of the demonstrator and assumed that they had to imitate the broader action of opening the box. Interestingly, they also assumed that they had to close the box after getting the chocolate inside, even though the demonstrator did not close the box. Thus, it was unclear to the adult subjects what part of the demonstration they had to copy and only interviews allowed the researchers to determine what led to the misunderstanding.

A recent series of experiments tested captive chimpanzees' ability to imitate other chimpanzees.[40] By testing individuals within their social group with group members

[39] Horowitz (2003).

[40] Andrew Whiten and collaborators have pioneered this approach by keeping the tested subjects in their social group and systematically training a chimpanzee to act as model (Whiten *et al.* 2007, 2009, Horner *et al.* 2006, Hopper *et al.* 2007).

demonstrating new feeding techniques, captive chimpanzees were seen to faithfully copy a group-specific technique, even if an alternative method was possible. Furthermore, these new traditions were seen to transfer to neighboring groups. This method was successful because it considered the social dimension of social learning, which clearly shows that internal validity must be taken seriously if we want to study cultural processes in animals. Therefore, when asked if "chimpanzees imitate chimpanzees," the answer from captivity models is yes for some tasks.[41]

The social dimension of imitation was reinforced by the observation that imitation disappears if the model is a machine performing exactly the same thing as the chimpanzee model was. This clearly shows that naïve chimpanzees gather information not only about perceptive aspects – like movements or objects – but also about the model's intentions.[42] Interestingly, in a study using a different type of task with the Leipzig chimpanzees, the individuals were less successful at imitating the behavior of a chimpanzee demonstrator. This result mirrors their known social deficits mentioned above.

Moreover, individuals copied aspects of the tasks as a function of previous experience and knowledge; if they can understand the challenge a task represents, they will tend to solve it in a manner they are used to or that they know will work, while if they cannot understand the technical challenges – for example, with an opaque box instead of a transparent box – they are much more likely to imitate the successful behavior they see being performed.[43] Thus, imitation is a dynamic social process and used by chimpanzees only when needed, when understanding of the task is partial, and when performed by a trustworthy model. This is reminiscent of what has been shown in humans, where children "need to balance prior knowledge about causal relations, the new evidence that is presented to them by the adult, and knowledge of the adult's intentions. Moreover, there is often no single 'right answer' to the question of what to imitate."[44]

Wild culture impact on cognition

As we previously saw for humans, we have now seen that upbringing conditions can have an important effect on cognitive development in captive chimpanzees. To see if the same effect occurs in the wild, I will now turn to cognitive abilities in the natural

[41] By considering the social dimension of imitation and therefore asking about imitative skills of a chimpanzee model, it was possible to convincingly show that imitation with a high level of conformity was observed (Whiten *et al.* 2005, Hopper *et al.* 2008). The consideration for the social component of imitation allowed them to make a cultural transmission experiment whereby, after training one individual to do a task in one possible specific way and releasing her into the group, the whole group will learn it and faithfully copy the model. If instead of a chimpanzee model you place a machine doing exactly the same movement, chimpanzees do not copy anything, which again emphasizes the social dimension of imitation.

[42] For the experiment with a machine, see Hopper (2010). For the second study on imitating a conspecific, see Tennie *et al.* (2006).

[43] For the study about imitation with a transparent or opaque box, see Horner and Whiten (2005).

[44] Buchsbaum *et al.* (2011, p. 332) reviewed the multiple sources of information approach adopted by children to imitate daily tasks.

context, i.e. with higher ecological validity, and thereby complete our comparison of the effect of culture. We should remember, however, that this higher ecological validity makes direct comparisons with captive settings more difficult.

Altruism and sharing food

All wild chimpanzees regularly share food with close kin (e.g. mothers share large amounts of food with their infant and juvenile offspring), as well as with non-related adult group members (e.g. meat, a highly prized food, is shared extensively in all populations studied) (see Chapters 4 and 6). Chimpanzees also share the important social service of grooming. Furthermore, this kind of sharing within the group is not done randomly: Taï chimpanzees share grooming interactions with high levels of symmetry within dyads over periods of many months. In other words, some individuals may exchange grooming only rarely, while other partners groom each other very frequently. Furthermore, social support in aggressive encounters is traded for meat and meat is traded for sex.[45] So in Taï chimpanzees, a complex network of sharing and exchange of food and social behavior is observed in a social group, which allows individuals to access resources that they might not be able to obtain alone.

A similar network of exchange has been seen in the Ngogo chimpanzees, where grooming is traded for meat and support. In other populations, meat-sharing seems to result more from the harassing pressure from beggars and it does not seem to be traded for sex or support. The different socio-ecological conditions prevailing in each group with different levels of competition between individuals (animal density was quite different), and the different availability of resources (quantities of meat shared differs largely between the different populations) seem to lead to different dynamics of exchanges. The observations from the wild are thus dramatically different from what has been seen with captive individuals. Therefore, it should come as no surprise that food-sharing and providing help in the wild is different from what is found in captive chimpanzees. For most of the studies, sharing in captivity was supposed to take place through barriers and with the help of complex apparatus. A recent study done in one captive group using a simpler experimental approach showed that captive chimpanzees can spontaneously share with others (see Table 8.5).

Social learning, teaching, and culture

Learning to nut-crack is very complex and takes many years, with the youngsters copying their mothers and the mothers teaching their youngsters many aspects of the technique (see Chapter 6). Clearly the complexity of the technique plays a decisive role, as learning the simpler task of termite fishing (acquired by youngsters 2–3 years earlier than nut-cracking) is not accompanied by teaching interactions

[45] See Gomes and Boesch (2009, 2011). Because chimpanzees live in fission–fusion societies where some individuals do not see other group members for weeks, such reciprocity studies need to include different longer time intervals.

Figure 8.9 **Sartre eating *Sacoglottis* while looking at reflections in the water**. These fruit are very abundant and a major food source during the rainy season in the Taï forest. Chimpanzees carry these fruit to water pools that accumulate in the forest during these months and wash the dirt off them before eating them. At the same time, chimpanzees are regularly confronted with their reflections as well as the reflections of the trees in the water. Each time Sartre dips the fruit into the water, he sees that the reflection trembles until the water quiets down. What is it that Sartre is looking at? Is he seeing how the reflection chews on the fruit just like he does?

from the mother. Imitation or simply copying aspects of the mother's technique is readily done by young chimpanzees in natural situations, as has been shown for termite fishing in Gombe, for nut-cracking in Taï and Bossou, and for different foraging techniques in Mahale.[46] Like young children, young chimpanzees in the wild use all possible sources of information so that they can learn the different complex tasks used in their cultures.

The complexity of the natural tasks has not been paralleled in any of the captive studies, which have all used much simpler tasks. Furthermore, intraspecific teaching has never been studied, but instead the question studied was, "what do captive chimpanzees understand about humans?"[47] However, this question has little

[46] See our discussion in Chapter 6, as well as Inoue-Nakamura and Matsuzawa (1997), Corp and Byrne (2002).

[47] Until recently, all studies about theory of mind, mind reading, or the understanding of others in chimpanzees have been done with humans performing the role of the "social partner" (Premack and Woodruff 1978, Povinelli 1996, 1999, 2000, Povinelli *et al.* 1990, Povinelli and Eddy 1996, Tomasello *et al.* 1993, Call and Tomasello 2008, Kaminski *et al.* 2004, Brauer *et al.* 2007, Hare *et al.* 2006). A lot of time, money, and energy have been invested in exploring chimpanzee's understanding of humans

relevance to our comprehension of cultural abilities in chimpanzees. Only recently have two captive studies addressed the question of "chimpanzees' understanding of other chimpanzees." As expected, the captive chimpanzees did better when asked about a "chimpanzee other" than when asked about a "human other";[48] captive chimpanzees knew what another chimpanzee knows through seeing something and used this attribution of knowledge to the other to their own benefit. In support of the importance of appropriate early environmental experience, the same experiment repeated with the New Iberia chimpanzees failed to show any attribution of knowledge to others.

Contribution to the culture debate

Our hardcore experimental psychologist undoubtedly reacted negatively to my suggestion that each population of captive chimpanzees should be considered a population on its own and should not be taken as representative of the whole species. If he agrees that the ecological and social conditions of captivity are very different from the natural conditions, he totally disagrees that it has such an important impact on the development of cognition in chimpanzees. He believes chimpanzees to be much more resilient than humans, as, contrary to humans, young chimpanzees have been shown to have much less interactive, communicative, and teaching exchanges with their mothers. Furthermore, according to him, the culture contribution to daily life being dramatically less important in chimpanzees than in humans, he thinks that we cannot reasonably expect culture to have a similar effect on chimpanzee lives and certainly not on chimpanzee cognition. In a sense, he agrees that studies specifically addressing the effects of different lifestyles would be important, but as long as such studies have not supported such a strong effect, we should consider captive experimental work on cognition to be the main contributor to our knowledge of intelligence in animals.[49]

and what in the human face, eyes, attitude, and posture is most important for this ability. If cross-species attribution is a fascinating topic, it is not clear how relevant it is for studying within-species attributions of others.

[48] Despite the fact that this study was the first to ask what "chimpanzees know about other chimpanzees' knowledge," in a twisted logic, the authors neglected this important novelty but claimed that only the competitive nature of the test explained the better performance (Hare *et al.* 2000, 2001). They suggested that previous failures of captive chimpanzees who had to beg from humans or gather information from humans were not ecologically relevant, as wild chimpanzees would not beg for food (contrary to all the begging behaviors seen during nut- and meat-sharing, for example). They nevertheless conclude that, "It is therefore likely that primate social-cognitive abilities evolved to a large degree to allow individuals to outcompete competitors, and so it is in these kinds of settings that we are most likely to see these abilities expressed" (p. 149). Competition would be more natural than cooperation to chimpanzees. For the New Iberia chimpanzee's attribution of knowledge to others, see Karin-d'Arcy and Povinelli (2002).

[49] Some readers may feel I caricaturized the opinion of the hardcore psychologist to nicely illustrate my opinions. Some examples could clear this up. David Premack (2007) and Celia Heyes (1998), two of the most famous psychologist skeptics of mental abilities in chimpanzees, want more highly controlled laboratory work. Premack says that, "… cognition is not exclusively a field phenomenon; it can

This brings us to the heart of the problem, as we may well face a personal inclination issue here. In my mind, as long as we do not have all the data to confirm the amplitude of the detrimental effect of captive conditions on the development of cognition, we need to adopt a parsimonious approach and consider that the same dramatic detrimental effects that deprivation has on humans might also occur in those animal species that are closely related to us and have similar learning and developmental phases. The impressive differences in cognitive development between the New Iberia chimpanzees and other captive chimpanzees, however, confirm the validity of such a Darwinian parsimony. Contrary to the hardcore psychologist's suggestion, as the chimpanzees of Taï and Gombe have shown us, mother–infant interactions are abundant, constant over many years, and of a very different nature in different contexts and populations. Such a situation is very similar in humans, where mother–infant interactions can be quite different in different cultures and the way that western mothers actively stimulate their children is absent in many human societies.

The underestimation of population differences is a thorny issue when it comes to discussions about culture, as we naturally always refer to our own culture, which can sadly be very misleading when trying to understand what "humans" are. In this light, our hardcore psychologist needs to remember that not all humans are westerners and that the socially and contextually integrative, less-individualistic African and Asian way of educating children is observed in very large segments of humanity, as we have discussed in the teaching section of Chapter 6.

The review of this chapter shows that in the few cases in which exact comparisons could be done (see Table 8.6), upbringing conditions have strong effects on cognition. In the more ecologically valid conditions, comparisons are less systematic, but all evidence tends to indicate that cognitive abilities differ between wild populations, between wild and captive populations, and between captive populations. More studies are needed to directly address this issue. However, for the time being, all show that we need to adopt a population-sensitive approach in chimpanzees if we want to answer questions about how flexible they are, how culture affects their cognition, and how this compares to humans.

At the start of this chapter, I provided a comparison of chimpanzees and Hadza digging for underground resources and proposed an ecologically valid comparison that stresses the similarities of solutions found by different species when solving similar natural tasks. As seen in Tables 8.1 and 8.2, this in no way precludes identifying important differences. By incorporating a higher level of ecological validity, such comparisons could directly answer questions about specificity in each species.

Throughout this chapter, we have seen that, in both chimpanzees and humans, the cognitive capacities demonstrated in one population are not only a function of the daily challenges they face, but also an expression of the specific culture of that

take place in the laboratory. Indeed, in the case of chimpanzees, advanced cognition would appear to be largely a laboratory phenomenon. For only the chimpanzee who has been specially trained – exposed to the culture of a species more evolved than itself – shows advanced cognition" (Heyes 1998, pp. 171–172).

given population. The more divergent the cultural conditions between two populations, the more important the cognitive differences will be (Figure 8.7). In addition, culture has a snowball effect as it influences many different aspects of cognition due to the impact it produces in the early learning environment. In this sense, culture produces a cognitive specialization that develops from different aspects in the social and physical environment and shapes the individual experience during development. The resulting cultural cognitive specialization will, in turn, open access to different aspects of the environment – for example, through more complex forms of tool use, more elaborate levels of cooperation, or more systematic cumulative innovations – and allow for more independence from the natural limits of the environment.

The second aspect that is emerging is that experience is important for the development of intelligence. In chimpanzees and humans, both species with extended learning abilities, it is the presence of daily challenges that selects for improved cognitive solutions. Intelligence should therefore be viewed as a very flexible and dynamic tool that adapts to the ecological and social challenges encountered and allows an individual to find the optimal solutions under such conditions. Thus, we see different levels of intelligence in different domains emerging in different social groups depending upon the living conditions. Culture molding both the ecology and social system has a direct influence on the intelligence developed and this contributes to the cultural specialization we have seen in chimpanzees and even more so in humans.

Synopsis

- The link between culture and cognition has been studied rarely, and even less so in animals in captive conditions that are removed from their culture.
- In humans, plenty of observations support the idea that culture affects the development and performance level of cognition.
- Human cultures favor a specialization of cognitive abilities that reflect the challenges of each of the cultures considered.
- When data are available, it appears that chimpanzees develop different levels of cognitive abilities depending on the culture and environment they experience.
- Absence of evidence of certain abilities in captive chimpanzees are likely to reflect the deprived conditions they experience or have experienced.

9 Uniquely chimpanzee – uniquely human

If [the intellectual and moral properties of man] were formerly of high importance to primeval man and to his ape-like progenitors, they would have been perfected or advanced through natural selection.

Charles Darwin, 1871: vol. 2: 153.

How could "natural selection," or survival of the fittest in the struggle for existence, at all favour the development of mental powers so entirely removed from the material necessities of savage men, and which even now, with our comparatively high civilization, are, in their farthest developments, in advance of the age …?

Alfred Wallace, 1870: 351–352

Hadzaland, Tanzania, 21 July 2011

Sophia and her mother Madi are searching for tubers, their staple food source, during this long dry season. Digging sticks are essential tools for Hadza women, just as bows and arrows are essential for Hadza men. Sophia has been digging for some time already and her stick is becoming very short after all the sharpening she has done to it in the last few days. As she walks by some saplings, she uses her knife to cut a long stick from a Cordia bush and shortens it to a length of 1.6 m. This is the main type of sapling the Hadza use for their digging tools as they are hard enough to access the underground tubers. She carries her stick with her as she and some other women move along. Once they have moved about 50 m, Sophia picks up a stone, places one end of the stick on an outcropping rock and turns it while pounding the bark all around it. In that way, she slits the bark so as to be easily and completely removed. After hitting it eight times, she removes all the bark from this part of the stick, then turns it over and does the same thing to the other half. After 3 minutes, the stick is entirely peeled (see Figure 9.1). She then catches up with her mother and joins her in digging for more tubers. Moda, another woman who is foraging with them, lights a fire nearby and the women harden their sticks in it after having sharpened one end with their knives.

What makes us human? The classical answer was simply "culture." After reading this book, I hope you will agree with me that we have learned so much about nonhuman cultures over the last 50 years, the answer is no longer so straightforward. True, humans build cars, cook their food, and play Chopin's nocturnes, but at the

Figure 9.1 **A Hadza woman pounds the bark on a new digging stick with a stone**. On other occasions, the Hadza women remove the bark with only their teeth and hands, but they regularly use stones to prepare the bark.

same time, so many cultural abilities have been shown to be present in different animal species that the idea of "human culture" needs to be qualified. It is my attempt in this chapter to integrate all these new observations on animal culture to qualify some of characteristics of "human culture" and "chimpanzee culture." These suggestions remain provisional due to the scanty nature of our knowledge of animal cultures.

Let me again share some of my field impressions of the chimpanzees with you. Salomé and Sartre were impressively successful at finding grubs hidden deep under the bark of large fallen trees and at skillfully using short sticks to extract them whole before eating them (see Chapter 3 and Figure 9.2). The only difficulty Salomé encountered when searching for grubs was in tearing the bark from the trunk to expose this tasty food source. The bark was still so well-attached to the wood in some places that it was difficult for her to get a good grip. I could not help but evaluate what they were doing through my human eyes. Personally, I would have used the stick to remove sawdust and other material from under the bark, which would undoubtedly help me to get a better grip. However, Salomé used the stick only for opening insect shells and extracting grubs. She was not the only one using her stick in such a way and I saw many other chimpanzees using the same technique in the same situation. This gave me the impression that the Taï chimpanzees tend

Figure 9.2 **Sartre using a stick to extract beetle grubs near his mother**. These *Cerambycid* beetle grubs live in the dead wood under the bark of fallen trees. Chimpanzees visit these trees when the grubs have matured and, after tearing off the bark, they use a stick to break open the egg shells protecting the grubs and then extract them from the resulting small hole.

to have a more rigid way of looking at the function of a stick and, therefore, they tend to use it for one single purpose. As a human, I much more readily saw tools as multi-purpose. We already noticed this tendency in our comparison of the Hadza tool use with those of the chimpanzees (see Table 8.3 for a comparison of the multi-functionality of tools in humans and chimpanzees).

Brutus was a very skillful hunter and from him I learned how to anticipate prey movement and how to consider the influence that other hunters have on the prey so as to place oneself in the best position. Brutus repeatedly surprised me by correctly anticipating future hunting movements. Hunting in chimpanzees can be a very smooth teamwork experience where each member of the team naturally takes up his position and encircles prey progressively and efficiently. Brutus and Falstaff would make anticipatory movements well in advance and take up positions ahead of the prey, waiting for the other hunters to push the monkeys towards the trees they were in. At other times, Brutus would climb alone in the trees and the other males would sit and watch him from the ground without joining in. The human observers keen to collect data on hunting behavior would start to get impatient. With my human glasses on, I would simply have called out, just like during a rugby game, "come on

Figure 9.3 **Brutus sharing meat with Kendo, while his friend Loukoum waits for her share**. Hunting provides precious meat that is shared between many adults in the group, including team hunters, like Kendo was on this day, and non-hunting group members, like his female friend seen here. Brutus, who was long one of the community's best hunters, was extremely generous with the meat he had secured.

guys, join me, let's do this … Falstaff, you go up there on my right and Snoopy, you push the monkeys in that direction." However, chimpanzees are not like this; the hunt is a voluntary undertaking by all team members and if everyone is not in the mood at the same time, then it becomes much more difficult to find enough hunters to participate. While humans try to actively recruit others to participate or even force them to join in, active recruitment during hunts is simply not a Taï chimpanzee behavior. Only one male, Rousseau, was heard giving loud hunting barks during some of the hunts in a way that suggested an attempt to enlist others, but he was one of the group's laziest hunters and would stop as soon as the others joined in.

What makes us human? Until now, I have refrained from directly answering some aspects of this question simply because, to answer this, you must first understand the nature of our closest relatives. From the similarities we learn about the differences.

Realizing that we are surely underestimating the extent of culture in chimpanzees, I nevertheless want to build on the knowledge we have discussed in previous chapters to propose some ideas about what I see as potential differences between the specific cultural abilities of chimpanzees and humans. Chimpanzee and human cultures are each unique in their own right, just as their morphologies and behavioral repertoires are unique. Each species on this planet is unique and distinct from all others and, therefore, they all have unique characteristics. At the same time, each species is the product of a long evolutionary history and, as a result, species with a longer shared evolutionary history share more features. Recent data show that humans and chimpanzees separated between 5 and 7 million years ago, while gorillas and chimpanzees separated about 9 million years ago and the first primate separated from the other mammals some 60 million years ago. Going back even further, the first mammals appeared some 220 million years ago. This means that, as mammals, we shared ancestors with chimpanzees for approximately 215 million years. This long common history should make it unsurprising that we share many features with chimpanzees.

Culture is about differences between social groups. Sartre would never have become who he was if he had not grown up with a very considerate mother, along with a gifted hunter – Snoopy – for a big brother, an irreplaceable playmate in his mother's best friend's daughter, and adult male models, like Brutus, Kendo, and Falstaff. But just observing Sartre and the other Taï chimpanzees is not enough. We would not notice that nut-cracking, knuckle-knocking, leaf-clipping, cooperative hunting, or tending to injured individuals have cultural components if we did not compare this group with the chimpanzees of Gombe, Mahale, and so on. Only a *multi-population cross-species approach* to the question of "what makes us humans" will do justice to the complexity of cultures and allow us to take off our ethnocentric and anthropocentric glasses.

Cultural diversity in humans and chimpanzees

Culture is a social phenomenon by which members of one social group can be distinguished from members of other social groups. The social dimension is central to both the products produced by a culture as well as to the acquisition of cultural traits by the individuals. Because one important element of the consensus definition of culture we mentioned in Chapter 1 is that culture is independent of genetic factors, we need to be very careful in our discussions to keep the conditions appropriate for culture to express itself. Genetic influences are best proven by removing animals from their natural context and seeing if the behavior of interest is still observed. In this sense, the captive experiments that have had so much difficulty finding evidence for imitation, teaching, cooperation, and other social cultural traits in chimpanzees have clearly proven that those learning mechanism are *not* genetically programmed and therefore can sustain cultural traits.[1] True, they

[1] Direct genetic testing remains very difficult as we do not yet know of any genes that would specifically code for the expression of any cultural traits (see also Box 2.3).

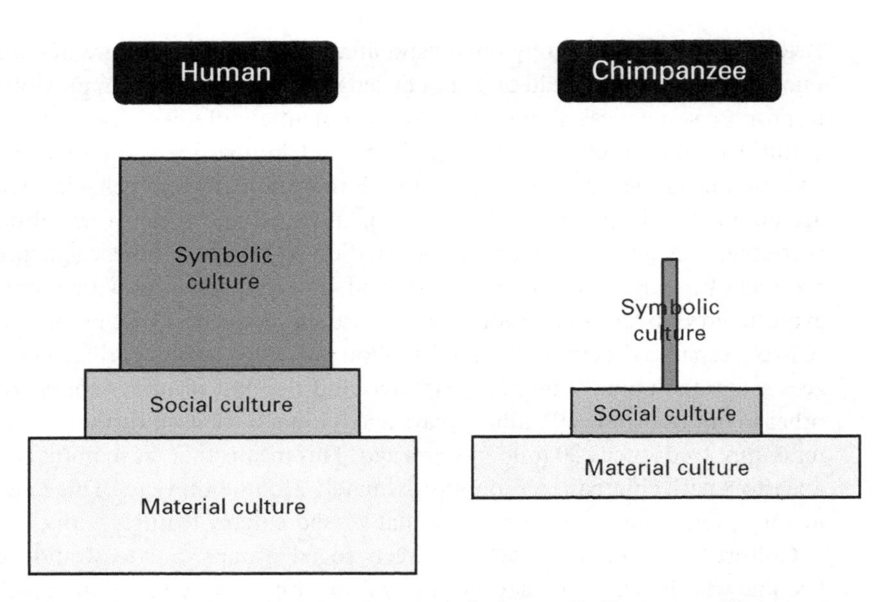

Figure 9.4 **Schematic comparison of cultural domains in humans and chimpanzees.** Here, we are comparing the relative surface of each domain of culture for the two species and the relative prevalence of each cultural domain is strikingly different. For humans, symbolic culture is very important in all known human societies while material culture is present everywhere, but its size here corresponds more to the extremes observed in modern western societies. Our knowledge about chimpanzee symbolic culture is limited to a few communicative traits, while material culture represents the largest proportion of their culture.

could still be purely environmental adaptations, but this cannot be tested in captivity. Much of the discussions in the previous chapters have shown that, for the most part, environmental influences can only partly explain the differences.

A qualitative comparison of human and chimpanzee culture reveals that both species possess traits in the three domains of culture (see Figure 9.4); as no systematic quantification of the number of cultural traits in each domain has been done for both species yet, I suggest that in humans, symbolic culture is predominant, as evidenced by all the myths, beliefs, stories, songs, and ideas that are important in all oral traditional cultures in all societies. In many human societies, material culture is less predominant than symbolic culture and it clearly varies extensively in magnitude from those human societies where all of an individual's material belongings can be transported by that individual alone to modern western societies where material belongings for some (including houses, furniture, etc.) are so plentiful that they are no longer transportable. I placed material culture at the base of the pyramid in Figure 9.4 because I assume that this is the first domain where culture evolves, as evidenced in some bird species, sea otters, and different primate species.[2] In comparison, symbolic culture in chimpanzees is still very limited and

[2] See Wilson (1975), Bonner (1980), Maynard-Smith and Szathmary (1995), Boesch (2012b). The main exceptions to this pattern are the marine mammals, such as dolphins and whales, including killer whales, which show most of their cultural differences in the domain of communication.

has been proven in only a few instances related to communication within the group. Social culture is present in both species and, relatively speaking, is more frequently observed in chimpanzees than in humans, partly because social life plays a larger role than symbolic life if you cannot speak. However, the mere presence of symbolic culture in chimpanzees shows that its existence is not dependent upon language, but that language is more an amplifier that has supported the explosion of symbolic culture in humans.

A great deal of cultural diversity has been documented in both humans and chimpanzees. In fact, there is so much diversity that once you know how a human or chimpanzee behaves, you can very reliably predict which well-studied social group that individual comes from. For example, a chimpanzee who hunts solitarily, dips for ants with long sticks, and leaf-grooms to smash lice undoubtedly comes from Gombe, while a chimpanzee who hunts mainly in groups could come from Ngogo or Taï, but if you also know that he knuckle-knocks to propose sex to females, then you know without a doubt that he comes from Taï (see Figure 9.5). The reliability

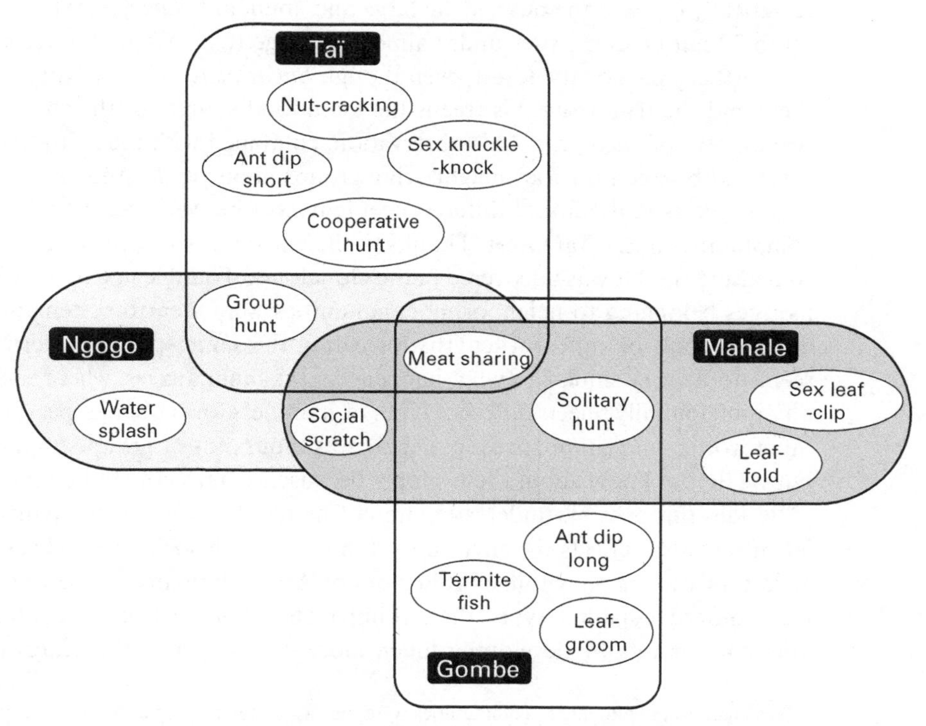

Figure 9.5 **Cultural diversity can be used to identify a chimpanzee's origin.** This schema illustrates that if you know a few group-specific cultural traits, you can clearly distinguish the origin of each individual, even with as few as four chimpanzee populations. Some traits are shared by several groups, such as "meat-sharing," and "solitary hunting," which are shared by the Mahale and Gombe chimpanzees, and "social scratch," which is shared by the Mahale and Ngogo chimpanzees, while others have so far only been seen in one group, like "leaf-fold" and "leaf-groom."

of such predictions serves as testimony to how strongly group members conform to the cultural solutions of their group. While some material cultural solutions are partly related to the prevailing environmental conditions, the cultural solutions for symbolic culture appear very arbitrary and removed from the environment. Thus, in both humans and chimpanzees, culture constitutes a "collective cast of mind" that sets each group apart from its neighbors and at the same time makes each group less dependent on the constraints imposed by nature.

There seems to be more cultural diversity in humans than in chimpanzees, but this is partly due to the fact that we know so much more about humans. Even a short time spent travelling in Africa will show you a wider variety of human cultures than science has discovered in chimpanzees in the last 50 years. We easily gain the impression that cultural diversity is larger in humans by just crossing Kenya or Côte d'Ivoire and experiencing all the diversity in the way people build houses and structure their villages, wear clothes and make axes. However, we should not underestimate chimpanzee cultural diversity. When Hedwige and I first explored Taï National Park, we noticed that the chimpanzees in the south of the park systematically cracked the nuts of the large and abundant *Sacoglottis* trees. We found up to 20 nut-cracking sites under almost all large trees. We had never seen this in the northern part of the forest, even though *Sacoglottis* trees are equally abundant there and the fruit from this tree is the main food source for the chimpanzees for two months of each year. This observation, combined with the cultural differences we found between the four neighboring groups in our main study area (see Figure 5.6), suggests that cultural differences might exist between almost all chimpanzee communities in the Taï forest. Thanks to the video traps I mentioned in Chapter 3, something similar was uncovered in the Goualougo Triangle in Congo, where chimpanzees belonging to neighboring communities were seen to systematically select different tools of different lengths for eating the same species of termites.[3] This seems to be very similar to what happens in Taï chimpanzees, where each community preferentially selects different types of hammers for the same species of nuts. In other words, wild chimpanzees could possess thousands of group-specific cultures,[4] but we do not know about them simply because we have not studied them.

Besides this possible underestimation of material culture in chimpanzees, material culture also reflects the environment in which each social group lives. Since our ancestors' expansion within and then out of Africa, humans have adapted to many more different habitat types than chimpanzees. Humans have adapted to these different habitats by developing much more diverse material cultures than those

[3] Six neighboring chimpanzee communities were monitored for their selection of three different tool types. The Mosika community selected longer puncturing tools for underground termites than four other communities, while the Malasi community selected shorter fishing tools. The brush sticks for aerial termites were longer in the Mayele community compared to three others. However, the diameter of the tools was homogeneous in all six communities (Sanz *et al.* 2004).

[4] This value is a minimal estimate. If we pessimistically estimate there to be 100 000 chimpanzees left in Africa, about half would still be living in intact habitats and would, therefore, live in natural social groups. Counting an average of 50 chimpanzees per community, it gives us an estimate of 1000 chimpanzee communities, each of which will have some cultural differences with their neighbors.

seen in chimpanzees. Furthermore, material culture artifacts have been found to be much more numerous in human groups living in temperate or cold climates than in human groups living close to the equator.[5]

Beliefs, songs, and myths are the main examples of symbolic culture in humans and we have always presumed these to be absent in other species because they cannot talk. But is this really so? I, like everyone else, cannot enter the mind of a chimpanzee, but as someone who has followed them in the wild for decades, I would like to propose a few observations here to entertain such a possibility. One day, I was following a group of chimpanzees when they entered a small river valley on the eastern side of the territory and fed on *Sacoglottis* fruit. The four individuals I was following stumbled upon a recently fallen tree where a forest cobra had been trapped and appeared to have died from the injuries it sustained from the tree fall. Chimpanzees seem to be aware of potential dangers and give very loud alarm calls when they encounter large snakes like Gabon vipers, but they often do not even react to the presence of forest cobras. In this case, the four looked silently at the dead snake for about 5 minutes and then continued on their way. What fascinated me was when the group returned to the same valley 3 days later and two of the individuals who had previously seen the snake made a 200 m detour from their normal route to come back to the same place and look at the snake again (by then only the skeleton remained). They looked at it carefully without touching anything and then left to catch up with the others. Why did they come back? As they have never been seen to kill, eat, or use snakes in any way, they certainly did not come back there for this purpose. Were they simply so curious that they made such a detour 3 days later or did they wonder about whether the snake was really dead? I will never know the answer, but this intriguing anecdote still stays in my mind.

Another time, I observed a similar situation with a duiker that had been freshly killed by a leopard. These ambush killers seem to avoid unexpected encounters with chimpanzees and when five chimpanzees approached on this occasion, the leopard vanished into the forest, leaving his large, half-eaten prey behind. The chimpanzees made some alarm calls and scrutinized the surroundings and the duiker for several minutes. Some sat and watched while others inspected the corpse, smelling the ground around it and the leaves of small saplings nearby. On very rare occasions, chimpanzees kill duikers, but I have never seen them eat them. Shortly after, the chimpanzees continued on with their normal activities. The next time they were close to this area, some of the individuals who had been present on the previous occasion went 150 m out of their way to return to the location of the dead duiker. The leopard must have returned and finished his meal because no sign of the duiker remained, but the chimpanzees briefly checked the spot and then immediately continued on their way. What were they checking for? As in the case of the snake, the intriguing thing to me was that although I immediately knew what they were heading for, I did not see them communicate in any way about it. Although I only rarely witnessed the chimpanzees surprising a leopard in the middle of its meal, when this

[5] Collard and Foley (2004).

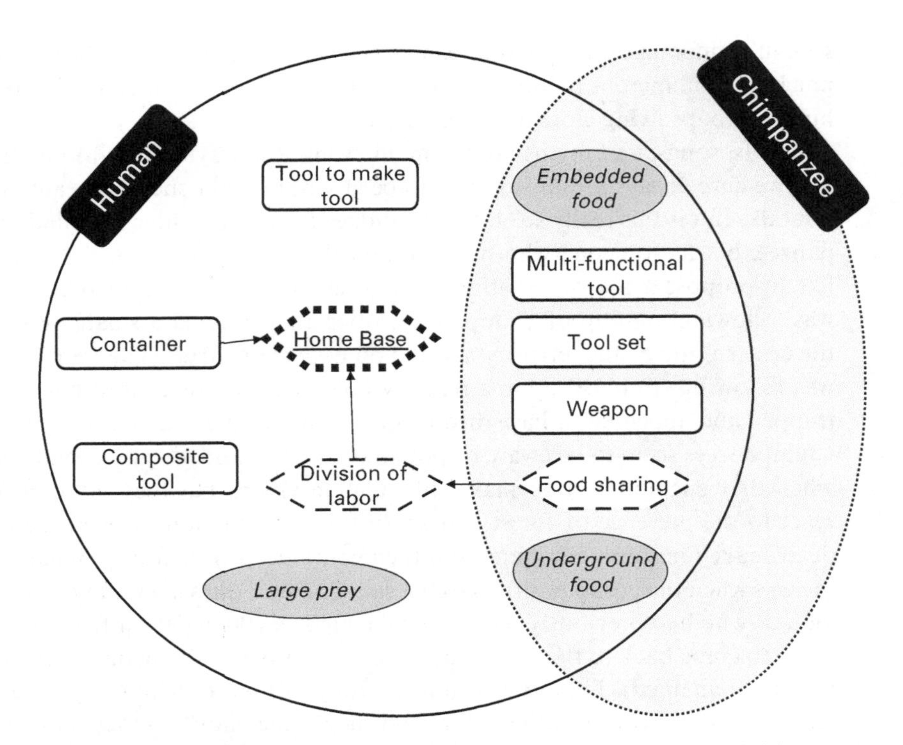

Figure 9.6 **Material culture characteristics in human and chimpanzee.** Human and chimpanzee material cultures share many similarities, such as the multi-functional tool, the tool set, and the weapon, and the profits from those traits in terms of access to hard-shelled and underground foods as well as small prey are reinforced by some specific social developments like food sharing and division of labor between male hunters and female gatherers. A key social characteristic, the home base, is unique to humans and seems to follow the emergence of a division of labor between the sexes and "containers," which allow for the feeding of individuals who remained in the home base. Two additional inventions unique to human material cultures include the ability to use tools to make tools and composite tools, which improve the chances of gaining food and allow access to large prey for the first time.

did happen, I regularly saw some of the chimpanzees return several days later to check on the spot. What were they expecting to see and why did they come back?

Language has undoubtedly had a major impact on the development of human social and symbolic cultural products and in many ways this has changed the type of communication possible between individuals. It allows us to communicate about past or future events, which has opened up a whole new world of possibilities. Symbolic culture in chimpanzees is seen mainly in the communicative domain and this is most likely where it also began in humans. Thus, language could have played an important role in symbolic culture becoming so prevalent in humans.

Another clear difference between human and chimpanzee culture is the complexity of the products humans make. Cars, computers, and axes all include many more elements than we have seen in chimpanzee technology. Composite tools, which are tools that include more than one material element, are unique to humans.[6] Even relatively simple composite tools like hammers and stone-tipped spears have not been observed in other animal species (see Figure 9.6). Much of our material culture is composed of composite tools. Combining different materials allows us to noticeably improve the properties of natural material in a way that was not possible before; for example, a composite hammer made of a heavy stone or iron head attached to a piece of wood can strike with much more force than only a stone or piece of wood alone can. Similarly, a spear made of a sharpened stone head attached to a stick pierces with much higher force and precision when thrown over long distances than a stone thrown alone. Chimpanzees have been seen to combine materials when they use sequential tools, but composite tools are more efficient because they produce this effect simultaneously. Why have chimpanzees not been seen to use composite tools? Axes and lances, which appear to be the first human composite tools, were used to hunt prey from a distance. This requires an open habitat and was proposed to be essential to the evolution of our ancestors.[7] So chimpanzees have either rarely encountered the conditions that necessitate the use of such tools – remember that the largest populations of chimpanzees live in the wetter parts of Africa – or they cannot cognitively conceive of such tools.

Cumulative cultural evolution has been viewed by some to be uniquely human, but the ethnography of cultural technology in chimpanzees suggests that it has happened regularly for both material and symbolic cultural domains (see Chapters 3 and 5). Once individuals in a species are able to copy others, there is no real qualitative difference between learning a simple movement and a more complex one. For example, there is no difference between learning to hit someone with a stick and learning to hit a nut that has been placed on an anvil. From the cultural point of view, the second case involves learning something that was acquired thanks to two preceding innovations. Cumulative cultural evolution is an example of how small quantitative changes can produce large effects; whether what I am copying has gone through 3 accumulations or 15 will mean the difference between a stone hammer – as is used for nut-cracking – or a lead hammer – as is used for working iron. Accumulation of innovations will be required much more frequently in a species that faces many different types of environmental conditions and this will increase the more this species invades new, unknown habitats and expands its range.

Cultural transmission mechanisms in chimpanzees and humans

Social learning, as we have seen in Chapters 2 through 6, is an essential component for the acquisition of cultural traits and has been amply documented in both

[6] Ambrose (2001, 2010), Boesch (2012a).

[7] The "Savanna Hypothesis" was proposed by Raymond Dart (1925) and even if it has been modified and adapted to our new knowledge, the fact remains that the majority of our ancestors known to produce axes and lances during the Middle Paleolithic and Middle Stone Age were found mostly in open regions in East and South Africa.

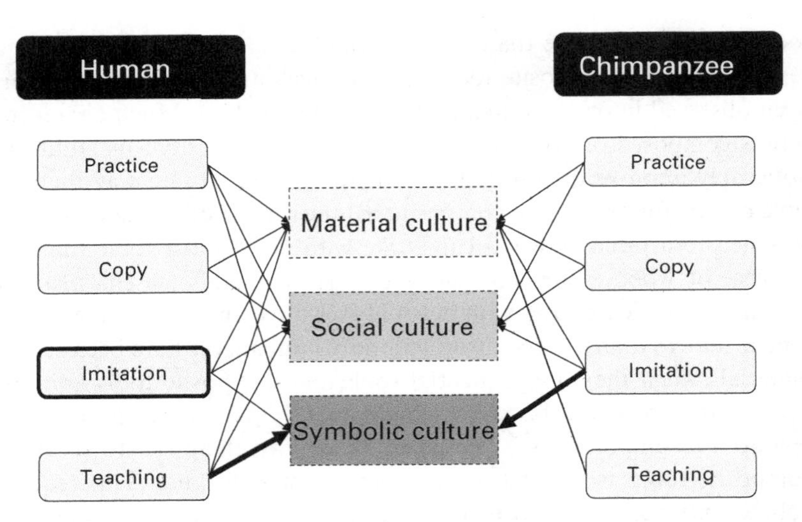

Figure 9.7 **Cultural learning mechanisms in humans and chimpanzees.** An arrow links two boxes when the mechanism exists and the thickness of the arrow indicates its relative frequency within each species. While practice is performed by humans for all three cultural domains, we have not yet observed practice for symbolic culture in chimpanzees. Different copying mechanisms have been observed between the two species, first for material and second for social culture. Similarly, imitation was seen in all three domains in both species, but has been suggested to play a more important role for symbolic cultural trait acquisition in chimpanzees. Finally, while teaching has been seen to occur in all three domains in humans, until now, it has only been seen for material culture in chimpanzees. The bold box around imitation for humans represents the over-imitation (see below) that is only seen in children.

chimpanzees and humans to support the acquisition of cultural behaviors. As we discussed for material, social, and symbolic culture, we expect both to use different learning mechanisms to acquire these traits since individual learning is less reliable as we move from material to symbolic culture, and the understanding of the task itself relies more on an understanding of the other. In other words, it makes a difference if one is learning about an object, a social interaction, or communication. If my propositions in Figure 9.7 were confirmed, the differential use of different social learning mechanisms when comparing humans with chimpanzees suggests important differences between the two species.

As I have already mentioned, the discussion about the importance of imitation versus other forms of social learning is part of a long-lasting debate in psychology. In my opinion, some of the misunderstanding comes from a lack of clarity about when either process should be used in different species. As can be seen in Figure 9.7, the different social learning mechanisms used will be adapted to the specific traits that are being learned. For example, copying how objects are used, which psychologists call emulation, will be observed regularly in material culture but simply cannot be used for elements of symbolic culture. Imitation, on the other hand, will be much less frequently seen for traits for which practice importantly contributes to the understanding of the technical aspects of the tasks. Therefore, imitation should be expected much more readily for complex material cultural traits. Thus, a better understanding of the complexity of a hammer or of a stick's physical properties, such as its shape, weight, and hardness, can only be gained through practice and not imitation. This has also been shown for stone knapping in humans (see Chapter 6). For symbolic communication elements, the advantage of practice is often limited as it is essential to be quickly understood in competitive situations – like when competing for sexual access or social partners – and furthermore, there is no clear feedback when miscommunications occur compared to when trying to crack nuts with a bad hammer (see Chapter 5). The main exception to this is language acquisition, where young children start to practice several years before they face a situation where there is a need to be error-free.

Imitation in human children differs from imitation in chimpanzees and adult humans in one striking aspect: children *over-imitate* by blindly copying everything that the model is doing, even if they know or have been told that aspects are wrong or useless.[8] Such strict mimicking at first does not seem to be environmentally sensible and we would expect over-imitation to be detrimental for learning material tasks in natural populations. Therefore, it is not surprising that over-imitation has not been observed in animal species and it would be intriguing to see if it is present in human societies that are more dependent on nature. However, over-imitation is reminiscent of how some bird and antelope species copy mate preference; some females in those species prefer males that they have seen paired with other females independent of any of this male's other characteristics.[9] The argument here is that it would be too time-consuming to directly test the quality of all males in an area, so females might simply use the success males have with other females as an approximation of his quality. In a similar way, over-imitation could be the best learning technique for those unable to ascertain the best method for themselves or for those unable to understand the associated technical challenges. This would suggest that over-imitation appeared rather late in the course of human evolution, when the technological complexity of our material culture became such that understanding the technical details became too time-consuming to individuals.

Teaching in chimpanzees has been seen mainly for traits in the material culture domain and mostly for traits that require a long acquisition phase – in other words, for the more complex material culture traits (see Figure 9.7). Chimpanzee life history plays such an important role in explaining the presence of teaching that we observe an interaction between the complexity of tasks and the reproductive agenda of the mothers. Thus far, this interaction has not been considered in other species.

[8] Alexandra Horowitz's study of imitation in adult humans (2003) showed that children copy a model much more faithfully than adults, and subsequent studies have dubbed this tendency "over-imitation" (Horner and Whiten 2005, Lyons *et al.* 2007, McGuigan *et al.* 2007). The tendency of children to reproduce useless or wrong elements while imitating has been attributed to a distortion of the causal understanding of the tasks (Lyons *et al.* 2007).

[9] Copying the sexual preference of other females has been clearly demonstrated in different species of antelopes, like fallow deer, and birds, like the black and sage grouse (Gibson and Hoglund 1992).

However, it is well known that human mothers invest in their children for much longer time periods than chimpanzees while also having relatively shorter interbirth intervals. Thus, in general, the younger children arrive when the older ones are not yet fully independent, as is rarely the case with chimpanzees. This reproductive schedule puts more pressure on human parents to make sure their children learn as much and as quickly as possible before their new siblings arrive. As a result, active and passive teaching is expected to be more systematically employed by human parents.

Nevertheless, active teaching in humans has often been overestimated as it has only become systematic with the recent introduction of schooling, which not only affects how children learn but also how schooled parents themselves teach their own children. As we have seen, in the material culture domain, learning of cultural techniques in traditional human societies involves a mixture of practice, passive observation, copying, and some limited passive and active teaching interventions.

Cultural transmission is first and foremost a social process, so much so that the performer or inventor's identity plays a decisive role in whether a behavior is adopted by group members. For species, like chimpanzees that live in small, closed social groups, a trustworthy partner from whom you will copy will be an individual you have known for years and with whom you have interacted with for long enough to know that he or she is reliable and a good social-role model (see Chapter 6). Human societies, particularly western ones, are much larger and more open than chimpanzee societies and this may lead us to more easily accept less familiar models. However, this should not distract us from the fact that young children are still much more likely to imitate trustworthy individuals than unknown or untrustworthy individuals.

Uniquely chimpanzee culture

Ecology of culture

Chimpanzees are tropical forest dwellers and no chimpanzee population has ever been found where there is not at least gallery forest available. However, compared to the many forest primates that are restricted to dense forest regions, chimpanzees have been very successful at adapting to open forests and even some dry savanna areas, but they remain totally dependent on trees in which they build their night nests.[10] Therefore, chimpanzee material culture reflects a specialization to forested regions and expresses itself mainly through improved foraging tools and skills that help to improve diet. Such flexible cultural foraging skills buffer groups against uncertainties about food production and therefore contribute to their survival in

[10] Night ground nests have been found in only a few chimpanzee populations and this is puzzling as the chimpanzee's main predator, the leopard, has a larger range than chimpanzees, and, therefore, they risk higher predation by sleeping on the ground (Boesch 2009). So we do not yet understand what determines range limits in chimpanzees.

a more diverse set of environmental conditions. Furthermore, chimpanzees spend about 40% of their daily activities looking for and eating food and an additional 35% of their time traveling between food sources. They spend an average of only 25% of their time socializing and resting. This illustrates how important food is to them and I think this partly explains why many cultural traits have a direct connection with food in this species.

Chimpanzees are social animals with a specific and flexible fission–fusion system. Such a social system is observed in only a minority of animal species as social animals generally live in stable, clearly defined social groups in which social interactions between group members can be directly monitored by others. In fact, almost all monkeys live in groups in which most members remain in visual contact with one another for most of the day. In contrast, in a fission–fusion system, individual members are associated with only a few group members at a time and can be out of sight of the other individuals for days or weeks. In Taï, we found that some pairs of individuals were seen together only once every 20 days on average. Thus, much about what happens between group members cannot be directly observed and individuals have to infer it when they meet again. However, Brutus was regularly seen associated with his preferred social partners, like Falstaff, Salomé, and Ondine, and this made cooperation between these pairs much more natural and likely. In contrast, Brutus was less frequently seen with his rivals and so he would not know what was happening between them most of the time. Social groups in chimpanzees are usually made up of between 50 and 80 individuals, half of which are adults, and this sets the context of within-group solidarity we have been talking about. In other words, the social environment consists of a relatively small but flexible group of individuals with a high proportion of trusted partners that have lived together for many years or decades.

Subjective reality and culture

Sartre and the other young Taï chimpanzees see their learning environment as not only what they directly perceive, but also what they see group members doing with it. We saw in Chapter 6 that the way Salomé interacts with stones and clubs and Brutus interacts with other social partners and monkeys influences the way Sartre and his age-mates see and interact with those objects themselves. The Taï chimpanzees' "cooperative hunting/meat-sharing/nut-cracking" learning environment will mold infants in a different way than the "solitary hunting/meat-sharing/termite-fishing" learning environment that prevails in the Gombe chimpanzees or the "reduced group activity/no meat/few tool/restricted movement" learning environment of many captive chimpanzees. Growing up with such different subjective realities will unconsciously shape infants' knowledge about objects and, accordingly, they will be prone to perform different activities with them and discover different properties about them. Thus, we should expect that wild chimpanzees will possess a stronger propensity to act in groups and use tools compared to many of their captive counterparts.

As we saw in Chapter 6, the mental transformations that are necessary to select tools will become more flexible as more tools are being used. In other words, the more anticipation needed to transform a raw material into a tool, the more complex transformations can be anticipated mentally, which in turn will open the door to more complex tool use. We see a complex interaction develop here as the complexity of the subjective reality experienced by individuals will affect the complexity of the cultural technique they invent. As we have seen, the more time required to learn a technique, the more frequently different forms of teaching are observed.

Symbolic culture might be one of the more surprising components of culture in chimpanzees. In Taï, Brutus expresses "I want to have sex with you" through a symbolic gesture that has no direct connection with this intended meaning, and in the Mahale Mountains, Alofu says the same thing in a different way that also has no direct connection to his intended meaning (see Figure 5.3). Indeed, it is intriguing to see that, even without speech, individuals within a social group can develop a communication system that is so analogous to human sign language, where signs stand for whole words. As far as I can tell, chimpanzee sign-code has no syntax – whereby the meaning of signs would be altered when combined with different signs – and so it remains a very rudimentary means of communication compared to human language. It is tempting here to imagine that our ancestors, with their increasing social complexity and larger groups and territories that would have led to long periods of separation between group members, would have profited from developing such a rudimentary sign-code system to communicate. Such precursors of language in another species open a window to a new understanding of the interplay of symbolism and language in humans.

It is too early to talk about the importance of symbolic culture in wild chimpanzee populations because of the difficulty we encounter in deciphering it. However, this difficulty with deciphering it means that we might be dramatically underestimating it. At the same time, symbolic culture in chimpanzees as proposed in this book illustrates what form symbolism can take in non-speaking animals.

Transmission mechanisms

Group-specific teaching styles are apparent in chimpanzees. Salomé, Héra, and other females have been shown to actively influence the learning of cultural traits in their youngest offspring for techniques that take a long time to learn and where conflicts with their future reproduction can occur. Gombe chimpanzees do not seem to acquire such techniques and therefore their teaching style is much more passive, with infants basically acquiring most of their skills on their own through stimulation within the learning environment provided by the group.

In contrast, Sartre never saw his mother Salomé leaf-clip before drumming on tree buttresses or knuckle-knock on tree trunks to propose mating to males, but nevertheless, Sartre learned those signs. The presence of group members other than the mother with diverse experience to serve as models is very essential for Sartre to acquire the skills of his social group. Some behaviors that are simply not observed in mothers are very important to their offsprings' adult lives and learning from others in the right environment is essential. This is especially true for social and symbolic culture, which is concentrated on adult social interactions in chimpanzees. This strengthens the fact that cultural learning is complex – because it involves different types of models at different ages for different traits – and specific – as different mechanisms of learning may be used for different traits.

Niche construction, in the sense that an animal is able to alter some of the constraints of the environment so as to access resources not otherwise accessible, is one of the main effects of culture in chimpanzees. Their ability to access water in very dry habitats by constructing wells has opened a new type of environment to them and explains their ability to live in surprisingly dry regions like those in Senegal, Mali, and Uganda (see Chapter 3). Furthermore, in rainforests, their cultural technology has allowed them to access some very rich resources that are inaccessible to most animal species, such as very hard nuts, underground honey, and termites. Such resources buffer them from food shortages during lean food periods and thus have positive effects on survival and reproduction.

Cumulative cultural evolution has been proposed to be very rare in animals, basically because animals were thought to be limited in their ability to learn faithfully through imitation and teaching. We have now seen that this was based on premature generalizations drawn from the "culture outside of culture" approach, and that both imitation and teaching can be very important in wild chimpanzees depending on which cultural traits are being learned.

If teaching, imitation, cooperation, and many other sophisticated social skills are present in wild chimpanzees, why have they not developed more complex cultures and why have they not colonized more and varied environmental niches? Why is cumulative cultural evolution limited in its expression to relatively simple technological traits? In other words, why are chimpanzees not humans? My answer is simply ecology. I suggest that the common ancestor of chimpanzees and humans had already acquired most of the ability to sustain culture. We should not expect any animal species to develop abilities that are not directly relevant to the environment they live in and as long as chimpanzees have remained in mainly forested habitats, they have faced a very stable and similar set of ecological challenges. It does not make sense to expect them to adopt more solutions than the ones that have allowed them to successfully conquer all of the forested regions throughout tropical Africa. While we know that the humidity levels have changed dramatically across the continent many times in the last 5 million years, with the forest cover shrinking and expanding with these variations, the chimpanzee's ancestors have seemingly tracked such changes and remained in the ecologically stable forests. Those trapped outside forested regions either became extinct or adapted and possibly mixed with our ancestors.[11] At the same time, human ancestors conquered the open regions of the African continent and were confronted with new ecological changes.

[11] A recent genetic study of human divergence from chimpanzees showed that such interbreeding occurred as recently as 1.5 million years ago (Patterson *et al.* 2006).

Uniquely human culture

There has been a long history of proposing what makes human culture unique and in Table 9.1, I have tried to summarize the most influential ones. Interestingly, it is one of the oldest hypotheses of human evolution, one that stresses the importance of the "home base" in our past that seems to have stood up best to the test of time. As seen in Figure 9.6, the home base in humans is also associated with two key inventions: the container and the sexual division of labor.

Ecology of culture

Humans, like chimpanzees, have successfully adapted to forest habitats, but we also inhabit the driest regions in Africa, as well as all parts of the temperate climate zones and some of the coldest polar regions. In addition, we have colonized coastal regions along rivers, lakes, seas, and oceans. This has required vastly different environmental adaptations in humans compared to chimpanzees and this has directly influenced the breadth of our material culture. In addition, as already mentioned, these more open habitats have been subject to extreme climate variability with significant shifts in seasonality that has forced vegetation changes and restructured mammalian communities and the paleo-diet of our ancestors, sometimes as often as every 23 000 years.[12] There seems to be a strong correspondence between these extreme climate periods and the appearance of some of our ancestors, such as *Homo habilis*, *Homo rudolfensis*, *Homo erectus*, and *Homo ergaster*. It can be said for certain that the diversity and breadth of human material culture are what sets it apart from animal culture (Figure 9.4). However, this cultural explosion is based on many cultural abilities that we share with chimpanzees. Especially important for this rapid human adaptation to so many different conditions were the pre-existing abilities for cumulative cultural evolution, as seen in chimpanzees, and faithful cultural transmission mechanisms for complex material traits, such as teaching and imitation. No qualitative change seemed necessary to allow this explosion, simply important changes in the frequencies of innovation and transmission.

Transmission mechanisms

One fascinating aspect about social learning mechanisms in humans is that studies have shown children to be "over-imitators" in the sense that they imitate the fine details of what they see a demonstrator doing, even if those details are obviously wrong or useless.[13] This seems to be the result of a very malleable understanding of causality, where the models' mistakes result in a distortion in the naïve child's causal belief about the effect of specific actions, and this persists even when the

[12] Maslin and Trauth (2009, pp. 153–154).
[13] Lyons *et al.* (2007, 2011), Horner and Whiten (2005).

Table 9.1 Alternative abilities presented as decisive in explaining human cultural uniqueness. Each ability is presented alongside evidence for and against it in humans and others species and some key references

Ability	Support in humans	Contradiction in humans	Contradiction in other species	References
Rational imitation	Daily life and laboratory evidence	Mixed learning mechanisms used in all human learning studies	Present also in dogs, chimpanzees, marmosets	Gergely *et al.* 2002; Range *et al.* 2007; Voelkl and Huber 2000; Horner and Whiten 2005
Teaching	Schools, daily life, and laboratory experiments	Much rarer in cross-cultural studies on learning of daily life skills	Present in chimpanzees, cheetahs, meerkats, etc.	Rogoff 2003; Greenfield *et al.* 2003b; Csibra and Gergely 2011; Thornton and Raihani 2008
Cumulative evolution (ratchet effect)	Archaeological remains and present complex technological design	None	Present in chimpanzees	Boyd and Richerson 1985; Tomasello *et al.* 1993; Boesch 1995, 2012a
Cooperation	Large group cooperation between unrelated individuals	Lower prevalence in cross-cultural studies with little group cooperation and no third-party punishment	Little group cooperation present in different species. Warfare and hunting in chimpanzees	Nowak and Sigmund 1992; Fehr and Gächter 2002; Marlowe *et al.* 2011; Boesch 2009; Mitani *et al.* 2002
Sharing	Daily life and laboratory evidence	Cross-cultural studies with limited sharing seen	Present in wild and captive chimpanzees	Henrich *et al.* 2006; Boesch 2009; Horner *et al.* 2010
Home base	Seen in archaeological times and in cross-cultural studies	None	Absent in great apes and most primates	Isaac 1978, Potts 1984, Boesch 2009, Marlowe 2010

child is made aware of the irrelevance of their actions. Intriguingly, this tendency to "over-imitate" decreases with age and older individuals think more rationally when copying different tasks. In the course of human evolution, what critical information needed to be faithfully acquired and was too complex to be understood? My suggestion would be language. Throughout the course of human evolution, tools remained relatively simple for an extremely long period of time; as an example, stone knapping appeared 2 million years ago but refined flake productions appeared only 200 000 years ago and composite tools appeared only 300 000 years ago. For children who had seen their parents fashion such tools since they were born, it might have been within the limits of their natural causal understanding to learn the tasks with practice, attentive observation, and some stimulations, as chimpanzees do today. The explosion of complex composite tools appeared only recently in our past. In contrast, words and sentences had no visible causal effects on the outside world and their learning would have been so opaque that faithful to blind copying would have been the most rapid and efficient way to acquire it.

Adult humans evaluate the tasks they are asked to copy and this leads them to imitate some aspects but not others in a way similar to chimpanzees. Adult humans do not faithfully copy and this leads to rather low conformity when copying. It has been argued that strict imitation can be detrimental, as innovation and flexibility are required to adapt to the ever-changing conditions one encounters when working with natural objects.

Cumulative cultural evolution is certainly a very essential part of our modern lifestyles. The invention of agriculture and the industrial revolution have dramatically altered many aspects of the way culture has expressed itself in modern humans. To understand why cumulative cultural evolution became so prevalent, we need to understand our cultural changes before this time. It is puzzling that our technological achievements remained rather rudimentary and extremely stable for hundreds of thousands of years – the shape of hand axes, for example, remained the same throughout most of the world for over a million years – which certainly does not give the impression that innovations accumulated very rapidly.[14] However, over this same time period, our ancestors were invading the tropical and temperate zones in Africa, Asia, and Europe and were forced to innovate to adapt to these new environments. My point is that, in archaeology, we do not really see signs of an acceleration of cumulative cultural evolution until quite late in human evolution. Thus, the explosion of cumulative evolution might have coincided with the emergence of language.[15]

Evolution of human culture and biological changes

Why do humans have more developed cultural niches than any other species? What happened in our past that led to such a cultural revolution? Why do we also see

[14] Iovita and McPherron (2011).

[15] Mellars and Stringer (1989), McBrearty and Brooks (2000).

such large cultural differences between human populations? What I suggest here is a quantitative jump in the use of pre-existing abilities during the course of human evolution and not a qualitative change in abilities. As we saw in Table 9.1, some time ago, Glynn Isaac proposed the home-base theory, whereby humans would center their activities around a fixed home base, where old, very young, and sick group members would remain while fit group members would gather and hunt for food, which they would bring back to share. Food sharing would benefit all individuals in the group and allow a division of labor to develop. It is possible that the home base was a solution adopted following some important reorganization in the reproductive schedule of our female ancestors, when they started to wean their infants before they were independent and able to care alone for them.[16]

It also seems that the home base corresponds with the invention of two key objects that are totally missing from chimpanzee technology: first, "containers," which would have allowed individuals to bring large quantities of food items back to the dependent group members who had been unable to walk the distances required to gather them, and second the "composite tools" including shelters, which would protect the dependent weaned group members left behind from bad weather conditions (Figure 9.6). These inventions provide some of the clearest examples of simple to complex cumulative cultural improvements in humans. In addition, the development of systematic cooperation and altruism within family groups would have been the only possible way to make such groups more reproductively successful.

The second key biological change in our ancestors was the colonization of totally new habitats compared to those used by other great apes. Humans are nearly alone among primates in colonizing coastal regions, which has allowed us to systematically exploit water resources and mountainous regions for the first time before dispersing to the coldest parts of our planet.[17] This required more flexibility in expanding our food niche as many of these new habitats did not contain as much easily accessible food items as the tropical rainforests. Savannas are known to harbor higher densities of large mammals but less fruit than forests, and thus humans needed to acquire devices to make new forms of hunting possible as well as develop technological devices to expand access to underground food resources to compensate for the savanna's lower fruit productivity. Acquiring food from water also requires different techniques than acquiring food in the forest and savanna. Thus, technology suddenly became more important and innovation became the key to

[16] Boesch (2009, Chapter 7) expands on this model of human evolution based on recent findings of *Homo erectus* and early *Homo sapiens* fossil remains (Cronk *et al.* 2000, Hawkes and Paine 2006). As it is difficult to determine reproductive events from bones, we are still unclear about why such changes in interbirth intervals happened in the first place. This acceleration of reproduction in modern humans, which led to children being less developed when born and being weaned earlier, has long been stressed as one of our most important characteristics (e.g. Passingham 1982, Potts 1984, Hart and Sussman 2005).

[17] Recent studies using the latest technology have shown that, by 2 million years ago, representatives of *Homo erectus* were living in many different environmental contexts in Africa and that this was strikingly different than for their ancestors, the australopithecines (Plummer *et al.* 2009); "Combined evidence suggests that early *Homo* was flexible in its habitat use, and that the capacity to extract resources from a range of open and more wooded environments was a vital component of its adaptation."

successfully adapting to all of these new conditions. As a result, flexible material culture developed rapidly in the face of all these new challenges.

What led to such changes in the first place? The classic "Savanna Hypothesis" posits that groups of the common ancestor of humans and chimpanzees became trapped in the East African forests approximately 10 million years ago by a shift in the tectonic plates east of the Rift Valley that led to the surface becoming elevated by several hundreds of meters and resulted in the drying and cooling of the entire region.[18] Our ancestors were forced to either adapt to more open conditions or die. This resulted in the birth of the australopithecines, who started to build on their chimpanzee-like abilities in their newly imposed environment. The impressive amount of time it took for *Homo erectus* to improve upon the hand axe illustrates how slow such a process was and how much time it took for humans to build upon already existing cognitive abilities.

In other words, on top of the biological and cultural skills humans inherited from their great ape ancestor, humans developed a vast cultural domain and this has required a tremendous amount of memory capacity to be able to store all our material and symbolic knowledge and solutions. This was not a new uniquely human ability, but stemmed from the abilities we inherited from our ancestors being placed in a totally new context, which led to the evolution of human culture. In my view, it was primarily this change that led to our increased brain size compared to other species. By conquering so many different habitats, many of which would have been uninhabitable to us without our cultural innovations, we found it necessary to store so many new and diverse types of information that we needed a much larger brain.[19] In addition, memorization is not only about storage space but also about being able to rapidly access information and all this contributed to the development of our larger, restructured brain.

From symbolic communication to language

The evolution of language still remains one of the big mysteries in human evolution. How and why did we acquire language? If we consider what we now know about chimpanzee symbolic communication, we can make a proposition. The chimpanzee social system has led to the development of some symbolic forms of communication – for example, sign-codes to communicate shared meanings within the group and symbolic tree-drumming to inform others about changes in travel direction or

[18] The "Savanna Hypothesis" was first proposed by Raymond Dart (1925) and was then elaborated on by many anthropologists (e.g. Isaac 1978, Johanson and Edey 1981, Leakey and Lewin 1977, Lovejoy 1981). In its original form, it proposed that this sudden need to adapt to drier habitats forced our ancestors to acquire tools, learn to hunt for meat, cooperate with one another, share meat and food, and return to a home base regularly – all behavior patterns proposed at the time to be uniquely human adaptations within the primate lineage. Revisions of the hypothesis have been made as some of our ancestors, including *Sahelanthropus tchadensis* (Brunet *et al.* 1995), have been found in dry regions west of the Rift Valley.

[19] Much importance has recently been placed on memory, especially working memory, and it has been proposed that this became very important in the course of human evolution and much more developed than in other animal species (see Fitch *et al.* 2010, Frith 2007).

resting times – and we can imagine that more will be discovered in the near future. Faced with the increase in their foraging range and a much more extended fission–fusion system related to the existence of a home base, our human ancestors would have needed to expand upon such a rudimentary symbolic system, especially to share travel goals or to inform one another about rich food sources.[20]

Ethnography of culture

With the evolution of more extended learning capacities in animal species, it has become more challenging to do cross-species comparisons, for we first need to understand within-species variability before we can compare between species. This in turn has allowed these species to adapt to more varied ecological conditions, which further increased the flexibility of behavior. This has opened up a fascinating new aspect of the biology of certain animal species, but at the same time it has made comparisons within and between species much more challenging. The temptation of some to restrict such comparisons to artificial or simplistic ecological conditions will result in many problems when confronted with the naturally found diversity. Thus, for a sensible study of culture, an ethnographic approach is required that takes into account all the different conditions experienced by the species under study. The explosion of fieldwork with different animal species, including humans, in the last 50 years has placed science in the unique situation to directly address the question of cultural diversity within and between species.

New discoveries are being made about the natural behavior of wild animals every day and evidence of behavioral variations between different populations has been increasing for chimpanzees, orangutans, gorillas, capuchin monkeys, New Caledonian crows, killer whales, Californian otters, and other species. While it is wonderful to learn so much about culture in animals, patience and care are required when one wants to study differences between species and what makes humans unique. This produces a situation where conclusions about culture or other abilities are only temporary and where claims of cultural limitations might only reflect our limited knowledge about animals.

The realization that teaching styles differ between different chimpanzee and human groups, and that cumulative cultural evolution and symbolic culture in communication are seen in chimpanzees all results from extensive and detailed long-term ethnographic studies of wild chimpanzee populations. Such discoveries contradict what has been found with captive chimpanzees and were unexpected by psychologists unfamiliar with how flexible and innovative this species is in the wild. This justifies increasing field research and observations of wild chimpanzee populations. If we also consider that the presence of researchers has been proven to be

[20] It is very revealing in this context how many languages use words that directly refer to body parts for numbers. This makes it seem as if counting was first done with signs and only later was integrated into the language.

one of the best ways to protect wild animal populations, then such efforts would be the best possible way to invest research money.

Human and chimpanzee cultures have much in common but a comparison suggests that there might be important quantitative differences as well, including more cumulative culture in humans which leads to more complex cultural artifacts, a larger reliance on imitation in humans in the three domains of culture, and more complex forms of symbolic culture in humans. However, the one thing that is unique to human culture is language, which acts to amplify cultural transmission and development. This language-based amplification not only makes it easier to learn complex techniques of material culture, but it also makes active teaching possible for symbolic culture, which would otherwise be especially difficult. Furthermore, language-based teaching can be aimed at larger groups of individuals and can happen outside of the context, which therefore increases the value of teaching for cultural integration and dissemination.

The ethnographic approach to culture shows that culture exists in different animal species and not only in humans, as has too often been presumed. At the same time, it also allows us to progress in our understanding of the similarities and differences between different species. It is through more detailed studies of nonhuman animal species in the wild that we have been able to increase our understanding of what connects culture in humans to those of other species and at the same time what makes human cultures special.

Synopsis

- Human culture is characterized by a diverse amount of material culture products that are dependent on the society, elaborate social culture, and extend symbolic culture.
- Chimpanzee culture, on the other hand, has aspects of the same three cultural domains but to a lesser degree and this is especially clear for symbolic culture, which has thus far only rarely been documented.
- Acquisition of cultural behavior in chimpanzees seems more specialized, with teaching mostly occurring for material culture and imitation dominating for symbolic culture.
- Acquisition of cultural behavior in humans seems to rely on an extended mixture of learning mechanisms for all three domains of culture.
- Composite tools and containers are the two types of material artifacts that are uniquely human and the former presents some of the highest levels of elaboration and complexity.
- Language acts as an especially strong amplifier to apply our cultural abilities to many novel and new situations.
- Humans and chimpanzees have the same basic abilities, but ecological changes encountered in the expanding range of humans have been the driving force behind the cultural explosion seen in human evolution.

Epilogue: Will we have the time to study chimpanzee culture?

Margot gave us the most lovely and charming surprise when she reappeared in the group at the end of November 2000 carrying her two newborn babies. Twins! This is extremely rare in chimpanzees and this was the first time we had ever seen a mother with twins at Taï. The sight of the two tiny faces peering out at the world from the safety of their mother's arms was very, very delightful (Figure 10.1). From the few examples I had heard of chimpanzees giving birth to twins, I knew it could be very challenging for a mother to invest in two constantly hungry, demanding babies and to prevent them from falling when she is feeding high up in the trees. However, Margot seemed at ease with the two of them sucking eagerly on her breasts and while she was resting on the ground, she played "airplane" tenderly with them by placing one infant at a time on her foot while reassuring them with one hand. Mustapha, her older son, was very curious about the babies and Margot let him touch them when he was quiet but refused to let him play with them as they were still so small. The babies, who we named Miriam and Makeba, were very active and curious about the world around them and they grew as two very playful infants.

To my horror, as the twins were about to turn 2 years old, I learned that a poacher had come into the forest and killed Margot and young Miriam. He carried the three of them – two dead, one alive – back to the village, where he ate the two dead chimpanzees and sold Makeba as a pet. According to our sources, she was most likely sold to someone in Abidjan, the capital, but we were unable to track her down and it is unknown whether she is still alive today. Shockingly, we also learned that the same poacher shot two other adult females, who then vanished into the thick forest, where they most likely died in awful agony. What happened to Margot, Miriam, and Makeba is just one example of what is happening every day throughout the pristine forests remaining in Africa as new roads are opened and new blocks of forest are made accessible to hunters. Sadly, the process repeats itself again and again because animals in the newly open stretches of forest are killed and many chimpanzee cultures disappear forever from our planet.

When we went looking for cultural differences between chimpanzee groups, we explored fairly remote areas where chimpanzees were known to live. On one of our trips in 1990, we visited the forests surrounding Mount Kopé, some 60 km south of Taï National Park. Mount Kopé is a large, rocky inselberg that emerges out of the forest with an abundance of stones and we were curious to see how chimpanzees would select hammers with such a different variety of available raw material

Figure 10.1 **Margot with her twin babies, Miriam and Makeba.**

compared to Taï forest. We had to camp out on top of the mountain and could hear the chimpanzees calling at dusk below us on the steep slopes of the mountain. As the chimpanzees were not habituated to humans, we did not see them. The slopes of the mountain were covered with forest and we found plenty of *Coula* trees with traces of nut-cracking by chimpanzees. Most likely due to the over-abundance of stones, the Mount Kopé chimpanzees use stones exclusively as anvils and hammers to crack the nuts. In contrast, the chimpanzees of the North group in Taï use stones as anvils less than 10% of the time. Thus, this nicely illustrates the behavioral flexibility of chimpanzees.

In 2009, 19 years later, I had to take the same isolated dirt road that runs alongside Mount Kopé to reach the Taï research camp and was astonished to see what remained of this formerly great mountain. The small village of Grabo that had previously rested at the foot of Mount Kopé now stretched over more than 5 km and I really needed to look many times to confirm to myself that I was really looking at the same mountain as I could not see any trees left and everything was covered by fields. Were all the chimpanzees with their unique culture gone?

In the Monogaga classified forest, which stretches along the Atlantic Ocean in the south of Taï, *Coula* trees were almost absent and so chimpanzees concentrated on the hard *Panda* nuts. We were intrigued to find anvils alongside many broken Achatine snail shells. While these approximately 12 cm long Achatine snails are typical of all African forests, the Taï chimpanzees have never been seen to show any interest whatsoever in them, let alone crack and eat them. Had the Monogaga chimpanzees invented another new cultural habit that we uncovered by chance? Unfortunately, there will be no answer to this question as the chimpanzees and the forest are now gone.

Côte d'Ivoire is one of the most active and developed African countries but has one of the highest deforestation rates. And this comes at a cost. In 2007, we revisited 11 of the sites that we had explored back in 1990 to confirm the role of the Sassandra River as the boundary for the Taï chimpanzee nut-cracking behavior (see Chapter 3), and we discovered a shocking reality. Most sites that were situated within protected forests were now fields as far as the eye could see, even to the point that in many cases, it was simply impossible for us to find any of the forests.[1] They were gone! Even the forests in some national parks had been replaced by huge stretches of plantations. Marahoué National Park in the middle of the country had had the highest density of chimpanzees and a large population of elephants in 1990, but they had all disappeared over the last 17 years. Only Taï National Park remains a relatively safe haven for the chimpanzees and this is most likely because this park is the only one that profited from uninterrupted international support that kept active law enforcement operational as well as a long-term research presence near Taï village and an eco-tourist project in the southernmost area of the park.[2]

Chimpanzee culture is disappearing as the forests are cut and we are in a mad race to save both the chimpanzees and their habitat. One can certainly become pessimistic towards nature conservation as deforestation, climate change, illegal hunting, as well as poverty and water access shortages are continually increasing, in part due to inadequate economic and political authorities. On the other hand, more precise monitoring tools are being developed that have helped us to realize the amplitude of the problems and evaluate the efficiency of different conservation activities. Effective and uninterrupted law enforcement in Africa has proven essential to protecting national parks,[3] and combined with low impact eco-tourism and long-term research, we have a set of proven lasting conservation measures that should be implemented. Added to the effective management of sustainable forests

[1] Genevieve Campbell invested much effort to reach all of the same sites that we had visited 17 years earlier and she had to face the depressing truth that most of the forests still existed only on paper and in reality were gone (Campbell *et al.* 2008). She estimated that the chimpanzee population in Côte d'Ivoire declined by 90% over a period of 17 years!

[2] The important positive effects of research and tourism on chimpanzee and other animal conservation were clearly shown in Taï National Park (Campbell *et al.* 2011). We found a very clear sign over long distances related to the presence of researchers in that monkey population densities were much higher closer to our research camps and this effect was still visible as far as 40 km away (Köndgen *et al.* 2008).

[3] The effective presence of conservation teams on the ground, as in the case of national park guards, has proven to be the most effective way of protecting great ape populations in more than 100 protected zones throughout Africa, and this was even more important than the very effective role of conservation non-governmental organizations or researcher presence in those protected zones (Tranquilli *et al.* 2011).

outside of protected areas, we have the tools to make nature conservation a reality and a success.[4]

For as long as humans have been thinking, we have loved the idea of being unique and different from other living beings. One of these differences is based on our unique intelligence. If these differences truly exist, then as humans, we should be able to understand and acknowledge the fact that cultural abilities are shared with different animal species, including chimpanzees. Our planet is also their planet and we should guarantee them the right to survive, since only we have the power to destroy it.

[4] More information can be found on the webpage www.apesmanifesto.org.

References

Alexander, R. 1974. The evolution of social behavior. *Annual Review of Ecology and Systematics*, **5**: 325–383.

Allen, C. 2004. Is anyone a cognitive ethologist? *Biology and Philosophy*, **19**: 587–607.

Allen, C. and Bekoff, M. 2007. Animal minds, cognitive ethology, and ethics. *The Journal of Ethics*, **11**: 299–317.

Allen, C. and Hauser, M. 1991. Concept attribution in nonhuman animals: theoretical and methodological problems in ascribing complex mental processes. *Philosophy of Science*, **58**: 221–240.

Altmann, S. and Altmann, J. 1970. *Baboon ecology: African field research*. Chicago, IL: University of Chicago Press.

Alvard, M. 2003. The adaptive nature of culture. *Evolutionary Anthropology*, **12**: 136–149.

Ambrose, S. 2001. Paleolithic technology and human evolution. *Science* **291**, 1748–1753.

2010. Coevolution of composite-tool technology, constructive memory and language. *Current Anthropology*, **51**: S135–S141.

Anderson, J. 2011. A primatological perspective on death. *American Journal of Primatology*, **73**: 410–414.

Anderson, J., Gillies, A. and Lock, L. 2010. *Pan* thanatology. *Current Biology*, **20**: R349–R351.

Arnett, J. 2008. The neglected 85%: why American psychology needs to become less American. *American Psychologist*, **63**: 602–614.

Assersohn, C., Whiten, A., Kiwede, Z., Tinka, J. and Karamagi, J. 2004. Use of leaves to inspect ectoparasites in wild chimpanzees: a third cultural variant? *Primates*, **45**: 255–258.

Astington, J. and Baird, J. 2004. Why language matters for theory of mind. *International Society for the Study of Behavioural Development Newsletter*, **45**: 7–9.

Atkinson, Q. 2011. Phonemic diversity supports a serial founder effect model of language expansion from Africa. *Science*, **332**: 346–349.

Atran, S., Medin, D. and Ross, N. 2005. The cultural mind: environmental decision making and cultural modeling within and across populations. *Psychological Review*, **112**: 744–776.

Aunger, R. 1994. Are food avoidances maladaptive in the Ituri forest of Zaire? *Journal of Anthropology Resources*, **50**: 277–310.

1996. Acculturalization and the persistence of indigenous food avoidances in the Ituri forest, Zaire. *Human Organization*, **55**: 206–218.

Aureli, F., Schaffner, C., Boesch, C. *et al.* 2008. Fission–fusion dynamics: new research frameworks. *Current Anthropology*, **49**: 627–642.

Avis, J. and Harris, P. 1991. Belief-desire reasoning among Baka Children: evidence for a universal conception of mind. *Child Development*, **62**: 460–467.

Bakermans-Kranenburg, M., Ijzendoorn, M. and Juffer, F. 2008. Earlier is better: a meta-analysis of 70 years of intervention improving cognitive development in institutionalized children. *Monographs of the Society for Research in Child Development*, **73**: 279–293.

Bania, A., Harris, S., Kinsley, H. and Boysen, S. 2009. Constructive and deconstructive tool modification by chimpanzees (*Pan troglodytes*). *Animal Cognition*, **12**: 85–95.

Bard, K., Myowa-Yamakoshi, M., Tomonaga, M. *et al.* 2005. Group differences in the mutual gaze of chimpanzees *(Pan troglodytes*). *Developmental Psychology*, **41**: 616–624.

Bard, K., Brent, L., Lester, B., Worobey, J. and Suomi, S. 2011. Neurobehavioral integrity of chimpanzee newborns: comparisons across groups and across species reveal gene–environment interaction effects. *Infant and Child Development*, **20**: 47–93.

Barnard, A. 2000. *History and theory in anthropology*. Cambridge: Cambridge University Press.

2011. *Social anthropology and human origins*. Cambridge: Cambridge University Press.

Baron-Cohen, S. 1999. The evolution of a theory of mind. In *The descent of mind* (Corballis, M. and Lea, S., eds.), pp. 259–277. Oxford: Oxford University Press.

Barret, C. and Behne, T. 2005. Children's understanding of death as the cessation of agency: a test using sleep versus death. *Cognition*, **96**: 93–108.

Barrett, L., Henzi, P. and Rendall, D. 2007. Social brains, simple minds: does social complexity really require cognitive complexity. *Philosophical Transactions of the Royal Society*, Series B, **362**: 561–575.

Barth, J., Reaux, J. and Povinelli, D. 2005. Chimpanzees' (*Pan troglodytes*) use of gaze cues in object-choice tasks: different methods yield different results. *Animal Cognition*, **8**: 84–92.

Bates, L. and Byrne, R. 2010. Imitation: what animal imitation tells use about animal cognition. *Cognitive Sciences*, **1**: 685–695.

Baumard, N. 2010. Has punishment played a role in the evolution of cooperation? A critical review. *Mind and Society*, **9**: 171–192.

Beaune, S. 2004. The invention of technology: prehistory and cognition. *Current Anthropology*, **45**: 139–151.

Bekoff, M. 2004. Wild justice and fair play: cooperation, forgiveness, and morality in animals. *Biology and Philosophy*, **19**: 489–520.

Bekoff, M., Townsend, S. E. and Jamieson, D. 1994. Beyond monkey minds: towards a richer cognitive ethology. *Behavioral and Brain Sciences*, **17**: 571–572.

Berry, J., Poortinga, Y., Segall, M. and Dasen, P. 2002. *Cross-cultural psychology: Research and applications*. 2nd edition. Cambridge: Cambridge University Press.

Biro, D., Inoue-Nakamura, N., Tonooka, R. *et al.* 2003. Cultural innovation and transmission of tool use in wild chimpanzees: evidence from field experiments. *Animal Cognition*, **6**: 213–223.

Biro, D., Humle, T., Koops, K. *et al.* 2010. Chimpanzee mothers at Bossou, Guinea, carry the mummified remains of their dead infants. *Current Biology*, **20**: R352.

Boesch, C. 1991a. Teaching in wild chimpanzees. *Animal Behaviour*, **41**: 530–532.

1991b. Symbolic communication in wild chimpanzees? *Human Evolution* **6**: 81–90.

1991c. The effect of leopard predation on grouping patterns in forest chimpanzees. *Behaviour*, **117**: 220–242.

1992. New elements of a theory of mind in wild chimpanzees. *Behavioral and Brain Sciences*, **15**: 149–150.

1993a. Aspects of transmission of tool use in wild chimpanzees. In *Tools, language and intelligence: Evolutionary implications* (Gibson, K. and Ingold, T., eds.), pp. 171–183. Cambridge: Cambridge University Press,

1993b. Toward a new image of culture in chimpanzees. *Behavioral and Brain Sciences*, **16**: 514–515.

1994a. Chimpanzees–red colobus: a predator–prey system. *Animal Behaviour*, **47**: 1135–1148.

1994b. Cooperative hunting in wild chimpanzees. *Animal Behaviour*, **48**: 653–667.

1995. Innovation in wild chimpanzees. *International Journal of Primatology*, **16**: 1–16.

1996. Three approaches for assessing chimpanzee culture. In *Reaching into thought* (Russon, A., Bard, K. and Parker S., eds.), pp. 404–429. Cambridge: Cambridge University Press.

2002. Cooperative hunting roles among Taï chimpanzees. *Human Nature*, **13**: 27–46.

2003. Is culture a golden barrier between human and chimpanzee? *Evolutionary Anthropology*, **12**: 26–32.

2007. What makes us human (*Homo sapiens*)? The challenge of cognitive cross-species comparison. *Journal of Comparative Psychology*, **121**: 227–240.

2008. Imaginative scrub-jays, causal rooks, and a liberal application of Occam's after-shave. *Behavioral and Brain Sciences*, **31**: 134–135.

2009. Comparative social cognition. *Annual Review of Psychology*, **60**: 87–113.

2009. *The real chimpanzees: Sex strategies in the forest*. Cambridge: Cambridge University Press.

2012a. Ecology and cognition of tool use in chimpanzees. In *Tool use in animals: Cognition and ecology* (Sanz, C., Call, J. and Boesch, C., eds.). Cambridge: Cambridge University Press.

2012b. From material to symbolic cultures: culture in primates. In *The Oxford handbook of culture and psychology* (Valsiner, J., ed.). New York, NY: Oxford University Press.

Boesch, C. and Boesch, H. 1981. Sex differences in the use of natural hammers by wild chimpanzees: a preliminary report. *Journal of Human Evolution*, **10**: 585–593.

1983. Optimization of nut-cracking with natural hammers by wild chimpanzees. *Behaviour*, **83**: 265–286.

1984a. Mental map in wild chimpanzees: an analysis of hammer transports for nut cracking. *Primates*, **25**: 160–170.

1984b. Possible causes of sex differences in the use of natural hammers by wild chimpanzees. *Journal of Human Evolution*, **13**: 415–440.

1989. Hunting behavior of wild chimpanzees in the Taï National Park. *American Journal of Physical Anthropology*, **78**: 547–573.

1990. Tool use and tool making in wild chimpanzees. *Folia Primatologica*, **54**: 86–99.

Boesch, C. and Boesch-Achermann, H. 2000. *The chimpanzees of the Taï Forest: Behavioural ecology and evolution*. Oxford: Oxford University Press.

Boesch, C. and Tomasello, M. 1998. Chimpanzee and human cultures. *Current Anthropology*, **39**: 591–614.

Boesch, C., Marchesi, P., Marchesi, N., Fruth, B. and Joulian, F. 1994. Is nut cracking in wild chimpanzees a cultural behaviour? *Journal of Human Evolution*, **26**: 325–338.

Boesch, C., Uehara, S. and Ihobe, H. 2002. Variations in chimpanzee–red colobus interactions. In *Behavioural diversity in chimpanzees and bonobos* (Boesch, C., Hohmann, G. and Marchant, L., eds.), pp. 221–230. Cambridge: Cambridge University Press.

Boesch, C., Crockford, C., Herbinger, I. *et al.* 2008. Intergroup conflicts among chimpanzees in Taï National Park: lethal violence and the female perspective. *American Journal of Primatology*, **70**: 1–14.

Boesch, C., Head, J. and Robbins, M. 2009. Complex tool sets for honey extraction among chimpanzees in Loango National Park, Gabon. *Journal of Human Evolution*, **56**: 560–569.

Boesch, C., Bolé, C., Eckhardt, N. and Boesch, H. 2010. Altruism in chimpanzees: the case of adoption. *PLoS One*, **5**(1): e8901.

Boesch, E. 1991. *Symbolic action theory and cultural psychology*. Berlin/New York: Springer.

2003. Why does Sally never call Bobby "I"? *Culture and Psychology*, **9**: 287–297.

Boesch-Achermann, H. and Boesch, C. 1994. Hominization in the rainforest: the chimpanzee's piece of the puzzle. *Evolutionary Anthropology*, **3**: 9–16.

Bogart, S. and Pruetz, J. 2008. Ecological context of savanna chimpanzee (*Pan troglodytes verus*) termite fishing at Fongoli, Senegal. *American Journal of Primatology*, **70**: 605–612.

2011. Insectivory of savanna chimpanzees *(Pan troglodytes verus*) at Fongoli, Senegal. *American Journal of Physical Anthropology*, **145**: 11–20.

Bonner, J. 1980. *The evolution of culture in animals*. Princeton, NJ: Princeton University Press.

Bowles, S. and Gintis, H. 2002. *Homo reciprocans*. *Nature*, **415**: 125–128.

Boyd, R. 2006. The puzzle of human sociality. *Science*, **314**: 1555–1556.

Boyd, R. and Richerson, P. 1985. *Culture and the evolutionary process*. Chicago, IL: The University of Chicago Press.

1996. Why culture is common and cultural evolution is rare. In *Evolution of social behaviour patterns in primates and man* (Runciman, W., Maynard-Smith, J. and Dunbar, R., eds.), pp. 77–94. London: British Academy.

Bradsley, N. 2008. Dictator game giving: altruism or artifact? *Experimental Economy*, **11**: 122–133.

Bratman, M. 1992. Shared cooperative activity. *The Philosophical Review*, **101**: 327–341.

2009. Modest sociality and the distinctiveness of intention. *Philosophical Studies*, **144**: 149–165.

Brauer, J., Call, J. and Tomasello, M. 2007. Chimpanzees really know what others can see in a competitive situation. *Animal Cognition*, **10**: 439–448.

Brent, L., Bloomsmith, M. and Fisher, S. 1995. Factors determining tool-using ability in two captive chimpanzee (*Pan troglodytes*) colonies. *Primates*, **36**: 265–274.

Brill, B. and Roux, V. 2003. Comments to Stout, D. 2003. Skill and cognition in stone tool production: an ethnographic case study from Irian Jaya. *Current Anthropology*, **43**: 715–716.

Broesch, T., Callaghan, T., Henrich, J., Murphy, C. and Rochat, P. 2011. Cultural variations in children's mirror self-recognition. *Journal of Cross-Cultural Psychology*, **42**: 1019–1031.

Brown, P. and Levinson, S. 2000. Frames of spatial reference and their acquisition in Tenejapan Tzeltan. In *Culture, thought and development* (Nucci, L., Saxe, G. and Turiel, E., eds.), pp. 167–197. New York, NY: Erlbaum.

Brunet, M., Beauvilain, A., Coppens, Y. *et al.* 1995. The first australopithecine 2500 kilometres west of the Rift Valley (Chad). *Nature*, **378**: 273–274.

Buchsbaum, D., Gopnik, A., Griffiths, T. and Shafto, P. 2011. Children's imitation of causal action sequences is influenced by statistical and pedagogical evidence. *Cognition*, **120**: 331–340.

Buffon, G. 1762. *Natural history: The history of man and quadrupeds*. Paris: Imprimerie Royale.

Bugnyar, T. and Heinrich, B. 2005. Ravens, *Corvus corax*, differentiate between knowledgeable and ignorant competitors. *Proceedings of the Royal Society, Series B*, **272**: 1641–1646.

2006. Pilfering ravens, *Corvus corax*, adjust their behaviour to social context and identity of competitors. *Animal Cognition*, **9**: 369–376.

Bugnyar, T. and Kotrschal, K. 2002. Observational learning and the raiding of food caches in ravens, *Corvus corax*: is it "tactical" deception? *Animal Behaviour*, **64**: 185–195.

Bugnyar, T., Schwab, C., Schloegl, C., Kotrschal, K. and Heinrich, B. 2007. Ravens judge competitors through experience with play caching. *Current Biology*, **17**: 1804–1808.

Bulloch, M., Boysen, S. and Furlong, E. 2008. Visual attention and its relation to knowledge states in chimpanzees, *Pan troglodytes. Animal Behaviour*, **76**: 1147–1155.

Busse, C. 1977. Chimpanzee predation as a possible factor in the evolution of red colobus monkey social organization. *Evolution*, **31**: 907–911.

1978. Do chimpanzees hunt cooperatively? *American Naturalist*, **112**: 767–770.

Butterfill, S. 2012. Joint action and development. *Philosophical Quarterly*, **62**: 23–47.

Byrne, R. 2007. Culture in great apes: using intricate complexity in feeding skills to trace the evolutionary origin of human technical prowess. *Philosophical Transactions of the Royal Society* B, **362:** 577–585.

Byrne, R. and Russon, A. 1998. Learning by imitation: a hierarchical approach. *Behavioral and Brain Sciences*, **21**: 667–721.

Caldwell, C. and Whiten, A. 2002. Evolutionary perspectives on imitation: is a comparative psychology of social learning possible? *Animal Cognition*, **5**: 193–208.

Call, J. 2009. Contrasting the social cognition of humans and nonhuman apes: the shared intentionality hypothesis. *Trends in Cognitive Science*, **1**: 368–379.

Call, J. and Tomasello, M. 2008. Does the chimpanzee have a theory of mind? 30 years later. *Trends in Cognitive Sciences*, **12**: 187–192.

Call, J., Hare, B. and Tomasello, M. 1998. Chimpanzee gaze following in an object choice task. *Animal Cognition*, **1**: 89–100.

Call, J., Carpenter, M. and Tomasello, M. 2005. Copying results and copying actions in the process of social learning: chimpanzees (*Pan troglodytes*) and human children (*Homo sapiens*). *Animal Cognition*, **8**: 151–163.

Callaghan, T., Rochat, P., Lillard, A., *et al.* 2005. Synchrony in the onset of mental-state reasoning. *Psychological Science*, **16**: 378–384.

Campbell, G., Kuehl, H., N'Goran, P. and Boesch, C. 2008. Alarming decline of West African chimpanzees in Côte d'Ivoire. *Current Biology*, **18**: R903–R904.

Campbell, G., Kuehl, H., Diarrassouba, A., N'Goran, P. and Boesch, C. 2011. Long-term research sites as refugia for threatened and over-harvested species. *Biology Letters*, **7**: 723–726.

Canale, G., Guidorizzi, C., Kierulff, M. and Gatto, C. 2009. First record of tool use by wild populations of the Yellow-Breasted Capuchin moneys (*Cebus xanthosternos*) and new records for the bearded capuchin (*Cebus libidinosus*). *American Journal of Primatology* **71**: 366–372.

Caro, T. and Hauser, M. 1992. Is there teaching in nonhuman animals? *Quaterly Review of Biology*, **67**: 151–174.

Carpendale, J. and Lewis, C. 2004. Constructing an understanding of mind: the development of children's social understanding within social interaction. *Behavioral and Brain Sciences*, **27**: 79–151.

Carpenter, M. 2009. Just how joint is joint action in infancy? *Topics in Cognitive Science*, **1**: 380–392.

Carruthers, P. and Smith, P. 1996. *Theories of theories of mind.* Cambridge: Cambridge University Press.

Cavalli-Sforza, L.L. and Feldman, M.W. 1981. *Cultural transmission and evolution: A quantitative approach.* Princeton, NJ: Princeton University Press.

Cavalli-Sforza, L., Feldman, M., Chen, K. and Dornbusch, S. 1982. Theory and observation in cultural transmission. *Science*, **218**: 19–27.

Chalmeau, R. 1994. Do chimpanzees cooperate in a learning task? *Primates*, **35**: 385–392.

Chalmeau, R. and Gallo, A. 1996. Cooperation in primates: critical analysis of behavioural criteria. *Behavioural Processes*, **35**: 101–111.

Cheney, D. and Seyfarth, R. 1990. *How monkeys see the world: Inside the mind of another species.* Chicago, IL: Chicago University Press.

2007. *Baboon metaphysics: The evolution of a social mind.* Chicago, IL: The University of Chicago Press.

Clark, M. 2002. *In search of human nature.* London: Routledge.

Collard, I. and Foley, R. 2004. Latitudinal patterns and environmental determinants of recent human cultural diversity: do humans follow biogeographical rules? *Evolutionary Ecology Research*, **4**: 371–383.

Collins, A. and McGrew, W. 1987. Termite fauna related to differences in tool-use between groups of chimpanzees (*Pan troglodytes*). *Primates*, **28**: 457–471.

Coppens Y. 1983. *Le singe, l'Afrique et l'Homme.* Paris: Fayard.

Corp, N. and Byrne, R. 2002. The ontogeny of manual skill in wild chimpanzees: evidence from feeding on the fruit of *Saba florida*. *Behaviour*, **139**: 137–168.

Corp, N., Hayaki, H., Matsusaka, T., *et al.* 2009. Prevalence of muzzle-rubbing and hand-rubbing behavior in wild chimpanzees in Mahale Mountains National Park, Tanzania. *Primates*, **50**: 184–189.

Correla, S., Dickinson, A. and Clayton, N. 2007. Western scrub-jays anticipate future needs independently of their current motivational state. *Current Biology*, **17**: 856–861.

Corriveau, K. and Harris, P. 2009. Choosing your informant: weighing familiarity and recent accuracy. *Developmental Science*, **12**: 426–437.

Crawford, M. P. 1937. The cooperative solving of problems by young chimpanzees. *Comparative Psychology Monographs*, **14**: 1–88.

Cronk, L., Chagnon, N. and Irons, W. 2000. *Adaptation and human behavior: An anthropological perspective.* New York, NY: Aldine De Gruyter.

Csibra, G. 2007. Teachers in the wild. *Trends in Cognitive Sciences*, **11**: 95–96.

Csibra, G. and Gergely, G. 2009. Natural pedagogy. *Trends in Cognitive Sciences*, **13**: 148–153.

2011. Natural pedagogy as evolutionary adaptation. *Philosophical Transactions of the Royal Society B*, **366**: 1149–1157.

Cutting, A. and Dunn, J. 1999. Theory of mind, emotion understanding, language and family background: individual differences and interrelations. *Child Development*, **70**: 853–865.

Dally, J., Emery, N. and Clayton, N. 2006. Food-caching scrub-jays keep track of who was watching when. *Science*, **312**: 1662–1665.

Dana, J., Cain, D. and Dawes, R. 2006. What you don't know won't hurt me: costly (but quiet) exit in dictator games. *Organizational Behavior and Human Decision Processes*, **100**: 193–201.

Dart, R. 1925. *Australopithecus africanus*, the man-ape of South Africa. *Nature*, **115**: 195–199.

Dasen, P. R. 1982. Cross-cultural data on operational development: asymptotic development curves. In *Regressions in mental development: Basic phenomena and theories* (Bever, T., ed.), pp. 221–232. Hillsdale, NJ: Lawrence Erlbaum Associates.

Dasen, P. R., Inhelder, B., Lavallée, M. and Reitschitzki, J. 1978. *La naissance de l'intelligence chez Penfant Baoule de Côte d'Ivoire*. Berne: Hans Huber.

Davenport, R. and Rogers, C. 1970. Differential rearing of the chimpanzee: a project survey. In *The chimpanzee* (Bourne, G., ed.), Vol. 3, pp. 337–360. Baltimore, MD: University Park Press.

Davenport, R., Rogers, C. and Rumbaugh, D. 1973. Long-term cognitive deficits in chimpanzees associated with early impoverished rearing. *Developmental Psychology*, **9**: 343–347.

Deblauwe, I., Guislan, P., Dupain, J. and van Elsacker, L., 2006. Use of a tool-set by *Pan troglodytes troglodytes* to obtain termites (*Macrotermes*) in the periphery of the Dja Biosphere Reserve, Southeast Cameroon. *American Journal of Primatology*, **68**: 1191–1196.

de León, L. 1994. Exploration in the acquisition of geocentric location by Tzotzil children. *Linguistics*, **32**: 857–884.

Douglas-Hamilton, I. and Douglas-Hamilton, O. 1975. *Life among the elephants*. New York, NY: Wythan Publications Ltd.

Draper, P. 1976. Social and economic constraints on child life among the !Kung. In *Kalahari hunter–gatherers: Studies of the !Kung San and their neighbors* (Lee, R.B. and DeVore, I., eds.), pp. 199–217. Cambridge, MA: Harvard University Press.

Duffy, K. 1984. *The children of the forest: Africa's Mbuti pygmies.* Prospect Heights, NY: Waveland Press.

Durham, W. 1991. *Coevolution: Genes, culture and human diversity*. Stanford, CA: Stanford University Press.

2001. Effects of experience and social context on prospective caching strategies by scrub jays. *Nature*, **414**: 443–446.

Fashing, P., Nguyen, N., Barry, T., *et al.* 2011. Death among Geladas (*Theropithecus gelada*): a broader perspective on mummified infants and primate thanatology. *American Journal of Primatology*, **73**: 405–409.

Fay, M. and Carroll, R. 1994. Chimpanzee tool use for honey and termite extraction in Central Africa. *American Journal of Primatology*, **34**: 309–317.

Fehr, E. and Fischbacher, U. 2003. The nature of human altruism. *Nature*, **425**: 785–791.

Fehr, E. and Gächter, S. 2002. Altruistic punishment in humans. *Nature*, **415**: 137–140.

Fitch, T., Huber, L. and Bugnyar, T. 2010. Social cognition and the evolution of language: constructing cognitive phylogenies. *Neuron*, **65**: 795–814.

Ford, J. 1962. *A quantitative method for deriving cultural chronology*. Pan American Union, Technical Bulletin No. 1.

Fragaszy, D. 2003. Making space for traditions. *Evolutionary Anthropology*, **12**: 61–70.

Fragaszy, D., Kennedy, E., Murnane, A., *et al.* 2009. Navigating two-dimensional mazes: chimpanzees (*Pan troglodytes*) and capuchins (*Cebus apella* sp.) profit from experience differently. *Animal Cognition*, **12**: 491–504.

Franks, N. and Richardson, T. 2006. Teaching in tandem-running ants. *Nature*, **439**: 153.

Frith, C. 2007. *Making up the mind: How the brain creates our mental world.* Oxford: Blackwell Publishing.

Furlong, E., Boose, K. and Boysen, S. 2008. Raking it in: the impact of enculturation on chimpanzee tool use. *Animal Cognition*, **11**: 83–97.

Gächter, S., Herrmann, B. and Thöni, C. 2010. Culture and cooperation. *Philosophical Transactions of the Royal Society*, B, **365**: 2651–2661.

Galat-Luong, A. and Galat, G. 2000. Chimpanzees and baboons drink filtrated water. *Folia Primatologica* **71**: 258.

Galef, B. 1988. Imitation in animals: history, definition, and interpretation of data from the psychological laboratory. In *Social learning: Psychological and biological perspectives* (Zentall, T.R. and Galef, B.G. Jr, eds.), pp 3–28. Hillsdale, NJ: Lawrence Erlbaum.

1992. The question of animal culture. *Human Nature*, **3**: 157–178.

2004. Approaches to the study of traditional behaviors of free-living animals. *Learning and Behavior*, **32**: 53–61.

Gardner, B. and Gardner, R. 1989. Prelinguistic development of children and chimpanzees. *Human Evolution*, **4**: 433–460.

Gardner, R. 2005. Animal cognition meets evo-devo. *Behavior and Brain Sciences*, **28**: 699–700.

2008. Comparative intelligence and intelligent comparisons. *Behavior and Brain Sciences*, **31**: 135–136.

Gardner, R. and Gardner, B. 1969. Teaching sign language to a chimpanzee. *Science*, **165**: 664–672.

Gergely, G., Bekkering, H. and Kiraly, I. 2002. Rational imitation in preverbal infants. *Nature*, **415**: 755.

Gibson, R.M. and Hoglund, J. 1992. Copying and sexual selection. *Trends in Ecology and Evolution*, **7**: 229–231.

Gilby, I. 2006. Meat sharing among the Gombe chimpanzees: harassment and reciprocal exchange. *Animal Behaviour*, **71**: 953–963.

Gilby, I. and Wrangham, R. 2007. Risk-prone hunting by chimpanzees (*Pan troglodytes schweinfurthii*) increases during periods of high diet quality. *Behavior Ecology and Sociobiology*, **61**: 1771–1779.

Gilby, I., Thompson, E., Ruane, J. and Wrangham, R. 2010. No evidence of short-term exchange of meat for sex among chimpanzees. *Journal of Human Evolution*, **59**: 44–53.

Gintis, H., Bowles, S., Boyd, R. and Fehr, E. 2003. Explaining altruistic behavior in humans. *Evolution and Human Behavior*, **24**: 153–172.

Gomes, C. and Boesch, C. 2009. Wild chimpanzees exchange meat for sex on a long-term basis. *PLoS One*, **4**(4): e5116.

2011. Reciprocity and trades in wild West African chimpanzees. *Behavioral Ecology and Sociobiology*, **65**: 2183–2196.

Goodall, J. 1963. Feeding behaviour of wild chimpanzees: a preliminary report. *Symposium of the Zoological Society, London*, **10**: 39–48.

1964. Tool-using and aimed throwing in a community of free-living chimpanzees. *Nature*, **201**: 1264–1266.

1968. Behaviour of free-living chimpanzees of the Gombe Stream area. *Animal Behaviour Monograph*, **1**: 163–311.

1986. *The chimpanzees of Gombe: Patterns of behavior*. Cambridge: The Belknap Press of Havard University Press.

Goodrich, G. and Allen, C. 2007. Conditioned anti-anthropomorphism. *Comparative Cognition and Behavior Reviews*, **2**: 147–150.

Gopnik, A. and Meltzoff, A. 1994. Minds, bodies and persons: young children's understanding of the self and others as reflected in imitation and theory of mind research. In *Self-awareness in animals and humans* (Parker, S., Mitchell, R. and Boccia, M., eds.), pp. 166–186. Cambridge: Cambridge University Press.

Greenfield, P. 2009. Linking social change and developmental change: shifting pathways of human development. *Developmental Psychology*, **45**(2): 401–418.

Greenfield, P., Maynard, A. and Childs, C. 2003a. Historical change, cultural learning, and cognitive representation in Zinacantec Maya children. *Cognitive Development*, **18**: 455–487.

Greenfield, P., Keller, H., Fuligni, A. and Maynard, A. 2003b. Cultural pathways through universal development. *Annual Review of Psychology*, **54**:461–490.

Gruber, T., Muller, M., Strimling, P., Wrangham, R. and Zuberbuehler, K. 2009. Wild chimpanzees rely on cultural knowledge to solve an experimental honey acquisition task. *Current Biology*, **19**:1–5.

Guala, F. 2005. *The methodology of experimental economics*. Cambridge: Cambridge University Press.

2012. Reciprocity: weak or strong? What punishment experiments do (and do not) demonstrate. *Behavioral and Brain Sciences*, **35**: 1–59.

Hala, S. and Chandler, M. 1996. The role of strategic planning in accessing false-belief understanding. *Child Development*, **67**: 2948–2966.

Hanus, D. and Call, J. 2008. Chimpanzees infer the location of a reward on the basis of the effect of its weight. *Current Biology*, **18**: R370–372.

Hare, B., Call, J., Agnetta, B. and Tomasello, M. 2000. Chimpanzees know what conspecific do and do not see. *Animal Behaviour*, **59**: 771–785.

Hare, B., Call, J. and Tomasello, M. 2001. Do chimpanzees know what conspecifics know? *Animal Behaviour*, **61**: 139–151.

Hare, B., Brown, M., Williamson, C. and Tomasello, M. 2002. The domestication of social cognition in dogs. *Science*, **298**: 1634–1636.

Hare, B., Call, J. and Tomasello, M. 2006. Chimpanzees deceive a human competitor by hiding. *Cognition*, **101**: 495–514.

Harlow, H. and Harlow, M. 1962. Social deprivation in monkeys. *Scientific American*, **207**: 136–146.

Harris, P. and Corriveau, K. 2011. Young children's selective trust in informants. *Philosophical Transactions of the Royal Society, B*, **366**: 1179–1187.

Harrison, G. W. and List, J. A. 2004. Field experiments. *Journal of Economic Literature*, **42**: 1009–1055.

Hart, D. and Karmel, M. 1996. Self-awareness and self-knowledge in humans, apes, and monkeys. In: *Reaching into thought: The minds of the great apes* (Russon, A., Bard, K. and Parker, S., eds.), pp: 325–347. Cambridge: Cambridge University Press.

Hart, D. and Sussman, R. 2005. *Man the hunted. Primates, predators and human evolution*. New York, NY: Westview Press.

Hatley, T. and Kappelman, J. 1980. Bears, pigs, and Plio-Pleistocene hominids: a case for the exploitation of belowground food resources. *Human Ecology*, **8**: 371–387.

Haun, D., Rapold, C., Call, J., Janzen, G. and Levinson, S. 2006. Evolutionary psychology of spatial representations in the Hominidae. *Current Biology*, **16**: 1736–1740.

Hauser, M. 2000. *Wild minds: What animals really think*. New York, NY: Henry Holt.

2006. *Moral minds: The nature of right and wrong*. New York, NY: HarperCollins.
2009. The possibility of impossible cultures. *Nature*, **460**: 190–196.

Hawkes, K. and Paine, R. 2006. *The evolution of human life history*. Santa Fe, NM: School of American Research Press.

Henrich, J., Boyd, R., Bowles, S., *et al*. 2001. In search of *Homo economicus*: behavioural experiments in 15 small-scale societies. *American Economics Review*, **91**: 73–78.

Henrich, J., Ensminger, J., Boyd, R., *et al.* 2005. "Economic man" in cross-cultural perspective: behavioral experiments in 15 small-scale societies. *Behavioral and Brain Sciences*, **28**: 795–855.

Henrich, J., McElreath, R., Barr, A., *et al.* 2006. Costly punishment across human societies. *Science*, **312**: 1767–1770.

Henrich, J., Ensminger, J., McElreath, R., *et al.* 2010a. Markets, religion, community size and the evolution of fairness and punishment. *Science*, **327**: 1480–1484.

Henrich, J., Heine, S. and Norenzayan, A. 2010b. The weirdest people in the world? *Behavioral and Brain Sciences*, **33**: 61–135.

Hernandez-Aguilar, A., Moore, J. and Pickering, T. 2007. Savanna chimpanzees use tools to harvest the underground storage organs of plants. *Proceedings of the National Academy of Sciences*, **104**(49): 19 210–19 213.

Herrmann, E., Call, J., Hernandez-Lloreda, M., Hare, B. and Tomasello, M. 2007. Humans have evolved specialized skills of social cognition: the cultural intelligence hypothesis. *Science* **317**: 1360–1366.

Herrmann, E., Wobber, V. and Call, J. 2008. Great Apes' (*Pan troglodytes*, *Pan paniscus*, *Gorilla gorilla*, *Pongo pygmaeus*) understanding of tool functional properties after limited experience. *Journal of Comparative Psychology*, **122**(2): 220–230.

Hewlett, B. and Cavalli-Sforza, L. 1986. Cultural transmission among Aka pygmies. *American Anthropologist*, **88**: 922–934.

Hewlett, B., Fouts, H., Boyette A. and Hewlett, B. 2011. Social learning among Congo Basin hunter–gatherers. *Philosophical Transactions of the Royal Society B*, **366**: 1168–1178.

Heyes, C. 1993. Imitation, culture and cognition. *Animal Behavior*, **46**: 999–1010.
1998. Theory of mind in nonhuman primates. *Behavior and Brain Sciences*, **21**:101–148.

Hicks, T. 2010. *A chimpanzee mega-culture? Exploring behavioral continuity in* Pan troglodytes schweinfurthii *across northern DR Congo*. PhD thesis. Faculteit der Natuurwetenschappen, University of Amsterdam.

Hicks, T., Fouts, R. and Fouts, D. 2005. Chimpanzees (*Pan troglodytes troglodytes*) tool use in the Ngotto Forest, Central African Republic. *American Journal of Primatology*, **65**: 221–237.

Hill, K. and Hawkes, K. 1983. Neotropical hunting among the Ache of eastern Paraguay. In *Adaptive responses of native Amazonians* (Hames, R. and Vickers, W., eds.), pp. 139–188. New York, NY: Academic Press.

Hill, K. and Hurtado, M. 1996. *Ache life history: The ecology and demography of a foraging people*. New York, NY: Walter de Gruyter.

Hohmann, G. and Fruth, B. 2003. Culture in Bonobos? Between-species and within-species variation in behavior. *Current Anthropology*, **44**(4): 563–571.

Holas, B. 1980. *Traditions Krou*. Paris: Ferdinand Nathan.

Homes, H., Black, C. and Miller, S. 1996. A cross-task comparison of false belief understanding in a Head Start population. *Journal of Experimental Child Psychology*, **63**: 263–285.

Hopper, L. 2010. Ghost experiments and the dissection of social learning in humans and animals. *Biological Reviews*, **85**: 685–701.

Hopper, L., Spiteri, A., Lambeth, S. *et al.* 2007. Experimental studies of traditions and underlying transmission processes in chimpanzees. *Animal Behaviour*, **73**: 1021–1032.

Hopper, L., Lambeth, S., Schapiro, S. and Whiten, A. 2008. Observational learning in chimpanzees and children studied through "ghost" conditions. *Proceedings of the Royal Society, Series B*, **275**: 835–840.

Hoppitt, W., Brown, G., Kendal, R., *et al.* 2008. Lessons from animal teaching. *Trends in Ecology and Evolution*, **23**(9): 486–493.

Horner, V. and Whiten, A. 2005. Causal knowledge and imitation/emulation switching in chimpanzees (*Pan troglodytes*) and children (*Homo sapiens*). *Animal Cognition*, **8**: 164–181.

2007. Learning from others' mistakes? Limits on understanding a trap-tube task by young chimpanzees (*Pan troglodytes*) and children (*Homo sapiens*). *Journal of Comparative Psychology*, **121**(1): 12–21.

Horner, V., Whiten, A., Flynn, E. and de Waal, F. 2006. Faithful replication of foraging techniques along cultural transmission chains by chimpanzees and children. *Proceedings of the National Academy of Sciences*, **103**(37): 13 878–13 883.

Horner, V., Proctor, D., Whiten, A., Bonnie, K. and de Waal, F. 2010. Prestige affects cultural learning in chimpanzees. *PLoS ONE* **5**: e10625.

Horner, V., Carter, D., Suchak, M. and de Waal, F. 2011. Spontaneaous prosocial choice by chimpanzees. *PNAS*, **108**(33): 13 847–13 851.

Horowitz, A. 2003. Do humans ape? Or do apes human? Imitation and intention in humans (*Homo sapiens*) and other animals. *Journal of Comparative Psychology*, **117**: 325–336.

Hosaka, K., Matsumoto-Oda, A., Huffman, M. A. and Kawanaka, K. 2000. Reactions to dead bodies of conspecifics by wild chimpanzees in the Mahale Mountains, Tanzania. *Primate Research* **16**: 1–15 [in Japanese with English summary].

Howe, N., Rinaldi, C., Jennings, M. and Petrakos, H. 2002. "No! the lambs can stay out because they got cozies": constructive and destructive sibling conflict, pretend play and social understanding. *Child Development*, **73**: 1460–1473.

Hrdy, S. 2009. *Mothers and others: The evolutionary origins of mutual understanding*. Cambridge, MA: The Belknap Press of Harvard University Press.

Huber, L., Range, F., Voelkl, B. *et al.* 2009. The evolution of imitation: what do the capacities of non-human animals tell us about the mechanisms of imitation? *Philosophical Transactions of the Royal Society B*, **364**: 2299–2309.

Huffman, M. 1996. Acquisition of innovative cultural behaviors in nonhuman primates: a case study of stone handling, a socially transmitted behavior in Japanese macaques. In *Social learning in animals: The roots of culture* (Heyes, C. and Galef, B., eds.), pp. 267–289. Orlando, FL: Academic Press.

Humle, T. and Matsuzawa, T. 2002. Ant-dipping among the chimpanzees of Bossou, Guinea, and some comparisons with other sites. *American Journal of Primatology*, **58**: 133–148.

Hunt, G. 1996. Manufacture and use of hook-tools by New Caledonian crows. *Nature*, **379**: 249–251.

Hunt, G. and Gray, R. 2003. Diversification and cumulative evolution in New Caledonian crow tool manufacture. *Proceedings of the Royal Society, London, Series B*, **270**: 867–874.

Hunt, K., Cleminson, A., Latham, J., Weiss, R. and Grimmond, S. 1999. A partly habituated community of dry-habitat chimpanzees in the Semliki Valley Wildlife Reserve, Uganda. *American Journal of Physical Anthropology Supplement*, **28**: 157.

Imanishi, K. 1952. Evolution of humanity. In *Man* (Imanishi, K., ed.), pp. 36–94. Mainichi-Shinbunsha.

Inoue, S. and Matsuzawa, T. 2007. Working memory of numerals in chimpanzees. *Current Biology* **17**(23): R1004–R1005.

Inoue-Nakamura, N. and Matsuzawa, T. 1997. Development of stone tool use by wild chimpanzees (*Pan troglodytes*). *Journal of Comparative Psychology*, **111**(2): 159–173.

Iovita, R. and McPherron, S. 2011. The handaxe reloaded: a morphometric reassessment of Acheuliand and Middle Paleolithic handaxes. *Journal of Human Evolution*, **61**: 61–74.

Isaac, G. 1978. The food-sharing sharing behavior of protohuman hominids. *Scientific American*, **238**: 90–108.

Jaeggi, S., Buschkuehl, M., Jonides, J. and Perrig, W. 2008. Improving fluid intelligence with training on working memory. *Proceedings of the National Academy of Sciences USA*, **105**: 6829–6833.

Jensen, K., Call, J. and Tomasello, M. 2007. Chimpanzees are vengeful but not spiteful. *Proceedings of the National Academy of Science*, **104**: 13 046–13 050.

Johanson, D. and Edey, M. 1981. *Lucy: The beginnings of humankind*. New York, NY: Simon and Schuster.

Jordan, P. and Shennan, S. 2003. Cultural transmission, language, and basketry traditions amongst the California Indians. *Journal of Anthropological Archaeology*, **22**: 42–74.

Kaminski, J., Call, J. and Tomasello, M. 2004. Body orientation and face orientation: two factors controlling apes' begging behavior from humans. *Animal Cognition*, **7**: 216–223.

Karin-D'Arcy, M. and Povinelli, D. 2002. Do chimpanzees know what each other see? A closer look. *International Journal of Comparative Psychology*, **15**: 21–54.

Kawai, M. 1965. Newly acquired precultural behavior of the natural troop of Japanese monkeys on Koshima islet. *Primates*, **6**: 1–30.

Kawamura, S. 1959. The process of sub-culture propagation among Japanese macaques. *Primates*, **2**: 43–60.

1965. Sub-culture among Japanese macaques. In *Monkeys and apes – Sociological studies* (Kawamura, S. and Itani, J., eds), pp. 237–289. Tokyo: Chuokoronsha.

Keeley, B. 2004. Anthropomorphism, primatomorphism, mammalomorphism: understanding cross-species comparisons. *Biology and Philosophy*, **19**: 521–540.

Keller, H., Lohaus, A., Kuensemueller, P., *et al.* 2004a. The bio-culture of parenting: evidence from five cultural communities. *Parenting: Science and Practice*, **4**: 25–50.

Keller, H., Yovsi, R., Borke, J. *et al.* 2004b. Developmental consequences of early parenting experiences: self-recognition and self-regulation in three cultural communities. *Child Development*, **75**(6): 1745–1760.

Keller, H., Joscha Kärtner, J., Borke, J., Yovsi, R. and Kleis, A. 2005. Parenting styles and the development of the categorical self: a longitudinal study on mirror self-recognition in Cameroonian Nso and German families. *International Journal of Behavioral Development*, **29**(6): 496–504.

Kenward, B., Weir, A., Rutz, C. and Kacelnik, A. 2005. Tool manufacture by naive juvenile crows. *Nature*, **433:** 121.

Knight, N., Sousa, P., Barrett, J. and Atran, S. 2004. Children's attribution of belief to humans and God: cross-cultural evidence. *Cognitive Science*, **28**: 117–126.

Koehler, W. 1959. *The mentality of apes*. New York, NY: Vintage Books.
Köndgen, S., Kühl, H., N'Goran, P., *et al*. 2008. Pandemic human viruses cause decline of endangered great apes. *Current Biology*, **18**: 260–264.
Kooriyama, T. 2009. The death of a newborn chimpanzee at Mahale: reactions of its mother and other individuals to the body. *Pan Africa News*, **16**(2): 4.
Kroeber, A. and Kluckhohn, C. 1952. *Culture: A critical review of concepts and definitions*. New York, NY: Random House.
Kubinyi, E., Viranyi, Z. and Miklosi, A. 2007. Comparative social cognition: from wolf and dog to humans. *Comparative Cognition and Behavior Reviews*, **2**: 26–46.
Kummer, H. 1971. *Primate societies*. Chicago, IL: Aldine.
Kuper, A. 1999. *Culture: The anthropologist's account*. Cambridge, MA: Harvard University Press.
Kutsukake, N. and Matsusaka, T. 2002. Incident of intense aggression by chimpanzees against an infant from another group of Mahale Mountains National Park, Tanzania. *American Journal of Primatology*, **58**: 175–180.
Laland, K. and Hoppitt, W. 2003. Do animals have culture? *Evolutionary Anthropology*, **12**: 150–159.
Laland, K. and Janik, V. 2006. The animal cultures debate. *Trends in Ecology and Evolution*, **21**(10): 542–547.
Langergraber, K., Boesch, C., Inoue, E., *et al*. 2011. Genetic and "cultural" similarity in wild chimpanzees. *Proceedings of the Royal Society B*, **278**: 408–416.
Lanjouw, A. 2002. Behavioural adaptations to water scarcity in Tongo chimpanzees. In *Behavioural diversity in chimpanzees and bonobos* (Boesch, C., Hohmann G. and Marchant, L., eds.), pp. 52–60. Cambridge: Cambridge University Press.
Leakey, L. S. B. 1961. *The progress and evolution of man in Africa*. London: Oxford University Press.
Leakey, R. 1980. *The making of mankind*. London: Book Club Associates.
Leakey, R. and Lewin, R. 1977. *Origins*. London: McDonald and Jones.
Leavens, D. and Bard, K. 2011. Environmental influences on joint attention in great apes: implications for human cognition. *Journal of Cognitive Education and Psychology*, **10**(1): 9–31.
Leavens, D., Hopkins, W. and Bard, K. 2005. Understanding the point of chimpanzee pointing: epigenesis and ecological validity. *Current Directions in Psychological Science*, **14**: 185–189.
Leavens, D., Bard, K. and Hopkins, W. 2010. BIZARRE chimpanzees do not represent "The Chimpanzee". Commentary on target article by Heinrich, Heine & Norensayan, "The weirdest people in the world?" *Behavioral and Brain Sciences*, **33**(1): 100–101.
Lee, P. and Moura. 2005. Response to terrestriality and tool use by Visalberghi *et al.* 2005. *Science*, **308:** 951–952.
Lee, R. B. and DeVore, I. 1968. *Man the hunter*. Chicago, IL: Aldine-Atherton.
Leendertz, F., Ellerbock, H., Boesch, C., *et al*. 2004. Anthrax kills wild chimpanzees in a tropical rainforest. *Nature* **430**: 451–452.
Lefebvre, L., Nicolakakis, N. and Boire, B. 2002. Tools and brains in birds. *Behaviour* **139**: 939–973.
Lefebvre, L., Reader, S. M. and Sol, D. 2004. Brains, innovations and evolution in birds and primates. *Brain and Behavioral Evolution*, **63**: 233–246.
Leslie, A., Friedman, O. and German, T. 2004. Core mechanisms in "theory of mind". *Trends in Cognitive Science*, **8**: 528–533.

Levitt, S. and List, J. 2007. What do laboratory experiments measuring social preferences reveal about the real world? *Journal of Economic Perspectives*, **21**: 153–174.

2008. *Homo economicus* evolves. *Science*, **319**: 909–910.

Limongelli, L., Visalberghi, E. and Boysen, S. 1995. Comprehension of cause–effect relationships in a tool-using task by common chimpanzees (*Pan troglodytes*). *Journal of Comparative Psychology*, **109**: 18–26.

List, J. 2006. The behavioralist meets the market: measuring social preferences and reputation effects in actual transactions. *Journal of Political Economy*, **114**(1): 1–37.

Liszkowski, U., Schafer, M., Carpenter, M. and Tomasello, M. 2009. Prelinguistic infants, but not chimpanzees, communicate about absent entities. *Psychological Science*, **20**(5): 654–660.

Lonsdorf, E. 2005. Sex differences in the development of termite-fishing skills in the wild chimpanzees, *Pan troglodytes schweinfurthii*, of Gombe National Park, Tanzania. *Animal Behaviour*, **70**: 673–683.

2006. What is the role of mothers in the acquisition of termite-fishing behaviors in wild chimpanzees (*Pan troglodytes schweinfurthii*)? *Animal Cognition*, **9**: 36–46.

Lonsdorf, E., Pusey, A. and Eberly, L. 2004. Sex differences in learning in chimpanzees. *Nature*, **428**: 715–716.

Lorenz, K. 1961. *King Solomon's ring*. London: Methuen.

Lovejoy, C.O. 1981. The origin of man. *Science*, **211**: 341–350.

Lumsden, C. and Wilson, E. 1981. *Genes, mind and culture*. Cambridge, MA: World Scientific.

Luncz, L., Mundry, R. and Boesch, C. 2012. Evidence for cultural differences between neighboring chimpanzee (*Pan troglodytes verus*) communities. *Current Biology*, **22:** 922–926.

Lycett, S., Collard, M. and McGrew, W. 2007. Phylogenetic analyses of behavior support existence of culture among wild chimpanzees. *Proceedings of the National Academy of Sciences USA*, **104**(45): 17 588–17 592.

2010. Are behavioral differences among wild chimpanzee communities genetic or cultural? An assessment using tool-use data and phylogenetic methods. *American Journal of Physical Anthropology*, **142**(3): 461–467.

Lyn, H., Russell, J. and Hopkins, W. 2010. The impact of environment on the comprehension of declarative communication in apes. *Psychological Science*, **21**(3): 360–365.

Lyons, D., Young, A. and Keil, F. 2007. The hidden structure of overimitation. *Proceedings of the National Academy of Sciences USA*, **104**(50): 19 751–19 756.

Lyons, D., Damrosch, D., Lin, J., Macris, D. and Keil, F. 2011. The scope and limits of overimitation in the transmission of artifact culture. *Philosophical Transactions of the Royal Society, Series B*, **366**: 1158–1167.

Majid, A., Bowerman, M., Kita, S., Haun, D. and Levinson, S. 2004. Can language restructure cognition? The case for space. *Trends in Cognitive Sciences*, **8**: 108–114.

Maki, S., Fritz, J. and England, N. 1993. An assessment of early differential rearing conditions on later behavioral development in captive chimpanzees. *Infant Behavior and Development*, **16**: 373–381.

Marlowe, F. 2010. *The Hadza hunter-gatherers of Tanzania*. Berkeley, CA: University of California Press.

Marlowe, F., Berbesque, J., Barr, A., *et al.* 2008. More "altruistic" punishment in larger societies. *Proceedings of the Royal Society B*, **275**: 587–590.

Marlowe, F., Berbesque, C., Barrett, C. *et al.* 2011. The spiteful origins of human cooperation. *Proceedings of the Royal Society, Series B*, **278**: 2159–2164.

Martin-Ordas, G. and Call, J. 2009. Assessing generalization within and between trap tasks in the great apes. *International Journal of Comparative Psychology*, **22**: 43–60.

Maslin, M. and Trauth, M. 2009. Plio-Pleistocene East African pulsed climate variability and its influence on early human evolution. In *The first humans – Origin and early evolution of the genus* Homo (Grine, F., Fleagle, J. and Leakey, R., eds.), pp. 151–158. New York, NY: Springer.

Matsumoto-Oda, A. and Tomonaga, M. 2005. "Intentional" control of sound production found in leaf-clipping display of Mahale chimpanzees. *Journal of Ethology*, **23**:109–112.

Matsuzawa, T. 2001. *Primate origins of human cognition and behavior*. Tokyo: Springer-Verlag.

2006. Sociocognitive development in chimpanzees: a synthesis of laboratory work and fieldwork. In *Cognitive development in chimpanzees* (Matsuzawa, T., Tomonaga, M. and Tanaka, M., eds.), pp: 3–33. Tokyo: Springer Verlag.

2007. Comparative cognitive development. *Developmental Science* **10**(1): 97–103.

Matsuzawa, T., Biro, D., Humle, T. *et al.* 2001. Emergence of culture in wild chimpanzees: Education by master-apprenticeship. In: *Primate origins of human cognition and behavior* (Matsuzawa, T., ed.), pp. 557–574. Tokyo: Springer-Verlag.

Maynard, A. and Greenfield, P. 2003. Implicit cognitive development in cultural tools and children: lessons from Maya Mexico. *Cognitive Development*, **18**: 489–510.

Maynard-Smith, J. and Szathmary, E. 1995. *The major transitions in evolution*. Oxford: Freeman.

McBrearty, S. and Brooks, A. 2000. The revolution that wasn't: a new interpretation of the origin of modern human behavior. *Journal of Human Evolution*, **29**: 453–563.

McCall, R. B. 2011. Research, practice and policy perspectives on issues of children without permanent parental care. *Monographs of the Society for Research in Child Development*, **76**: 223–272.

McGrew, W. 1987. Tools to get food: the subsistants of Tasmanian aboriginees and Tanzanian chimpanzees compared. *Journal of Anthropological Research*, **43**: 247–258.

1992. *Chimpanzee material culture: Implications for human evolution.* Cambridge: Cambridge University Press.

2004. *The cultured chimpanzee: Reflections on cultural primatology.* Cambridge: Cambridge University Press.

McGrew, W., Turin, C. and Baldwin, P. 1979. Chimpanzees, tools and termites: cross-cultural comparisons of Senegal, Tanzania, and Rio Muni. *Man*, **14**: 185–214.

McGuigan, N., Whiten, A., Flynn, E. and Horner, V. 2007. Imitation of causally opaque versus causally transparent tool use by 3- and 5-year-old children. *Cognitive Development*, **22**: 353–364.

Mech, D. L. 1970. *The wolf*. New York, NY: Natural History Press.

Medin, D. and Atran, S. 2004. The native mind: biological categorization and reasoning in development and across cultures. *Psychological Review*, **111**(4): 960–983.

Mellars, P. and Stringer, C. 1989. *The human revolution: Behavioural and biological perspectives on the origins of modern humans.* Edinburgh: Edinburgh University Press.

Meltzoff, A. 1995. Understanding the intentions of others: re-enactment of intended acts by 18-month-old children. *Developmental Psychology*, **31**: 838–850.

1996. The human infant as imitative generalist: a 20-year progress report on infant imitation with implications for comparative psychology. In *Social learning in animals: The roots of culture* (Galef, B. and Heyes, C., eds.), pp. 347–370. New York, NY: Academic Press.

Menzel, E., Davenport, R. and Rogers, C. 1970. The development of tool use in wild-born and restriction-reared chimpanzees. *Folia Primatologica*, **12**: 273–283.

Mercader, J., Panger, M. and Boesch, C. 2002. Excavation of a chimpanzee stone tool site in the African rainforest. *Science*, **296**: 1452–1455.

Mercader, J., Barton, H., Gillespie, J., *et al.* 2007. 4,300-year-old chimpanzee sites and the origins of percussive stone technology. *Proceedings of the National Academy of Sciences USA*, **104**(9), 3043–3048.

Medin, D. and Atran, S. 2004. The native mind: Biological categorization and reasoning in development and across cultures. *Psychological Review*, **111**: 960–983.

Melis, A., Hare, B. and Tomasello, M. 2006. Engineering cooperation in chimpanzees: tolerance constraints on cooperation. *Animal Behaviour*, **72**: 275–286.

Mills, M. G. L. 1990. *Kalahari hyenas: The comparative behavioural ecology of two species.* London: Chapman and Hall.

Mishra, R., Dasen, P. and Niraula, S. 2003. Ecology, language and performance on spatial cognitive tasks. *International Journal of Psychology*, **38**: 366–383.

Mitani, J. 2009. Cooperation and competition in chimpanzees: current understanding and future challenges. *Evolutionary Anthropology*, **18**: 215–227.

Mitani, J. and Watts, D. 1999. Demographic influences on the hunting behavior of chimpanzees. *American Journal of Physical Anthropology*, **109**: 439–454.

2001. Why do chimpanzees hunt and share meat? *Animal Behaviour*, **61**: 1–10.

2005. Correlates of territorial boundary patrol behaviour in wild chimpanzees. *Animal Behaviour*, **70**: 1079–1086.

Mitani, J., Watts, D. and Muller, M. 2002. Recent development in the study of wild chimpanzee behaviour. *Evolutionary Anthropology*, **11**(1): 9–25.

Mithen, S. 1996. *The prehistory of the mind.* London: Thames and Hudson.

Möbius, Y., Boesch, C., Koops, K., Matsuzawa, T. and Humle, T. 2008. Cultural differences in army ant predation by West African chimpanzees? A comparative study of microecological variables. *Animal Behaviour*, **76**: 37–45.

Moll, H. and Tomasello, M. 2007. Cooperation and human cognition: the Vygotskian intelligence hypothesis. *Philosophical Transactions of the Royal Society, B*, **362**: 639–648.

Morgan, B. and Abwe, E. 2006. Chimpanzees use stone hammers in Cameroon. *Current Biology*, **16**(16): R632–R633.

Morris, D. 1967. *The naked ape*. New York, NY: Random House.

Morris, M. and Peng, K. 1994. Culture and cause: American and Chinese attributions for social and physical events. *Journal of Personality and Social Psychology*, **67**: 949–971.

Moss, C. 1988. *Elephant memories: Thirteen years in the life of an elephant family*. New York, NY: William Morrow.

Moura, A. C. de A. and Lee, P. C. 2004. Capuchin stone tool use in Caatinga dry forest. *Science*, **306**: 1909.

Mulcahy, N. and Call, J. 2006. How great apes perform on a modified trap-tube task. *Animal Cognition*, **9**: 193–199.

Murdock, G. 1940. The cross-cultured survey. *American Sociological Review*, **5**: 361–370.

Nagell, K., Olguin, K., & Tomasello, M. 1993. Processes of social learning in the tool use of chimpanzees and human children. *Journal of Comparative Psychology*, **107**: 174–186.

Nakamichi, M., Koyama, N., and Jolly, A. 1996. Maternal responses to dead and dying infants in wild troops of ring-tailed lemurs at the Berenty Reserve, Madagascar. *International Journal of Primatology*, **17**: 505–523.

Nakamura, M. and Nishida, T. 2006. Subtle behavioral variation in wild chimpanzees, with special reference to Imanishi's concept of kaluchua. *Primates*, **47**: 35–42.

Nakamura, M., McGrew, W., Marchant, L. and Nishida, T. 2000. Social scratch: another custom in wild chimpanzees? *Primates*, **41**(3): 237–248.

Nelson, C., Zeanah, C., Fox, N. *et al.* 2007. Cognitive recovery in socially deprived young children: the Bucharest early intervention project. *Science*, **318**: 1937–1940.

Nishida, T. 1968. The social group of wild chimpanzees in the Mahali Mountains. *Primates*, **9**: 167–224.

1987. Local traditions and cultural transmission. In *Primate societies* (Smuts, S.S., Cheney, D.L., Seyfarth, R.M., Wrangham, R.W. and Strusaker, T.T., eds.), pp. 462–474. Chicago, IL: University of Chicago Press.

1990. *The chimpanzees of the Mahale mountains: Sexual and life history strategies*. Tokyo: University of Tokyo Press.

Nishida, T. and Hiraiwa, M. 1982. Natural history of a tool-using behaviour by wild chimpanzees in feeding upon wood-boring ants. *Journal of Human Evolution*, **11**: 73–99.

Nishida, T. and Uehara, U. 1980. Chimpanzees, tools and termites: another example from Tanzania. *Current Anthropology*, **21**: 671–672.

Nishida, T., Uehara, S. and Nyondo, R. 1983. Predatory behavior among wild chimpanzees of the Mahale Mountains. *Primates*, **20**: 1–20.

Nishida, T., Hiraiwa-Hasegawa, M., Hasegawa, T. and Takahata, Y. 1985. Group extinction and female transfer in wild chimpanzees in the Mahale National Park, Tanzania. *Zeitschrift für Tierpsychologie*, **67**: 284–301.

Nishida, T., Hasegawa, T., Hayaki, H., Takahata, Y. and Uehara, S. 1992. Meat-sharing as a coalition strategy by an alpha male chimpanzee? In *Topics in primatology: vol. 1. Human origins* (Nishida, T., McGrew, W. C., Marler, P., Pickford, M. and de Waal, F., eds.), pp. 159–174. Basel: Karger AG.

Nishida, T., Matsusaka, T. and McGrew, W. 2009. Emergence, propagation or disappearance of novel behavioral patterns in the habituated chimpanzees of Mahale: a review. *Primates*, **50**:23–36.

Noad, M., Cato, D., Bryden, M., Jenner, M. and Jenner, C. 2000. Cultural revolution in whale songs. *Nature*, **408**: 537.

Normand, E. and Boesch, C. 2009. Sophisticated Euclidian maps in forest chimpanzees. *Animal Behaviour*, **77**: 1195–1201.

Normand, E., Ban, S. and Boesch, C. 2009. Forest chimpanzees (*Pan troglodytes verus*) remember the location of numerous fruit trees. *Animal Cognition*, **12**: 797–807.

Normand, E., Caillaud, D., Ban, S. and Boesch, C. In press. Fruiting tree diversity selects for long-lasting memory in forest chimpanzees. *Proceedings of the Royal Society, London, Series B*.

Nowak, S. and Sigmund, K. 1992. Tit for tat in heterogeneous populations. *Nature*, **355**: 250–253.

O'Brien, M. J., Darwent, J. and Lyman, L. 2001. Cladistics is useful for reconstructing archaeological phylogenies: Paleoindian points from the Southeastern United States. *Journal of Archaeological Science*, **28**: 1115–1136.

O'Brien, M., Lyman, L., Mesoudi, A. and Van Pool, T. 2010. Cultural traits as units of analysis. *Philosophical Transactions of the Royal Society B*, **365**: 3797–3806.

Oswalt, W., 1976. *An anthropological analysis of food-getting technology*. New York, NY: John Wiley and Sons.

Over, H. and Carpenter, M. 2009. Priming third-party ostracism increases affiliative imitation in children. *Developmental Science*, **12**(3): F1–F8.

Packer, C., Scheel, D. and Pusey, A. E. 1990. Why lions form groups: food is not enough. *American Naturalist*, **136**: 1–19.

Paradise, R. and Rogoff, B. 2009. Side by side: learning by observing and pitching in. *Ethos*, **37**(1): 102–138.

Passingham, R. 1982. *The human primate*. London: Freeman.

Patterson, N., Richter, D., Gnerre, S., Lander, E. and Reich, D. 2006. Genetic evidence for complex speciation of humans and chimpanzees. *Nature*, **441**: 1103–1108.

Payne, K. and Payne, R. 1985. Large-scale changes over 19 years in songs of humpback whales in Bermuda. *Zeitschrift für Tierpsychologie*, **68**: 89–114,

Pears, K. and Moses, L. 2003. Demographics, parenting and theory of mind in preschool children. *Social Development*, **12**: 1–20.

Penn, D. and Povinelli, D. 2007. On the lack of evidence that non-human animals possess anything remotely resembling a "theory of mind". *Philosophical Transactions of the Royal Society, Series B*, **362**: 731–744.

Penn, D., Holyoak, K. and Povinelli, D. 2008. Darwin's mistake: explaining the discontinuity between human and nonhuman minds. *Behavioral and Brain Sciences*, **31**: 109–178.

Perner, J., Ruffman, T. and Leekam, S. 1994. Theory of mind is contagious: you catch it from your sibs. *Child Development*, **65**: 1128–1238.

Perry, S. and Manson, J. 2003. Traditions in monkeys. *Evolutionary Anthropology*, **12**: 71–81.

Piaget, J. 1935. *La Naissance de l'Intelligence chez l'Enfant*. Neuchâtel: Delachaux et Niestlé.

1945. *La Formation du Symbole chez l'Enfant*. Neuchâtel: Delachaux et Niestlé.

Piaget, J. and Inhelder, B. 1947. *La Représentation de l'Espace chez l'Enfant*. Paris: Presses Universitaires de France.

Plummer, T., Ditchfield, P., Bishop, L., *et al*. 2009. Oldest evidence of toolmaking hominins in a grassland-dominated ecosystem. *PLoS One* **4**(9): e7199.

Poole, J. 1996. *Coming of age with elephants: A memoir*. New York, NY: Hyperion.

Potts, R. 1984. Home bases and early hominids. *American Scientific*, **72**: 338–347.

Povinelli, D. 1996. Chimpanzee theory of mind? The long road to strong inference. In *Theories of theories of mind* (Carruthers, P. and Smith, P., eds.), pp. 293–329. Cambridge: Cambridge University Press.

1999. Social understanding in chimpanzees: new evidence from a longitudinal approach. In *Developing theories of intention: Social understanding and selfcontrol* (Zelazo, P.D., Astington, J.W. and Olson, D.R., eds.), pp. 195–225. Mahwah, NJ: Lawrence Erlbaum Associates.

2000. *Folk physics for apes: The chimpanzee's theory of how the world works.* Oxford: Oxford University Press.

Povinelli, D. and Eddy, T. 1996. What young chimpanzees know about seeing. *Monograph of the Society for Research in Child Development*, **61**: 1–152.

Povinelli, D. J. and O'Neill, D. K. 2000. Do chimpanzees use their gestures to instruct each other? In: *Understanding other minds: Perspectives from developmental cognitive neuroscience* (Baron-Cohen, S., Tager-Flusberg, H. and Cohen, D. J., eds.), pp. 459–487. Oxford: Oxford University Press.

Povinelli, D. and Vonk, J. 2003. Chimpanzee minds: suspiciously human? *Trends in Cognitive Sciences*, **7**: 157–160.

Povinelli, D., Nelson, K. and Boysen, S. 1990. Inferences about guessing and knowing by chimpanzees (*Pan troglodytes*). *Journal of Comparative Psychology*, **104**: 203–210.

Povinelli, D., Gallup, G., Eddy, T., *et al.* 1997. Chimpanzees recognize themselves in mirrors. *Animal Behaviour*, **53**: 1083–1088.

Power, M. 1991. *The egalitarians: Human and chimpanzee*. Cambridge: Cambridge University Press.

Premack, D. 2007. Human and animal cognition: continuity and discontinuity. *Proceedings of the National Association of Sciences, USA*, **104**(35): 13 861–13 867.

Premack, D. and Woodruff, G. 1978. Does the chimpanzee have a theory of mind? *Behavioral and Brain Sciences*, **4**: 515–526.

Pruetz, J. 2007. Evidence of cave use by savanna chimpanzees (*Pan troglodytes verus*) at Fongoli, Senegal: implications for thermoregulatory behavior. *Primates*, **48**:316–319.

Pruetz, J. and Bertolani, P. 2007. Savanna chimpanzees, *Pan troglodytes verus*, hunt with tools. *Current Biology*, **17**: 1–6.

Quiatt, D., Reynolds, V. and Stokes, E. 2002. Snare injuries to chimpanzees (*Pan troglodytes*) at 10 study sites in east and west Africa. *African Journal of Ecology*, **40**: 303–305.

Raby, C., Alexis, D., Dickinson, A. and Clayton, N. 2007. Planning for the future by western scrub-jays. *Nature*, **445**: 919–921.

Radford, A. and Ridley, A. 2006. Recruitment calling: a novel form of extended parental care in an altricial species. *Current Biology*, **16**: 1700–1704.

Range, F., Varanyi, Z. and Huber, L. 2007. Selective imitation in domestic dogs. *Current Biology*, **17**: 866–872.

Rendell, L. and Whitehead, H. 2001. Culture in whales and dolphins. *Behavioral and Brain Sciences*, **24**(3): 309–324.

Rendell, L., Fogarty, L., Hoppitt, W. *et al.* 2011. Cognitive culture: theoretical and empirical insights into social learning strategies. *Trends in Cognitive Sciences*, **15**(2): 68–76.

Reaux, J. and Povinelli, D. 2000. The trap-tube problem. In *Folk physics for apes: The chimpanzee's theory of how the world works* (Povinelli, D., ed.), pp. 109–131. Oxford: Oxford University Press.

Rogoff, B. 1990. *Apprenticeship in thinking: Cognitive development in social context*. New York, NY: Oxford University Press.

2003. *The cultural nature of human development*. New York, NY: Oxford University Press.

Rogoff, B., Paradise, R., Arauz, R., Correa-Chavez, M. and Angelillo, C. 2003. Firsthand learning through intent participation. *Annual Review of Psychology*, **54**: 175–203.

Ruffman, T., Perner, J., Naito, M., Parkin, L. and Clements, W. 1998. Older (but not younger) siblings facilitate false belief understanding. *Developmental Psychology*, **34**: 161–174.

Rumbaugh, D., Washburn, D., King, J. *et al.* 2008. Why some apes imitate and/or emulate observed behavior and others do not: fact, theory and implication for our kind. *Journal of Cognitive Education and Psychology*, **7**: 101–110.

Sahlins, M. 1976. *The use and abuse of biology: An anthropological critique of sociobiology*. London: Tavistock Publications.

Santos, L., Miller, C. and Hauser, M. 2003. Representing tools: how two non-human primate species distinguish between the functionally relevant and irrelevant features of a tool. *Animal Cognition*, **6**: 269–281.

Sanz, C. and Morgan, D. 2007. Chimpanzee tool technology in the Goualougo Triangle, Republic of Congo. *Journal of Human Evolution*, **52**: 420–433.

2009. Flexible and persistent tool-using strategies in honey-gathering by wild chimpanzees. *International Journal of Primatology*, **30**: 411–427.

Sanz, C., Morgan, D. and Gulick, S. 2004. New insights into chimpanzees, tools, and termites from the Congo Basin. *American Naturalist*, **164**(5): 567–581.

Sanz, C., Call, J. and Morgan, D. 2009. Design complexity in termite-fishing tools of chimpanzees (*Pan troglodytes*). *Biology Letters*, **5**: 293–296.

Sapolsky, R. 2006. Social cultures among nonhuman primates. *Current Anthropology*, **47**(4): 641–648.

Saxe, G. and Moylan, T. 1982. The development of measurements operations among the Oksapmin of Papua New Guinea. *Child Development*, **53**: 1242–1248.

Schaik, van C.P. and Knott, C. 2001. Geographic variation in tool use on Neesia fruits in Orangutans. *American Journal of Physical Anthropology*, **114**: 331–342.

Schaik, van C. P., Ancrenaz, M., Brogen, G., *et al.* 2003. Orangutan cultures and the evolution of material culture. *Science*, **299**: 102–105.

Schaller, G. B. 1972. *The Serengeti lion*. Chicago, IL: University of Chicago Press.

Schloegl, C., Kotrschal, K. and Bugnyar, T. 2008. Do common ravens (*Corvus corax*) rely on human or conspecific gaze cues to detect hidden food? *Animal Cognition*, **11**: 231–241.

Schwartz, A. 1975. *La Vie Quotidienne Dans Un Village Guéré*. Abidjan: INADES.

Segall, M., Dasen, P., Berry, J. and Poortinga, Y. 1999. *Human behavior in global perspective: An introduction to cross-cultural psychology.* 2nd edition. New York, NY: Pergamon Press.

Shennan, S. 2000. Population, culture history and the dynamics of culture change. *Current Anthropology*, **41**(5): 811–835.

Shettleworth, S. 1998. *Cognition, evolution and behaviour*. New York, NY: Oxford University Press.

Silberberg, A. and Kearns, D. 2009. Memory for the order of briefly presented numerals in humans as a function of practice. *Animal Cognition*, **12**: 405–407.

Silk, J., Brosnan, S., Vonk, J., Henrich, J., Povinelli, D., Richardson, A., *et al.* 2005. Chimpanzees are indifferent to the welfare of unrelated group members. *Nature*, **437** 1357–1359.

Silva, F. and Silva, K. 2006. Humans' folk physics is not enough to explain variations in their tool-using behavior. *Psychonomic Bulletin and Review*, **13**(4): 689–693.

Silva, F., Page, D. and Silva, K. 2005. Methodological–conceptual problems in the study of chimpanzees' folk physics: how studies with adult humans can help. *Learning and Behavior*, **33**(1): 47–58.

Slaughter, V. and Griffiths, M. 2007. Death understanding and fear of death in young children. *Clinical Child Psychology and Psychiatry*, **12**(4): 525–535.

Slaughter, V. and Lyons, M. 2003. Learning about life and death in early childhood. *Cognitive Psychology*, **46**: 1–30.

Smith, P. and Silberberg, A. 2010. Rational maximizing by humans *(Homo sapiens)* in an ultimate game. *Animal Cognition*, **13**(4): 671–677.

Speece, M. and Brent, S. 1984. Children's understanding of death: a review of three components of a death concept. *Child Development*, **55**(5): 1671–1686.

Stanford, C. 1998. *Chimpanzee and red colobus: The ecology of predator and prey*. Cambridge, MA: Harvard University Press.

Stanford, C., Wallis, J., Matama, H. and Goodall, J. 1994a. Patterns of predation by chimpanzees on red colobus monkeys in Gombe National Park, Tanzania, 1982–1991. *American Journal of Physical Anthropology*, **94**: 213–229.

Stanford, C., Wallis, J., Mpongo, E. and Goodall, J. 1994b. Hunting decisions in wild chimpanzees. *Behaviour*, **131**: 1–20.

Sternberg, R. 2008. Increasing fluid intelligence is possible after all. *Proceedings of the National Academy of Sciences USA*, **105**(19): 6791–6792.

Stout, D. 2003. Skill and cognition in stone tool production: an ethnographic case study from Irian Jaya. *Current Anthropology*, **43**(5): 693–715.

Sugiyama, Y. 1981. Observations on the population dynamics and behavior of wild chimpanzees of Bossou, Guinea, 1979–1980. *Primates*, **22**: 435–444.

1984. Population dynamics of wild chimpanzees at Bossou, Guinea, between 1976 and 1983. *Primates*, **25**: 391–400.

1994. Tool use by wild chimpanzees. *Science*, **367**: 327.

Sugiyama, Y. and Koman, J. 1979. Tool-using and -making behavior in wild chimpanzees at Bossou, Guinea. *Primates*, **20**(4): 513–524

Tanon, F. 1994. *A cultural view on planning: The case of weaving in Ivory Coast*. Tilburg, the Netherlands: Tilburg University Press.

Tebbich, S., Seed, A., Emery, N. and Clayton, N. 2007. Non tool-using rooks, *Corvus frugilegus*, solve the trap-tube problem. *Animal Cognition*, **10**: 225–231.

Tehrani, J. and Collard, M. 2002. Investigating cultural evolution through biological phylogenetic analyses of Turkmen textiles. *Journal of Anthropological Archaeology*, **21**: 443–463.

Teleki, G. 1973a. *The predatory behavior of wild chimpanzees*. Brunswick: Bucknell University Press.

1973b. Group response to the accidental death of a chimpanzee in Gombe National Park, Tanzania. *Folia Primatologica*, **20**: 81–94.

Tempelmann, S., Kaminski, J. and Liebal, K. 2011. Focus on the essential: all great apes know when others are being attentive. *Animal Cognition*, **14**: 433–439.

Tennie, C., Call, J. and Tomasello, M. 2006. Push or pull: imitation vs. emulation in great apes and human children. *Ethology*, **112**: 1159–1169.

2009. Ratcheting up the ratchet: on the evolution of cumulative culture. *Philosophical Transactions of the Royal Society B*, **364**: 2405–2415.

Teschke, I. and Tebbich, S. 2011. Physical cognition and tool-use: performance of Darwin's finches in the two-trap tube task. *Animal Cognition*, **14**: 555–563.

Thomas, E., Murphy, M., Pitt, R., Rivers, A. and Leavens, D. 2008. Understanding of visual attention by adult humans (*Homo sapiens*): a partial replication of Povinelli, Bierschwale and Cech (1999). *Journal of Comparative Psychology*, **122**(4): 428–436.

Thornton, A. and McAuliffe, K. 2006. Teaching in wild meerkats. *Science*, **313**: 227–229.

Thornton, A. and Raihani, N. 2008. The evolution of teaching. *Animal Behaviour*, **75**(6): 1823–1832.

Tomasello, M. 1990. Cultural transmission in tool use and communicatory signaling of chimpanzees? In *Comparative developmental psychology of language and intelligence in primates* (Parker, S. and Gibson, K., eds.), pp. 274–311. Cambridge: Cambridge University Press.

1999. *The cultural origins of human cognition*. Cambridge, MA: Harvard University Press.

2009. *Why we co-operate*. Cambridge, MA: MIT Press.

Tomasello, M. and Call, J. 1997. *Primate cognition*. New York, NY: Oxford University Press.

2004. The role of humans in the cognitive development of apes revisited. *Animal Cognition*, **7**: 213–215.

Tomasello M., Davis-Dasilva M., Camak L. and Bard K. 1987. Observational learning of tool-use by young chimpanzees. *Journal of Human Evolution*, **2**: 175–183.

Tomasello, M., Savage-Rumbaugh, S. and Kruger, A. 1993. Imitative learning of actions on objects by children, chimpanzees, and enculturated chimpanzees. *Child Development*, **64**: 1688–1705.

Tomasello, M., Call, J. and Hare, B. 1998. Five primate species follow the gaze of conspecifics. *Animal Behaviour*, **55**: 1063–1069.

Tomasello, M., Carpenter, M., Call, J., Behne, T. and Moll, H. 2005. Understanding and sharing intentions: the origins of cultural cognition. *Behavioral and Brain Sciences*, **28**: 675–691.

Topál, J., Gergely, G., Erdohegyi, A., Csibra, G. and Miklosi, A. 2009. Differential sensitivity to human communication in dogs, wolves and human infants. *Science*, **325**: 1269–1272.

Tranquilli, S., Anedi-Lartey, M., Amsini, F., *et al.* 2011. Lack of conservation effort rapidly increases African great ape extinction risk. *Conservation Letters*, **5**: 1–8.

Tsukahara, T. 1993. Lions eat chimpanzees: the first evidence of predation by lions on wild chimpanzees. *American Journal of Primatology*, **29**: 1–11.

Turnbull, C. 1961. *The forest people*. London: Jonathan Cape.

Tuttle, R. 2001. On culture and traditional chimpanzees. *Current Anthroplogy*, **42**(3): 407–408.

Udell, M., Dorey, N. and Wynne, C. 2008. Wolves outperform dogs in following human social cues. *Animal Behaviour*, **76**: 1767–1773.

Uehara, S. 1982. Seasonal changes in the techniques employed by wild chimpanzees in the Mahale Mountains, Tanzania, to feed on termites. *Folia Primatologica*, **37**: 44–76.

Ueno, A. and Matsuzawa, T. 2004. Food transfer between chimpanzee mothers and their infants. *Primates*, **45**: 231–239.

van Ijzendoorn, M. H., Bard, K. A., Bakermans-Kranenburg, M. J. and Ivan, K. 2009. Enhancement of attachment and cognitive development of young nursery-reared chimpanzees in responsive versus standard care. *Developmental Psychobiology*, **51**: 173–185.

Vinden, P. 1996. Junin Quechua children's understanding of mind. *Child Development*, **67**: 1707–1716.

1999. Children's understanding of mind and emotion: a multi-culture study. *Cognition and Emotion*, **13**: 19–48.

Visalberghi, E. and Fragaszy, D. 1990. Do monkeys ape? In *"Language" and intelligence in monkeys and apes: Comparative developmental perspectives* (Parker, S. and Gibson, K., eds.), pp. 247–273. Cambridge: Cambridge University Press.

Visalberghi, E. and Limongelli, L. 1994. Lack of comprehension of cause–effect relations in tool-using capuchin monkeys (*Cebus apella*). *Journal of Comparative Psychology*, **108**: 15–22.

Visalberghi, E. and Trinca, L. 1989. Tool use in capuchin monkeys: distinguishing between performing and understanding. *Primates*, **30**: 511–521.

Visalberghi, E., Fragaszy, D., Ottoni, E. *et al.* 2007. Characteristics of hammer stones and anvils used by wild bearded capuchin monkeys (*Cebus libidinosus*) to crack open palm nuts. *American Journal of Physical Anthropology*, **132**: 426–444.

Visalberghi, E., Addessi, E., Truppa, V., *et al.* 2009. Selection of effective stone tools by wild bearded capuchin monkeys. *Current Biology*, **19**: 213–217.

Vlamings, P., Hare, B. and Call, J. 2010. Reaching around barriers: the performance of the great apes and 3–5-year-old children. *Animal Cognition*, **13**: 273–285.

Voelkl, B. and Huber, L. 2000. True imitation in marmosets. *Animal Behaviour*, **60**: 195–202.

Vonk, J., Brosnan, S., Silk, J., Henrich, J., Richardson, A., Lambeth, S., *et al.* 2008. Chimpanzees do not take advantage of very low cost opportunities to deliver food to unrealted group members. *Animal Behaviour*, **75**: 1757–1770.

Waal, de F. 1996. *Good natured: The origins of right and wrong in humans and other animals*. Cambridge, MA: Harvard University Press.

2001. *The ape and the sushi master: Cultural reflections of a primatologist*. New York, NY: Basic Books.

2009. *The age of empathy: Nature's lessons for a kinder society*. New York, NY: Harmony.

Waal, de F. and Bonnie, K. 2009. In tune with others: the social side of primate culture. In *The question of animal culture* (Laland, K. and Galef, G., eds.), pp. 19–39. Cambridge, MA: Harvard University Press.

Waal, de F. and Ferrari, P. 2010. Towards a bottom-up perspective on animal and human cognition. *Trends in Cognitive Sciences*, **14**(5): 201–207.

Waal, de F., Boesch, C., Horner, V. and Whiten, A. 2008. Comparing social skills of children and apes. *Science*, **319**: 569–570.

Warneken, F. and Tomasello, M. 2006. Altruistic helping in human infants and young chimpanzees. *Science*, **311**, 1301–1303.

Warren, Y. and Williamson, E. 2004. Transport of dead infant mountain gorillas by mothers and unrelated females. *Zoo Biology*, **23**(4): 375–378.

Wassmann, J. and Dasen, P. 1998. Balinese spatial orientation: some empirical evidence of moderate linguistic relativity. *Journal of the Royal Anthropological Institute (N.S.)*, **4**: 689–711.

Watts, D. 2008. Tool use by chimpanzees at Ngogo, Kibale National Park, Uganda. *International Journal of Primatology*, **29**: 83–94.

Watts, D. and Mitani, J. 2000. Hunting behavior of chimpanzees at Ngogo, Kibale National Park, Uganda. *International Journal of Primatology*, **23**(1): 1–28.

2002. Hunting and meat sharing by chimpanzees of Ngogo, Kibale National Park, Uganda. In *Behavioural diversity in chimpanzees and bonobos* (Boesch, C., Hohmann, G. and Marchant, L., eds.), pp. 244–255. Cambridge: Cambridge University Press.

Watts, D., Muller, M., Amsler, S., Mbabazi, G. and Mitani, J. 2006. Lethal intergroup aggression by chimpanzees in the Kibale National Park, Uganda. *American Journal of Primatology*, **68**: 161–180.

Wellman, H., Cross, D. and Watson, J. 2001. Meta-analysis of theory of mind development: the truth about false-belief. *Child Development*, **72**: 655–684.

Whitehead, H. 1998. Cultural selection and genetic diversity in matrilineal whales. *Science*, **282**: 1708–1711.

Whiten, A. 1989. Transmission mechanisms in primate cultural evolution. *TREE* **4**(3): 61–62.

Whiten, A. and Custance, D. 1996. Studies of imitation in chimpanzees and children. In *Social learning in animals: The roots of culture* (Heyes, C. and Galef, B. Jr, eds.), pp. 291–318. New York, NY: Academic Press.

Whiten, A. and Ham, R. 1992. On the nature and evolution of imitation in the animal kingdom: reappraisal of a century of research. *Advance in the Study of Behavior*, **21**: 239–283.

Whiten, A. and Schaik, van C. 2007. The evolution of animal "cultures" and social intelligence. *Philosophical Transactions of the Royal Society B*, **362**: 603–620.

Whiten, A., Custance, D., Gomez, J., Texidor, P. and Bard, K. 1996. Imitative learning of artificial fruit processing in children (*Homo sapiens*) and chimpanzees (*Pan troglodytes*). *Journal of Comparative Psychology*, **110**(1): 3–14.

Whiten, A., Goodall, J., McGrew, W., *et al.* 1999. Cultures in chimpanzees. *Nature*, **399**: 682–685.

Whiten, A., Horner, V. and Marshall-Pescini, S. 2003. Cultural panthropology. *Evolutionary Anthropology*, **12**: 92–105.

Whiten, A., Horner, V., Litchfield, C. and Marshall-Pescini, S. 2004. How do apes ape? *Learning and Hormones*, **32**(1): 36–52.

Whiten, A., Horner, V. and de Waal, F. 2005. Conformity to cultural norms of tool use in chimpanzees. *Nature*, **437**: 737–740.

Whiten, A., Spiteri, A., Horner, V., *et al.* 2007. Transmission of multiple traditions within and between chimpanzee groups. *Current Biology*, **17**: 1038–1043.

Whiten, A., McGuigan, N., Marshall-Pescini, S. and Hopper, L. 2009. Emulation, imitation, over-imitation and the scope of culture for child and chimpanzee. *Philosophical Transactions of the Royal Society B*, **364**: 2417–2428.

Williams, J., Oehlert, G., Carlis, J. and Pusey, A. 2004. Why do male chimpanzees defend a group range? *Animal Behaviour*, **68**: 523–532.

Wilson, E. O. 1975. *Sociobiology*. Cambridge, MA: Harvard University Press.

Wimmer, H., Hogrefe, G. and Perner, J. 1988. Children's understanding of informational access as source of knowledge. *Child Development*, **59**: 386–396.

Wolpert, L. 2007. Causal belief makes human. In *What makes us human?* (Pasternak, C., ed.), pp. 164–181. Oxford: Oneworld Publications.

Wood, D., Bruner, J. S. and Ross, G. 1976. The role of tutoring in problem solving. *Journal of Child Psychology and Psychiatry*, **17**: 89–100.

Wrangham, R. 1975. *The behavioural ecology of chimpanzees in the Gombe Stream National Park*. Ph.D. thesis. Cambridge University.

2006. Chimpanzees: the culture-zone concept becomes untidy. *Current Biology*, **16**(16): R634.

Wrangham, R., Wilson, M. and Muller, M. 2006. Comparative rates of violence in chimpanzees and humans. *Primates*, **47**: 14–26.

Wynne, C. 2007. What are animals? Why anthropomorphism is still not a scientific approach to behavior. *Comparative Cognition and Behavior Reviews*, **2**: 125–135.

Zamma, K. 2002. Leaf-grooming by a wild chimpanzee in Mahale. *Primates*, **43**(1): 87–90.

Zentall, T. 1996. An analysis of imitative learning in animals. In *Social learning in animals: The roots of culture* (Heyes, C. and Galef, B., eds.), pp. 221–243. San Diego, CA: Academic Press.

Zuckerman, S. 1981. *The social life of monkeys and apes*. 2nd edition. London: Routledge & Kegan Paul.

Index